ADVANCED ESR METHODS IN POLYMER RESEARCH

ADVANCED ESR METHODS IN POLYMER RESEARCH

Edited by

SHULAMITH SCHLICK
University of Detroit Mercy
Detroit, Michigan

WILEY-INTERSCIENCE
A John Wiley & Sons, Inc., Publication

Copyright © 2006 by John Wiley & Sons, Inc. All rights reserved

Published by John Wiley & Sons, Inc., Hoboken, New Jersey
Published simultaneously in Canada

No part of this publication may be reproduced, stored in a retrieval system, or transmitted in any form or by any means, electronic, mechanical, photocopying, recording, scanning, or otherwise, except as permitted under Section 107 or 108 of the 1976 United States Copyright Act, without either the prior written permission of the Publisher, or authorization through payment of the appropriate per-copy fee to the Copyright Clearance Center, Inc., 222 Rosewood Drive, Danvers, MA 01923, (978) 750-8400, fax (978) 750-4470, or on the web at www.copyright.com. Requests to the Publisher for permission should be addressed to the Permissions Department, John Wiley & Sons, Inc., 111 River Street, Hoboken, NJ 07030, (201) 748-6011, fax (201) 748-6008, or online at http://www.wiley.com/go/permission.

Limit of Liability/Disclaimer of Warranty: While the publisher and author have used their best efforts in preparing this book, they make no representations or warranties with respect to the accuracy or completeness of the contents of this book and specifically disclaim any implied warranties of merchantability or fitness for a particular purpose. No warranty may be created or extended by sales representatives or written sales materials. The advice and strategies contained herein may not be suitable for your situation. You should consult with a professional where appropriate. Neither the publisher nor author shall be liable for any loss of profit or any other commercial damages, including but not limited to special, incidental, consequential, or other damages.

For general information on our other products and services or for technical support, please contact our Customer Care Department within the United States at (877) 762-2974, outside the United States at (317) 572-3993 or fax (317) 572-4002.

Wiley also publishes its books in a variety of electronic formats. Some content that appears in print may not be available in electronic formats. For more information about Wiley products, visit our web site at www.wiley.com.

Library of Congress Cataloging-in-Publication Data:

Advanced ESR methods in polymer research/edited by Shulamith Schlick.
   p.cm.
   Includes bibliographical references and index.
   ISBN-13: 978-0-471-73189-4
   ISBN-10: 0-471-73189-7
   1. Electron paramagnetic resonance—Research. 2. Polymers—Research. I. Schlick, Shulamith.

   QC763.A32.2006
   547'.7046—dc22
                                                                                           2006044267

Printed in the United States of America
10 9 8 7 6 5 4 3 2 1

DEDICATION

My experience and understanding of ESR methodologies have benefited greatly from interactions with my co-workers, who joined my lab and shared with me their ambitions, knowledge, creativity, and technical skills. Over the years these co-workers became my professional family. To them this book is dedicated.

CONTENTS

PREFACE		ix
ABOUT THE EDITOR		xi
CONTRIBUTORS		xiii

PART I ESR FUNDAMENTALS	1
1 Continuous-Wave and Pulsed ESR Methods *Gunnar Jeschke and Shulamith Schlick*	3
2 Double Resonance ESR Methods *Gunnar Jeschke*	25
3 Calculating Slow-Motion ESR Spectra of Spin-Labeled Polymers *Keith A. Earle and David E. Budil*	53
4 ESR Imaging *Shulamith Schlick*	85

PART II ESR APPLICATIONS	99
5 ESR Study of Radicals in Conventional Radical Polymerization Using Radical Precursors Prepared by Atom Transfer Radical Polymerization *Atsushi Kajiwara and Krzysztof Matyjaszewski*	101
6 Local Dynamics of Polymers in Solution by Spin-Label ESR *Jan Pilař*	133

7	**Site-Specific Information on Macromolecular Materials by Combining CW and Pulsed ESR on Spin Probes** *Gunnar Jeschke*	165
8	**ESR Methods for Assessing the Stability of Polymer Membranes Used in Fuel Cells** *Emil Roduner and Shulamith Schlick*	197
9	**Spatially Resolved Degradation in Heterophasic Polymers From 1D and 2D Spectral–Spatial ESR Imaging Experiments** *Shulamith Schlick and Krzysztof Kruczala*	229
10	**ESR Studies of Photooxidation and Stabilization of Polymer Coatings** *David R. Bauer and John L. Gerlock*	255
11	**Characterization of Dendrimer Structures by ESR Techniques** *M. Francesca Ottaviani and Nicholas J. Turro*	279
12	**High-Field ESR Spectroscopy of Conductive Polymers** *Victor I. Krinichnyi*	307

INDEX   339

PREFACE

In May 1994, I visited Professor Bengt Rånby at the Royal Institute of Technology in Stockholm, Sweden. Professor Rånby, at that time Emeritus, was enthusiastic about his numerous projects, including collaborations with Chinese scientists. On that occasion, I mentioned to him how useful his 1977 book entitled *ESR Spectroscopy in Polymer Research*, which he wrote together with J.F. Rabek, had been to me and many of my colleagues over the years. Professor Rånby confided that he planned a sequel, which "would be published sometime soon." I was hopeful, and expectant, but this was not to be.

So, what to do with all the excitement in the electron spin resonance (ESR) community over the extraordinary advances in ESR techniques in the last 20 years, techniques that have been used in Polymer Science? The pulsed, high field, double resonance, and DEER experiments, ESR imaging, simulations? Someone must tell the story, and I took the challenge.

In the winter of 2004, I was on sabbatical at the Max Planck Institute for Polymer Research in Mainz, Germany, shared an office with Gunnar Jeschke, and worked with him on the ESR chapter for the *Encyclopedia of Polymer Science and Technology (EPST)*.* Jacqueline I. Kroschwitz, the editor of *EPST*, encouraged me to enlarge the chapter into a full volume. In all planning and writing stages, I benefited greatly from numerous discussions with Gunnar, who has enriched the book by the three chapters that he contributed.

The final content of this book evolved during many talks with students and co-workers at UDM and colleagues at other institutions, and during long walks in my neighborhood. It took the talent, dedication, and patience of the contributors to travel

---

* Schlick, S.; Jeschke, G. Electron Spin Resonance, In *Encyclopedia of Polymer Science and Engineering*, Kroschwitz, J.I., Ed.; Wiley-Interscience: New York, NY, 2004; Chap. 9, pp. 614–651 (web and hardcopy editions).

through the seemingly endless revisions and to arrive at the published volume. I am grateful to Arza Seidel and her team at Wiley for guidance during all stages of this project.

Part I of the present volume includes the fundamentals and developments of the ESR experimental and simulations techniques. This part could be a valuable introduction to students interested in ESR, or in the ESR of polymers. Part II describes the wide range of applications to polymeric systems, from living radical polymerization to block copolymers, polymer solutions, ion-containing polymers, polymer lattices, membranes in fuel cells, degradation, polymer coatings, dendrimers, and conductive polymers: a world of ESR cum polymers. It is my hope that the wide range of ESR techniques and applications will be of interest to students and mature polymer scientists and will encourage them to apply ESR methods more widely to polymeric materials. And I extend an invitation to ESR specialists, to apply their talents to polymers.

Shulamith Schlick

February 2006

ABOUT THE EDITOR

Shulamith Schlick, D.Sc., is Professor of Physical and Polymer Chemistry in the Department of Chemistry and Biochemistry, University of Detroit Mercy in Detroit, Michigan.

Dr. Schlick received her undergraduate degree in Chemical Engineering at the Technion, Israel Institute of Technology in Haifa, Israel. At the same institution, she also obtained her M.Sc. in Polymer Chemistry and her D.Sc. degree in Molecular Spectroscopy. She taught at the Technion, Wayne State University, and the University of Windsor. In 1983, she assumed her present position at UDM. In recent years, she held Visiting Professorships at the Department of Chemistry, University of Florence, Italy, at the Department of Chemistry, University of Bologna, Italy, and at the Max-Planck Institute for Polymer Research, Mainz, Germany. She spent sabbatical leaves at the Centre d'Études Nucléaires de Grenoble, in Grenoble, France; as Varon Visiting Professor at the Weizmann Institute of Science, Rehovot, Israel; at the Department of Polymer Chemistry, Tokyo Institute of Technology; at the University of Bologna; and at MPI, Mainz, Germany.

Current research interests of the editor are morphology, phase separation, and self-assembling in ionomers and nonionic polymeric surfactants; electron spin resonance imaging (ESRI) of transport processes in polymer solutions and swollen gels; dynamical processes in disordered systems using electron spin probes and ^{2}H NMR; ESR and ESRI of degradation and stabilization processes in thermally-treated and UV-irradiated polymers; study of the stability of polymeric membranes used in fuel cells; and DFT calculations of the geometry and electronic structure of organic radicals, with emphasis on fluorinated radicals. Her research has resulted in more than 200 publications and has been supported by NSF, DOD, PRF, NATO, AAUW, Ford Motor Company, Dow Chemical Company, and the Fuel Cell Activity Center of

General Motors. Dr. Schlick was the recipient of two Creativity Awards from the Polymer Program of the National Science Foundation, and of an Honorary Doctorate (Doctor Honoris Causa) from Linköping University, Sweden, in May 2003.

Dr. Schlick is a member of the American Chemical Society, American Physical Society, American Association for the Advancement of Science, American Association of University Women, and International ESR Society.

CONTRIBUTORS

David R. Bauer, Research and Advanced Engineering, Ford Motor Company, Dearborn, Michigan, *ESR Studies of Photooxidation and Stabilization of Polymer Coatings (Chapter 10).*

David E. Budil, Department of Chemistry, Northeastern University, Boston, Massachusetts, *Calculating Slow-Motion ESR Spectra of Spin-Labeled Polymers (Chapter 3).*

Keith A. Earle, Department of Physics, University of Albany (SUNY), Albany, New York, *Calculating Slow-Motion ESR Spectra of Spin-Labeled Polymers (Chapter 3).*

John L. Gerlock, Ford Motor Company (retired), *ESR Studies of Photooxidation and Stabilization of Polymer Coatings (Chapter 10).*

Gunnar Jeschke, MPI for Polymer Research, Mainz, Germany, *Continuous-Wave and Pulsed ESR Methods (Chapter 1), Double Resonance ESR Methods (Chapter 2), Site-Specific Information on Macromolecular Materials by Combining CW and Pulsed ESR on Spin Probes (Chapter 7).*

Astushi Kajiwara, Nara University of Education, Nara, Japan, *ESR Study of Radicals in Conventional Radical Polymerization Using Radical Precursors Prepared by Atom Transfer Radical Polymerization (Chapter 5).*

Victor I. Krinichnyi, Institute of Problems of Chemical Physics, Chernogolovka, Moscow Region, Russia, *High-Field ESR Spectroscopy of Conductive Polymers (Chapter 12).*

Krzysztof Kruczala, Faculty of Chemistry, Jagiellonian University, Cracow, Poland, *Spatially Resolved Degradation in Heterophasic Polymers From 1D and 2D Spectral–Spatial ESR Imaging Experiments (Chapter 9).*

Krzysztof Matyjaszewski, Department of Chemistry, Carnegie Mellon University, Pittsburgh, Pennsylvania, *ESR Study of Radicals in Conventional Radical Polymerization Using Radical Precursors Prepared by Atom Transfer Radical Polymerization (Chapter 5).*

M. Francesca Ottaviani, Institute of Chemical Sciences, University of Urbino, Urbino, Italy, *Characterization of Dendrimer Structures by ESR Techniques (Chapter 11).*

Jan Pilar, Institute of Macromolecular Chemistry, Academy of Sciences of the Czech Republic, Prague, Czech Republic, *Local Dynamics of Polymers in Solution by Spin-Label ESR (Chapter 6).*

Emil Roduner, Institute of Physical Chemistry, University of Stuttgart, Stuttgart, Germany, *ESR Methods for Assessing the Stability of Polymer Membranes Used in Fuel Cells (Chapter 8).*

Shulamith Schlick, Department of Chemistry and Biochemistry, University of Detroit Mercy, Detroit, Michigan, *Continuous-Wave and Pulsed ESR Methods (Chapter 1), ESR Imaging (Chapter 4), ESR Methods for Assessing the Stability of Polymer Membranes Used in Fuel Cells (Chapter 8), Spatially Resolved Degradation in Heterophasic Polymers From 1D and 2D Spectral–Spatial ESR Imaging Experiments (Chapter 9).*

Nicholas J. Turro, Department of Chemistry, Columbia University, New York, *Characterization of Dendrimer Structures by ESR Techniques (Chapter 11).*

PART I

ESR FUNDAMENTALS

1

CONTINUOUS-WAVE AND PULSED ESR METHODS

GUNNAR JESCHKE
Max Planck Institute for Polymer Research, Mainz, Germany

SHULAMITH SCHLICK
University of Detroit Mercy, Detroit, Michigan

Contents

1. Introduction — 3
2. Fundamentals of Electron Spin Resonance Spectroscopy — 4
   2.1. Basic Principles — 4
   2.2. Anisotropic Hyperfine Interaction and g-Tensor — 10
   2.3. Isotropic Hyperfine Analysis — 12
   2.4. Environmental Effects on g- and Hyperfine Interaction — 12
   2.5. Accessibility to Paramagnetic Quenchers — 13
   2.6. Line Shape Analysis for Tumbling Nitroxide Radicals — 15
3. Multifrequency and High-Field ESR — 16
4. Pulsed ESR Methods — 18
Acknowledgments — 22
References — 22

1. INTRODUCTION

Electron spin resonance (ESR) is a spectroscopic technique that detects the transitions induced by electromagnetic radiation between the energy levels of electron

Advanced ESR Methods in Polymer Research, edited by Shulamith Schlick.
Copyright © 2006 John Wiley & Sons, Inc.

spins in the presence of a static magnetic field. The method can be applied to the study of species containing one or more unpaired electron spins; examples include organic and inorganic radicals, triplet states, and complexes of paramagnetic ions. Spectral features, such as resonance frequencies, splittings, line shapes, and line widths, are sensitive to the electronic distribution, molecular orientations, nature of the environment, and molecular motions. Theoretical and experimental aspects of ESR have been covered in a number of books,[1-8] and reviewed regularly.[9-11]

Currently available textbooks and monographs are written for students and scientists that specialize in the development of ESR technique and its application to a broad range of samples. Nowadays, however, research groups are interested in a specific field of applications, such as polymer science, and apply more than one characterization method to the materials of interest. An introduction to ESR that targets such an audience needs to be shorter, less mathematical, and focused on application rather than methodological issues. This chapter is an attempt to provide such a short introduction on the application of ESR spectroscopy to problems in polymer science.

Organic radicals occur in polymers as intermediates in chain-growth and depolymerization reactions,[12-15] or as a result of high-energy irradiation (γ, electron beams).[13,14] Paramagnetic transition metal ions are present in a number of functional polymer materials, such as catalysts and photovoltaic devices.[16] However, much of the modern ESR work in polymer science focuses on diamagnetic materials that are either doped with stable radicals as "spin probes", or labeled by covalent attachment of such radicals as "spin labels" to polymer chains.[9,17-22] This chapter therefore treats the *basic* concepts that are required to understand ESR spectra of a broad range of organic radicals and transition metal ions, and describes more advanced concepts as applied to the most popular class of spin probes and labels: nitroxide radicals.

2. FUNDAMENTALS OF ELECTRON SPIN RESONANCE SPECTROSCOPY

2.1. Basic Principles

Spins are magnetic moments that are associated with angular momentum; they interact with external magnetic fields (Zeeman interaction) and with each other (couplings). In most cases, the Zeeman interaction of the electron spin is the largest interaction in the spin system (high–field limit). The electron Zeeman (EZ) interaction can generally be described by the Hamiltonian below,

$$\mathcal{H}_{EZ} = \beta_e \boldsymbol{B_0} \boldsymbol{g} \boldsymbol{S} \tag{1}$$

where S is the spin vector operator, $\boldsymbol{B_0}$ is the transposed magnetic field vector in gauss (G) or tesla (1 T = 10^4 G), β_e is the Bohr magneton equal to 9.274×10^{-21} ergG^{-1} (or 9.274×10^{-24} JT^{-1}), and $\boldsymbol{g}$ is the g tensor. For a free electron, g is simply the number $g_e = 2.002319$. The transition energy is then $\Delta E = h\nu_{mw} = g_e\beta_e B_0$, where B_0 is the magnitude of the magnetic field. Typical values are $B_0 \approx 0.34$ T (3400 G) corresponding to microwave (mw) frequencies of ≈9.6 GHz (X band), or $B_0 \approx 3.35$ T corresponding to mw frequencies of ≈94 GHz (W band).

The g-value of a bound electron generally exhibits some deviation from g_e that is mainly due to interaction of the spin with orbital angular momentum of the unpaired electron (spin–orbit coupling). Spin–orbit coupling is a relativistic effect that tends to increase with increasing atomic number of the nuclei that contribute atomic orbitals to the singly occupied molecular orbital. Therefore, g-values deviate more strongly from g_e for transition metal complexes than for organic radicals. As the orbital angular momentum is quenched in the ground state of molecules, spin–orbit coupling comes about only by admixture of excited orbitals. Such admixture is stronger for low-lying excited states, which are relevant, for example, if the unpaired electron has high density at an oxygen atom. Oxygen-centered organic radicals thus tend to have higher g-values than carbon-centered ones.

As the orbital angular momentum relates to a molecular coordinate frame and the spin is quantized along the magnetic field (z axis of the laboratory frame), the g-value depends on the orientation of the molecule with respect to the field. This anisotropy can be described by a second rank tensor with three principal values, g_x, g_y, and g_z. The corresponding principal axes define the molecular frame. In fluid solutions, molecules tumble with a rotational diffusion rate that is much higher than the differences of the electron Zeeman frequencies between different orientations. In this situation, the g-value is orientationally averaged and only its isotropic value $g_{iso} = (g_x + g_y + g_z)/3$ can be measured. A good overview of isotropic g-values of organic radicals can be found in Ref. 23; Ref. 5 collects information on g tensors for transition metal complexes.

The real power of ESR spectroscopy for structural studies is based on the interaction of the unpaired electron spin with nuclear spins. This hyperfine interaction splits each energy level into sublevels and often allows the determination of the atomic or molecular structure of species containing unpaired electrons, and of the ligation scheme around paramagnetic transition metal ions. For a system with m nuclear spins (identified by index k) and a single electron spin, which may be larger than one-half as explained below, the hyperfine Hamiltonian is given in Eq. 2,

$$\mathcal{H}_{hfi} = h\Sigma \, S \cdot A_k \cdot I_k \qquad (2)$$

where the I_k are nuclear spin vector operators and the A_k are hyperfine tensors in frequency units (Hz). Each hyperfine tensor is characterized by three principal values A_x, A_y, and A_z and by the relative orientation of its principal axes system with respect to the molecular frame defined by the g-tensor. This relative orientation is most easily defined by three Euler angles α, β, γ, which correspond to a sequence of rotations about the z axis (by angle α), the new y' axis (by angle β), and the final z" axis (by angle γ); these rotations transform the principal axes frame of the hyperfine tensor into that of the g-tensor. The relative orientation is often given as direction cosines, which are the coordinates of unit vectors along the directions of the hyperfine principal axes given in the coordinate frame of the g-tensor.

Only the isotropic value $A_{iso} = (A_x + A_y + A_z)/3$ can be measured in fluid solutions, and is due to the Fermi contact interactions of electrons that reside in an s orbital of the nucleus under consideration. The contribution of a single orbital is proportional to the spin population (spin density) in that orbital, to the probability density $|\psi_0|^2$ of the orbital wave function at its center (inside the nucleus), and to the nuclear g-value, g_n. To a very good approximation, the hyperfine couplings for different isotopes of the same element thus have the same ratio as the g_n values.

Purely anisotropic contributions ($A_x + A_y + A_z = 0$) to the hyperfine coupling result from spin density in p, d, or f orbitals on the nucleus and from the dipole–dipole interaction T between the electron and nuclear spin. If the electron spin is confined to a region that is much smaller than the electron–nuclear distance r_{en}, both spins can be treated as point dipoles and the magnitude of T is proportional to r_{en}^{-3}. In this case, T has axial symmetry and its principal values are given by $T_x = T_y = -T$ and $T_z = 2T$. Furthermore, if the spin density in p, d, and f orbitals on that nucleus is negligible, as is the case for protons (^{1}H), the measurement of the hyperfine anisotropy can provide the electron–nuclear distance r_{en}. Any spin density at the nucleus under consideration is negligible if this nucleus is located in a neighboring molecule and does not interact (by van der Waals or hydrogen bonding) with a nucleus on which much spin density is located. Intermolecular distances larger than ≈ 0.3 nm can thus be inferred from hyperfine couplings.

For nuclei with significant hyperfine interaction, the other interactions of the nuclear spin also need to be considered. The nuclear Zeeman (NZ) interaction of these spins with the external magnetic field is described in Eq. 3.

$$\mathcal{H}_{NZ} = -\Sigma\, g_{n,k}\, \beta_n B_0\, I_k \qquad (3)$$

Nuclear spins with $I > \frac{1}{2}$ have an electric quadrupole moment that interacts with the quadrupole moment of the charge distribution around the nucleus. The Hamiltonian for this nuclear quadrupole (NQ) interaction is given in Eq. 4,

$$\mathcal{H}_{NQ} = h\Sigma\, I_k\, Q_k\, I_k \qquad (4)$$

where Q_k are the traceless ($Q_x + Q_y + Q_z = 0$) nuclear quadrupole tensors. Because the tensor is traceless, this interaction is not detected in fluid media.

Both the nuclear Zeeman and nuclear quadrupole interaction do not depend on the magnetic quantum number m_S of the electron spin. As the selection rule for ESR transitions is given by Eq. 5,

$$\Delta m_S = \pm 1 \quad \text{and} \quad \Delta m_I = 0 \qquad (5)$$

where m_I is the nuclear spin quantum number, these interactions do not make a first-order contribution to the ESR spectrum. In many cases, they can thus be neglected in spectrum analysis. This situation is illustrated in Fig. 1 for a nitroxide in which the nuclear spin $I = 1$ of the ^{14}N atom is coupled to the electron spin $S = \frac{1}{2}$ that resides mainly in the p_z orbitals on the N and O atom. The hyperfine coupling causes a splitting of each of the electron spin levels ($m_S = -\frac{1}{2}$ and $m_S = +\frac{1}{2}$) into three sublevels. When a constant microwave frequency ν_{mw} is irradiated and the magnetic field is swept, three resonance transitions are observed (Fig. 1a). The nuclear Zeeman interaction shifts both $m_I = +1$ sublevels to lower and both $m_I = -1$ sublevels to higher energy, but does not influence the resonance fields where the splitting between the levels with different m_S and the same m_I matches the energy of the mw quantum (Fig. 1b).

More generally, the higher sensitivity of ESR experiments can be used for the detection of NMR frequencies by applying both resonant mw and resonant radio frequency (rf) irradiation to the spin system. Such electron nuclear double-resonance (ENDOR) experiments are discussed in Chapter 2.

Transition metal ions can have several unpaired electrons when they are in their high-spin state; examples are Cr(III) ($3d^3$ configuration, $S = \frac{3}{2}$), Mn(II) ($3d^5$, $S = \frac{5}{2}$),

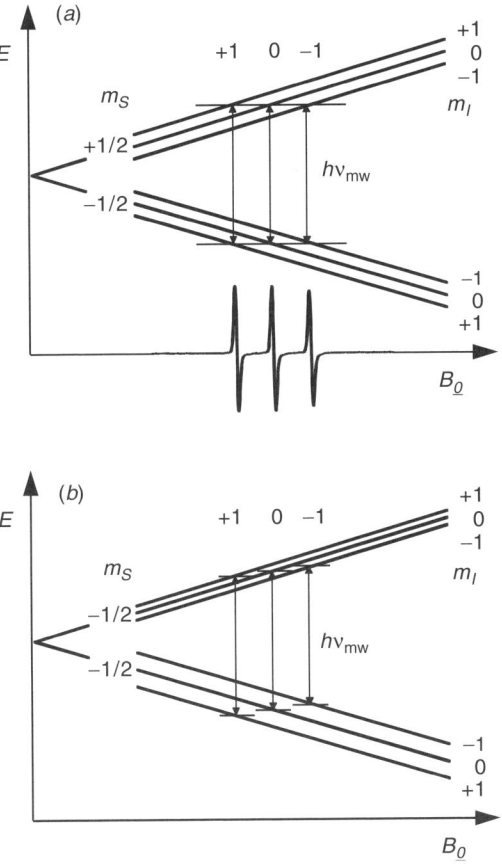

Fig. 1. Energy level schemes and ESR spectrum for a spin system of an electron spin $S = \frac{1}{2}$ coupled to a nuclear spin $I = 1$ (e.g., ^{14}N in a nitroxide). (*a*) Only the electron Zeeman and hyperfine interactions are considered. (*b*) The electron Zeeman, hyperfine, and nuclear Zeeman interactions are considered. Note that the splittings match the microwave quantum at the same resonance fields as in part a.

and Fe(III) ($3d^5$, $S = \frac{5}{2}$). The spins of these electrons are tightly coupled and have to be considered as a single group spin $S > \frac{1}{2}$. Such an electron group spin also has an electric quadrupole moment. For historical reasons, the electron spin analog of the nuclear quadrupole interaction is termed zero-field splitting (ZFS) and is described by Eq. 6,

$$\mathcal{H}_{ZFS} = h\,\mathbf{S}\,\mathbf{D}\,\mathbf{S} \qquad (6)$$

where $\mathbf{D}$ is a traceless tensor. Therefore, the ZFS can be characterized by two parameters, $D = 3D_z/2$ and $E = (D_x - D_y)/2$, rather than by giving all three principal values. For axial symmetry $E = 0$, and for maximum nonaxiality $E = D/3$.

With the exception of transition metal ions at a site with cubic symmetry, the ZFS often exceeds the electron Zeeman interaction at magnetic fields <1 T, sometimes even at the highest accessible fields (high-spin Fe(III)). In this situation, only the lowest lying doublet of spin states may be populated and only transitions within this doublet can be observed. It is convenient to describe such a doublet by an *effective spin* $S' = \frac{1}{2}$. The ZFS of the group spin $S > \frac{1}{2}$ then contributes to the effective g-tensor of the spin $S' = \frac{1}{2}$. For example, X-band ESR spectra of high-spin Fe(III) in a situation with maximum nonaxiality of the ZFS ($E = D/3$) exhibit a sharp feature at $g = 4.3$. Note that unlike the normal g-tensor, the effective g-tensor may depend on the applied magnetic field.

For low concentrations of the paramagnetic centers, the electron spins can be considered isolated from each other, and only a single electron spin S appears in the Hamiltonian. In systems with a high concentration of paramagnetic transition metal ions, this situation can be achieved by diamagnetic dilution with transition ions of the same charge and similar radius and coordination chemistry. However, there are a number of systems that feature coupled electron spins, for example, binuclear metal complexes and biradicals. Any pair of electron spins S_k and S_l in such a system interacts through space by dipole–dipole coupling, which is analogous to the dipolar part $\mathbf{T}$ of the hyperfine coupling. The Hamiltonian of the electronic dipole–dipole (DD) coupling is given by Eq. 7,

$$\mathcal{H}_{DD} = h\,\Sigma\,\mathbf{S}_k\,\mathbf{D}_{kl}\,\mathbf{S}_l \qquad (7)$$

where the $\mathbf{D}_{kl}$ are the traceless dipole–dipole tensors. If the two electron spins are far apart, the coupling can be described by a point-dipole approximation in which $\mathbf{D}_{kl}$ is an axial tensor with principal values $D_{z,kl} = 2d$ and $D_{x,kl} = D_{y,kl} = -d$. As d is inversely proportional to the cube of the distance r_{kl} between the two spins, a measurement of this coupling can thus yield the spin–spin distance. Such measurements are discussed in more detail in Chapter 2.

The two electrons can exchange if their wave functions overlap. Even for localized electrons, such an exchange is significant at a distance $r_{kl} < 1.5$ nm. For an antibonding overlap of the two orbitals, the exchange interaction J is negative and the triplet state of the pair has lower energy than the singlet state. This is called a ferromagnetic exchange coupling. Consequently, bonding overlap leads to a positive J, a lower lying singlet state, and antiferromagnetic coupling. The exchange coupling is not strictly isotropic, but except for electron spins at distances < 0.5 nm, the anisotropic contribution can usually be neglected. For a purely isotropic exchange coupling, the Hamiltonian is written in Eq. 8.

$$\mathcal{H}_{ex} = h \, \Sigma J_{kl} S_k S_l \tag{8}$$

Unlike the dipole–dipole coupling between the electron spins, the exchange coupling can thus be detected in fluid solutions.

The ESR spectra of monoradicals and mononuclear transition ion complexes can also be influenced by spin exchange, because the wave functions of the electrons overlap for a short time during diffusional collisions of paramagnetic species.[24] At moderate concentrations (1 M or larger), the collisions are so frequent that line broadening and a decrease of the hyperfine splitting can be observed. In macromolecular and supramolecular systems, this effect is sometimes perceptible at lower bulk concentrations, as diffusion may be restricted or local concentrations of some species strongly exceed their bulk concentration. Examples are discussed in Chapter 7.

When the various spin interactions can be separated experimentally or by spectral analysis, ESR spectra become a rich source of information not only on chemical structure of the paramagnetic species, but also on the structure and dynamics of their environment. Figure 2 provides an overview of time scales and length scales that can be accessed in this way. T_1 and T_2 are the longitudinal and transverse relaxation times, respectively.

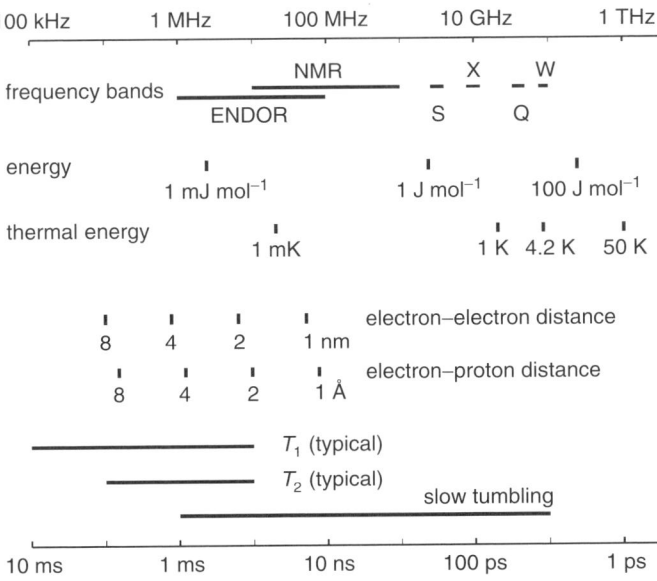

Fig. 2. Frequencies, time scales, energies, and length scales in ESR experiments.

2.2. Anisotropic Hyperfine Interaction and *g*-Tensor

Before considering the analysis of anisotropic solid-state ESR spectra in general, we discuss the orientation dependence of spin interactions of the nitroxide radical as an example. The ESR spectrum of a nitroxide is dominated by the hyperfine interaction of the electron spin with the nuclear spin of the ^{14}N atom and by *g*-shifts due to spin–orbit coupling mainly in the $2p_z$ orbital of the lone pair on the oxygen atom. The ^{14}N hyperfine coupling contains a sizeable isotropic contribution due to Fermi contact interaction in the 2*s* orbital on the nitrogen. An anisotropic contribution comes from the spin density in the nitrogen $2p_z$ orbital whose lobes are displayed in Fig. 3*a*. If the external magnetic field B_0 is parallel to these lobes (*z* axis of the molecular frame), the hyperfine interaction and thus the splitting within the triplet is large; if it is perpendicular to the lobes, the splitting is small. Conversely, *g*-shifts are small when the lobes of the orbital under consideration (here the $2p_z$ orbital on the oxygen) are parallel to the field and large when they are perpendicular. In the case of a nitroxide, the strongest shift is observed when the field is parallel to the N–O bond, which defines the *x* axis of the molecular frame. Hence, the triplets of lines at different orientations of the molecule with respect to the field do not only have different splittings, but their centers are also shifted with respect to each other.

In a macroscopically isotropic sample (all molecular orientations have the same probability), the spectrum consists of contributions from all orientations when the rotational motion is frozen on the time scale of the experiment. As ESR lines are derivative absorption lines, negative and positive contributions from neighboring orientations cancel. Powder spectra are thus dominated by contributions at the minimum and maximum resonance fields, and by contributions at resonance fields that are common to many spins. The latter contribution provides the center line in the nitroxide powder spectrum (Fig. 3*b*). It corresponds mainly to molecules with nuclear magnetic quantum number $m_I = 0$ (center line of all triplets, only *g*-shift). The detailed shape of this powder spectrum can be simulated, but interpretation is not easy, mainly because hyperfine and *g* anisotropy are of similar magnitude.

If one of the two interactions dominates, the spectra can be analyzed more easily. For dominating *g* anisotropy (Fig. 4*a*), signals in the CW ESR spectrum are observed at resonant fields corresponding to the principal values of the *g*- tensor: g_z (low-field edge), g_y, and g_x (high-field edge). For a *g*-tensor with axial symmetry (wave function of the unpaired electron has at least one symmetry axis C_n with $n \geq 3$), the intermediate feature coincides with one of the edges (Fig. 4*b*). For a dominating hyperfine interaction with a nuclear spin $I = \frac{1}{2}$ the spectrum consists of two of these powder patterns with mirror symmetry about the center of the spectrum (Fig. 4*c*).

When samples are available as single crystals, spectra corresponding to specific orientations of the paramagnetic center with respect to the external field can be measured separately. The orientation dependence of the spectrum can then be studied systematically and the principal axes frames of the *A*- and *g*-tensors can be related to the crystal frame. In polymer applications, samples are usually macroscopically isotropic, so that only the principal values of the interactions, and in favorable cases the *relative* orientations of their principal axes frames, can be obtained from spectral simulations. How these frames are related to the molecular geometry then needs to be

FUNDAMENTALS OF ELECTRON SPIN RESONANCE SPECTROSCOPY

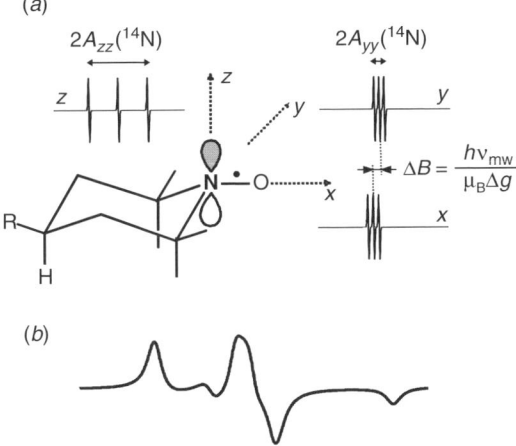

Fig. 3. Anisotropic interactions for a nitroxide radical. (*a*) Molecular frame of the nitroxide molecule and single-crystal ESR spectra along the principal axes of this frame. (*b*) Powder spectrum resulting from a superposition of the single-crystal spectra at all orientations of the molecule with respect to the external magnetic field.

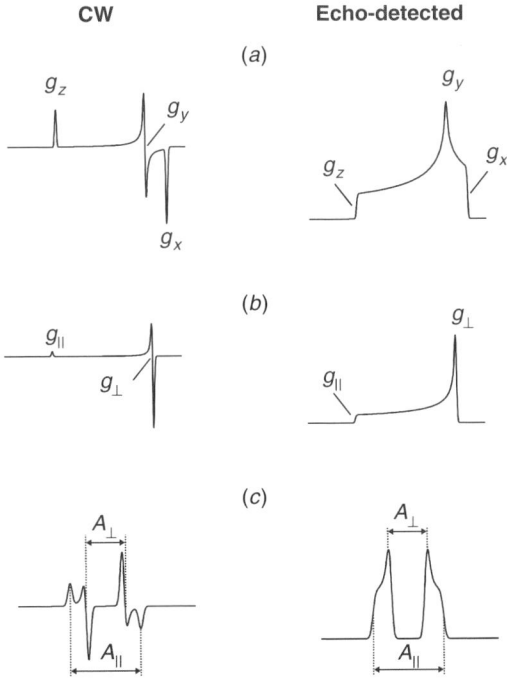

Fig. 4. Powder line shapes in continuous wave (CW) ESR (derivative absorption spectra) and echo-detected ESR (absorption spectra). (*a*) Rhombic g-tensor. (*b*) Axial g-tensor. (*c*) Axial hyperfine coupling tensor with dominating isotropic contribution.

established by theoretical considerations or by quantum chemical computations of the interaction tensors.

2.3. Isotropic Hyperfine Analysis

Anisotropic line broadening in solids often leads to a situation in which only one dominant hyperfine interaction is resolved, the one for the atom at which the spin is localized. In fluid media, however, anisotropic contributions average, lines are narrower, and a multitude of hyperfine interactions may be resolved. This situation is frequently observed for proton couplings in π radicals, where the electron spin is distributed throughout a network of conjugated bonds. Examples can be found in Ref. 23.

In isotropic ESR spectra, a single nucleus with spin I_k causes a splitting into $2I_k + 1$ lines corresponding to the magnetic quantum numbers $m_I = -I_k, -I_k + 1, \ldots I_k$. For a group of n_k equivalent nuclei (same isotropic hyperfine coupling), the number of lines is $2n_k I_k + 1$. For groups of nonequivalent spins, the number of lines (multiplicities) increases, and the total number of lines in the ESR spectrum is given in Eq. 9.

$$N_{\text{ESR}} = \prod (2n_k I_k + 1) \tag{9}$$

An example is shown in Fig. 5, where the spectrum for an electron spin coupled to four protons ($I = \frac{1}{2}$) exhibits a regular pattern of 16 lines. In complicated spectra consisting of multiple interacting nuclei, some of the smaller hyperfine couplings cannot be resolved. In such cases, ENDOR spectra are often easier to interpret, because each proton contributes only two lines; this technique is described in Chapter 2.

2.4. Environmental Effects on g- and Hyperfine Interaction

Self-assembly of polymer chains is due to noncovalent interactions: hydrogen bonding, π stacking, and electrostatic and van der Waals interactions. The high sensitivity of the NMR chemical shift of protons to π stacking (through ring currents) and hydrogen bonding provides one way for their characterization.[25] Since the magnetic

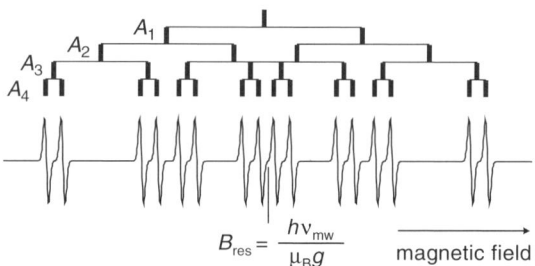

Fig. 5. Isotropic ESR spectrum for a system consisting of four nuclear spins $I_k = \frac{1}{2}$ coupled to a single electron spin $S = \frac{1}{2}$.

parameters of paramagnetic probes are also sensitive to such interactions, ESR spectroscopy can confirm and complement the information obtained by NMR.

The hyperfine interaction is influenced by any environmental effect that can perturb the spin density distribution. For example, in nitroxide radicals the unpaired electron is distributed between the nitrogen ($\approx 40\%$) and oxygen atom ($\approx 60\%$) in the polar N–O bond (Fig. 6). This distribution can change in the vicinity of a polar molecule (polar solvent or ion). Generally, a more polar solvent (higher dielectric constant) leads to a higher spin density ρ_N on the nitrogen atom and thus to a larger observed hyperfine coupling.[26] The spin density distribution is also influenced by hydrogen bonding to the oxygen atom, which also increases the hyperfine coupling.

The same interactions affect the deviation of g_x from the free electron value g_e, but in the opposite direction, since the extent of spin–orbit coupling is proportional to the spin density ρ_O on the oxygen atom. However, the effect on g_x also depends on the lone-pair energy, whose lowering causes stronger spin–orbit coupling. The lone-pair energy in turn is more affected by hydrogen bonding than by the local polarity, so that compared to A_z, g_x is more sensitive to hydrogen bonding than to polarity. Correlation of g_x to A_z thus enable the separation of polarity and hydrogen-bonding effects.[26] In principle, the same effects scaled by a factor of one-third can be seen in the isotropic values A_{iso} and g_{iso}, as the other principal values of the tensors are much less affected. As a rule, measurements of A_z and of g_x in solid samples at high field (W band) are much more precise than measurements of A_{iso} and g_{iso} at X-band frequencies.

2.5. Accessibility to Paramagnetic Quenchers

Spin exchange due to collision of paramagnetic species (see Section 2.1) can be used to check whether a spin-labeled site in a macromolecule is accessible by the solvent. To this end, a paramagnetic quencher is added to the solvent, and the effect on the spectrum or relaxation time of the spin label is measured. The quencher is a fast relaxing paramagnetic species, usually a molecule or transition ion complex with spin $S > \frac{1}{2}$. The situation is illustrated in Fig. 7 for oxygen as the quencher ($S = 1$, triplet ground state), which is soluble in nonpolar solvents and only moderately soluble in water. We can assume, without loss of generality, that at a certain time oxygen is in the T_{-1} triplet

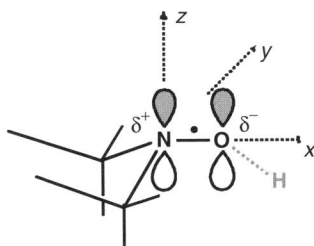

Fig. 6. Effects of the local polarity and hydrogen bonding on the nitroxide radical. The distribution of the unpaired electron between the two $2p_z$ orbitals on nitrogen and oxygen is affected.

substate and the nitroxide label is in the α state (spin up), which is the excited spin state for an electron (Fig. 7a). The two molecules diffuse and collide at a later time (Fig. 7b). Due to overlap of the wave functions, the three unpaired electrons become indistinguishable. Hence, when the two molecules separate again, there is a two-third's probability that the nitroxide is now with an unpaired electron in the β spin (spin down) and the oxygen molecule is in the T_0 state (Fig. 7c). Effectively, the collision with the quencher has thus relaxed the nitroxide from its spin excited state to the spin ground state. This corresponds to longitudinal relaxation. If longitudinal relaxation of the quencher is sufficiently fast and collisions are sufficiently frequent, the longitudinal relaxation time T_1 of the nitroxide is thus shortened. Indeed, the transverse relaxation time T_2 is also shortened, although this cannot be understood in such a simple picture. Collisions with a paramagnetic quencher thus lead to line broadening and faster longitudinal relaxation.

The shortening of T_1 is not directly visible in the ESR spectrum, but can be detected by saturation measurements with better sensitivity and higher precision than the shortening of T_2. In such CW ESR saturation measurements, the spectra are recorded as a function of mw power both in the presence and in the absence of the quencher. For nitroxides, a fit of the power dependence of the amplitude of the central line by a theoretical expression yields the parameter $P_{1/2}$, which is the power where the amplitude is reduced to one-half its value in the absence of saturation.[27] The difference of $\Delta P_{1/2}$ values in the presence and absence of quencher is a measure for the accessibility of the spin label by the quencher. Normalization to the width of the central line and to the half

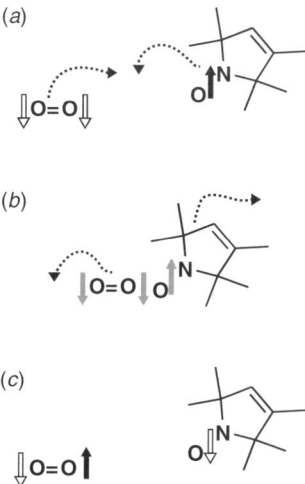

Fig. 7. Electron spin relaxation due to collision with a paramagnetic quencher. (a) An oxygen molecule in its T^{-1} state and a nitroxide with electron spin up are diffusing toward each other. (b) The two molecules collide and the three electrons are no longer distinguishable. (c) The two molecules have diffused apart after exchanging one electron. The oxygen molecule is now in its T^0 state, while the nitroxide has spin down.

saturation power of a standard sample, such as diphenyl picrylhydrazyl (DPPH), yields a dimensionless accessibility parameter. Accessibility to nonpolar solvents can be tested by saturating the solution with nitrogen (no quencher) and air (20% oxygen), while accessibility to polar solvents, such as water, can be tested with chromium(III)oxalate.

2.6. Line Shape Analysis for Tumbling Nitroxide Radicals

The mobility of a spin probe depends on the local viscosity (microviscosity) and on its connectivity to a larger, more immobile object. For spin labels, the mobility depends on the flexibility of the tether connecting it to the backbone, and on tumbling of the macromolecule as a whole. The mobility can be quantified by the rotational correlation time τ_r, which corresponds to the typical time during which a molecule maintains its spatial orientation. If the inverse of τ_r is of the same order of magnitude as the anisotropy of an interaction, this anisotropy is partially averaged and the ESR spectrum depends strongly on τ_r and on specific dynamics, such as the preference for a particular rotational axis or restrictions on the motion. For nitroxides at X-band, the ESR spectrum is dominated by the hyperfine anisotropy of ≈ 150 MHz. The largest effects are thus observed on time scales of a few nanoseconds, as illustrated in Fig. 8.

For rotational correlation times < 10 ps, the nitroxide spectrum consists of three lines with equal widths and amplitudes (fast limit), and no information on τ_r can be inferred from such spectra. For τ_r in the range 10 ps–1 ns, the transverse relaxation and thus the line width are dominated by effects of rotational motion.[28] The spectrum still consists of three derivative Lorentzian lines, but they now have different amplitudes and widths. In this regime, the rotational correlation time can be inferred from the ratio of the line amplitudes.[17] In the range 1–10 ns, spectra are best analyzed by simulations. At even longer rotational correlation times, the anisotropy is only moderately reduced by motion and the spectrum is basically a powder spectrum with slightly reduced outer extrema separation $2A'_{zz}$ (see spectrum at $\tau_r = 32$ ns in Fig. 8). If the outer-extrema separation $2A_{zz}$ in the rigid limit and the isotropic hyperfine coupling are known, for example, from measurements at very low and very high temperature, $\tau_r > 3$ ns can be estimated with good precision from the relative anisotropy, A_{rel}:

$$A_{rel} = (2A'_{zz} - 2A_{iso})/(2A_{zz} + 2A_{iso}) \qquad (10)$$

A test for linearity in an Arrhenius plot of $-\log(\tau_r)$ versus the inverse temperature reveals whether the dynamical process is an activated one.

For comparing dynamics in a series of materials, it is commonplace to plot the dependence of $2A'_{zz}$ versus T rather than computing τ_r. Such plots have a roughly sigmoidal shape (Fig. 9), with a maximum negative derivative close to $2A'_{zz} = 50$ G that corresponds to a rotational correlation time of ≈ 4 ns. The corresponding temperature T_{50G} (or T_{5mT}) is sometimes called ESR glass transition temperature (for a more detailed discussion, see Chapter 7).

Nitroxide radicals with $\tau_r < 4$ ns thus give a liquid-type spectrum and are considered mobile (or fast), while nitroxide radicals with $\tau_r > 4$ ns give a solid-type spectrum and are considered immobile (or slow). Polymers often exhibit distributions

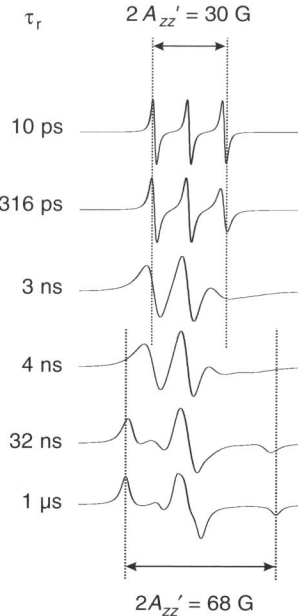

Fig. 8. Simulated nitroxide spectra at different rotational correlation times τ_r. In the fast limit ($\tau_r = 10$ ps) the outer-extrema splitting $2A'_{zz}$ is approximately twice the isotropic hyperfine coupling; in the rigid limit ($\tau_r = 10$ μs) it is twice the hyperfine coupling along the molecular z axis (see Fig. 3).

of correlation times, so that the spectrum may contain both fast and slow components. Simulations show that the presence of two components in the spectra can be observed even for broad monomodal distributions of τ_r, but in many cases it is due to genuinely bimodal distributions. This case is illustrated in Fig. 10 for a nitroxide radical in heterophasic poly(acrylonitrile–butadiene–styrene) (ABS); the fast and slow components in the ESR spectrum measured at 300 K are indicated, and represent radicals in butadiene-rich and acrylonitrile–styrene-rich domains, respectively; details will be described in Chapter 9.

3. MULTIFREQUENCY AND HIGH-FIELD ESR

Interpretation of solid-state ESR spectra may be difficult if several interactions in the Hamiltonian are of the same order of magnitude. Similarly, the spectrum of a tumbling nitroxide radical can often be reproduced by different motional models. In such cases, it may be impossible to analyze an ESR spectrum in an unambiguous way.

The problem can be overcome by measuring the spectrum not only at the standard frequency of ≈9.4 GHz (X band), where samples are most conveniently sized and spectrometers most available, but also at additional frequencies. For most organic

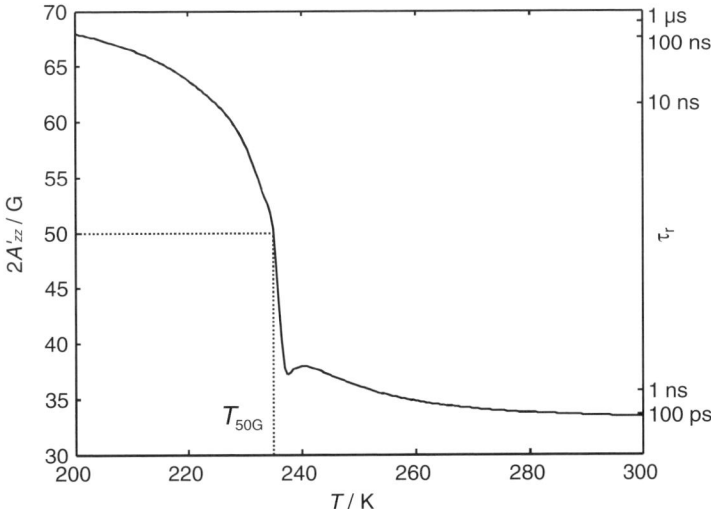

Fig. 9. Dependence of the outer-extrema splitting $2A'_{zz}$ in nitroxide spectra on temperature (simulation for an activated process with activation energy of 40 kJ mol^{-1}). At the temperature T_{50G}, where $2A'_{zz} = 50$ G, the correlation time matches the inverse anisotropy of the spectrum.

radicals, the g resolution is at best mediocre at X band, and measurements at higher frequencies, such as Q band (35 GHz) and W band (95 GHz) are advantageous. Increasing the frequency is also useful for studies on nitroxide dynamics, since the g-tensor has lower symmetry than the hyperfine tensor. High-field (high-frequency) spectra therefore discriminate more strongly between different motional models. Even for transition metal complexes, frequencies > 10 GHz may be advantageous if a small nonaxiality of the g-tensor has to be resolved. For spins $S > \frac{1}{2}$ with relatively small ZFS, lines may become narrower at higher fields, since second-order broadening of the $m_S = -\frac{1}{2} \leftrightarrow \frac{1}{2}$ transitions due to the ZFS decreases with increasing electron Zeeman interaction; this effect is prominent for Mn(II) complexes.

More advanced experiments, such as ENDOR, electron spin echo envelope modulation (ESEEM), or relaxation measurements by pulsed ESR rely on a selective excitation of spins close to the resonance field. Usually, the powder ESR spectrum is much broader than the excitation bandwidth of the pulses, which is in the range between 2 and 10 G. In cases where one anisotropic interaction dominates the spectrum, the experiments thus select contributions only from certain orientations of the molecule with respect to the external magnetic field. Such *orientation selection* is more efficient and easier to interpret at a field that is high enough for the g anisotropy to dominate. Finally, the size of mw resonators scales with wavelength and thus scales inversely with frequency. At higher frequency, spectra can thus be measured with much smaller sample volumes, yet the concentration does not need to be significantly increased.

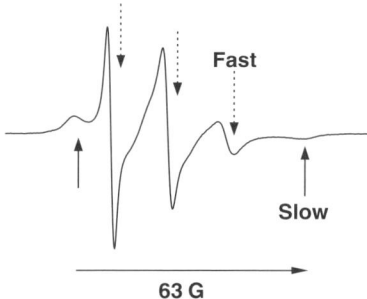

Fig. 10. X-band ESR spectrum at 300 K of a nitroxide radical derived from Tinuvin 770, a hindered amine stabilizer (HAS), in heterophasic ABS. Fast and slow components are indicated. The extreme separation of the slow component is 63 G.

In the case of transition metal complexes with large g anisotropy in disordered matrices, mw frequencies < 9.4 GHz are sometimes preferable, because local heterogeneities (strain) of the matrix lead to a distribution of the principal values of the g- and A-tensors (g- and A-strain) and thus to field-dependent line broadening. Such a situation is illustrated in Fig. 11 for ^{63}Cu(II) in Nafion perfluorinated ionomers swollen by acetonitrile:[29] the line width of the parallel components was measured at four mw frequencies in the range 1.2–9.4 GHz, and the narrowest line widths were detected for the two low-field lines of the parallel quartet at C band (4.7 GHz) and L band (1.2 GHz). In this way, clear superhyperfine splittings from ^{14}N nuclei were resolved, in addition of course to the hyperfine splittings from ^{63}Cu(II).

Solving a problem by ESR spectroscopy may thus sometimes require access to spectrometers at several different frequencies, and in particular, to a high-frequency spectrometer. That said, it is good practice to first gather as much information as possible with the simplest technique, which is CW ESR at X band. After this step, it should be decided whether more information is required and how it can best be obtained.

4. PULSED ESR METHODS

Continuous wave ESR is highly sensitive, applicable to most paramagnetic centers in a wide temperature range, and can be measured with relatively inexpensive spectrometers. However, quite often analysis of CW ESR spectra provides information only on one or two dominating interactions. Relaxation can be characterized to some extent by studying saturation of the spectrum at higher microwave power, but results are often only semiquantitative, as different contributions to spin relaxation cannot be separated. More information can be obtained by magnetic resonance experiments if pulsed instead of continuous irradiation is used, as demonstrated by the development of nuclear magnetic resonance (NMR) spectroscopy since the 1970s. The situation is somewhat less favorable in ESR spectroscopy, since in contrast to rf pulses in NMR,

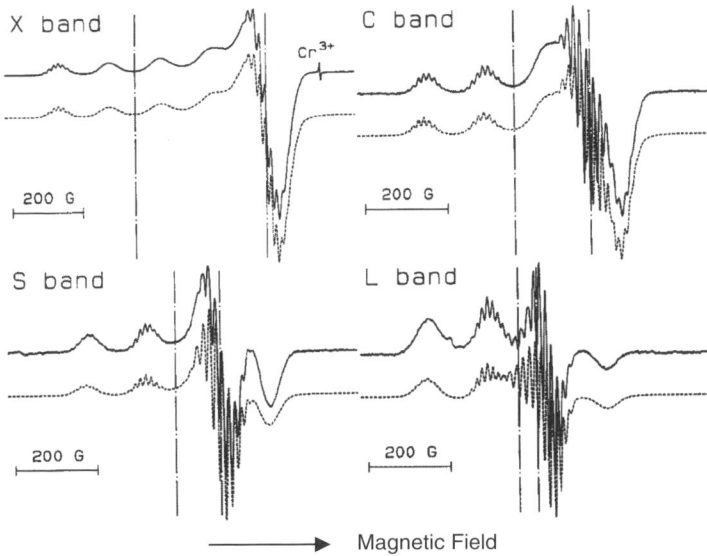

Fig. 11. Experimental (solid lines) and simulated (broken line) ESR spectra of ^{63}Cu(II) in Nafion perfluorinated membranes soaked by CH$_3$CN at the X band (9.36 GHz) and 110 K, and at the C band (4.7 GHz), S band (2.8 GHz), and L band (1.2 GHz) at 123. Vertical dashed lines indicate the position of g ∥. Note the clearly visible superhyperfine splittings from nitrogen ligands at the lower that X-band frequencies.

mw pulses cannot usually excite the entire spectrum at once. For this reason, pulsed ESR is somewhat less sensitive than CW ESR for many samples and manipulation of the spin dynamics is somewhat less effective than in pulsed NMR. Nevertheless, pulsed ESR can be applied to most samples of interest and allows for a better separation of different interactions in the spin Hamiltonian, or the detection of different types of spin relaxation mechanisms, compared with CW ESR.[8]

Separation of interactions allows for precise measurements of the small interactions of the observed electron spin with remote spins in the presence of line broadening due to larger contributions. Such techniques are therefore most useful for solid materials or soft matter, where ESR spectra are usually poorly resolved. The most selective techniques for isolating one type of interaction from all the others are pulsed double resonance experiments, such as ENDOR and electron–electron double resonance (ELDOR), which are discussed in more detail in Chapter 2. If the hyperfine couplings are of the same order of magnitude as the nuclear Zeeman frequency, ESEEM techniques may provide higher sensitivity than ENDOR techniques. In particular, the two-dimensional hyperfine sublevel correlation (HYSCORE) experiment provides additional information that aids in the assignment of ESEEM spectra. These experiments are also discussed in Chapter 2.

The separation of different contributions to spin relaxation relies on echo experiments.[30] Spin echoes are also the basis for almost all other pulsed ESR experiments in the solid-state and in soft matter, since the free induction signal induced by a single pulse usually decays within a time that is shorter than the receiver deadtime after that pulse. The simplest echo experiment is the two-pulse or Hahn echo experiment (Fig. 12), which consists of a first pulse with flip angle $\pi/2$, a delay τ, and a second pulse with flip angle π. The first pulse converts the longitudinal magnetization of the spins that exists in thermal equilibrium to transverse magnetization. Initially, the contributions by all spins are in phase (coherent), but as different spins have different resonance offsets Ω_S, they acquire a different phase $\phi = \Omega_S \tau$ during time τ and the signal thus vanishes. Additionally, magnetization within each packet of spins with equal resonance frequency decays by transverse relaxation with time constant T_2. The π pulse inverts the phase of each spin packet, which thus has the value $-\phi$ immediately after that pulse. Within another delay τ, each spin packet again acquires a phase ϕ. This exactly cancels the phase differences, so that at time 2τ all spin packets are again coherent. This coherence corresponds to observable transverse magnetization, which is called a spin–echo signal. After time 2τ, the signal is a replica of the unobservable free induction decay (FID) signal after the first pulse, except for an attenuation of the total amplitude by a factor $\exp(-2\tau/T_2)$. By measuring the echo amplitude as a function of τ (two-pulse echo decay), T_2 can be determined.

If the formally forbidden electron–nuclear transitions are weakly allowed, the two-pulse echo decay is modulated by the corresponding nuclear frequencies. For a spin system of two weakly coupled electron spins, it is modulated with the coupling between the two spins. Measurement of the echo amplitude as a function of the external magnetic field B_0 yields the absorption ESR line shape. This field-swept echo-detected ESR experiment is a useful alternative to CW ESR for systems with strong anisotropic line broadening. For example, in the situation in Fig. 4b the $g_\parallel$ feature can be easily missed, in particular if it is broadened by g strain. The strong anisotropy is then revealed more clearly in the absorption line.

The longitudinal relaxation time T_1 can be measured with the inversion recovery experiment that consists of a mw π pulse, a variable delay T, and a two-pulse echo

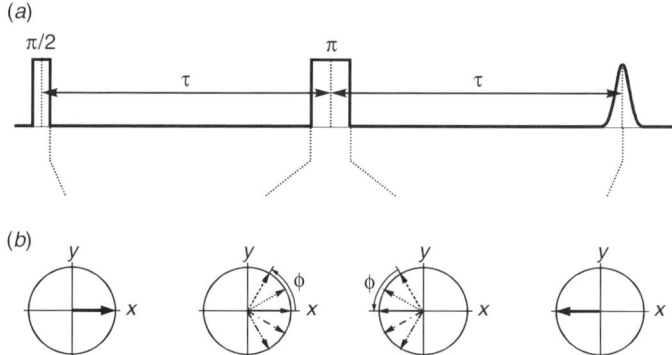

Fig. 12. Two-pulse echo experiment. (*a*) Pulse sequence. (*b*) Evolution of the magnetization vectors corresponding to spin packets with difference resonance offsets Ω_S.

sequence with fixed delay τ. The first π pulse inverts the longitudinal thermal equilibrium magnetization M_0 to $-M_0$. During time T the longitudinal magnetization again relaxes toward M_0 with time constant T_1. At the time of the $\pi/2$ pulse of the echo subsequence, the longitudinal magnetization is thus given by $[1-2\exp(-T/T_1)]M_0$. As only this longitudinal magnetization contributes to the echo experiment, the amplitude of the echo signal as a function of T is therefore proportional to $1-2\exp(-T/T_1)$. The inversion recovery experiment may be affected by spectral diffusion: changes in the resonance frequency of the observed spins during delay time T. Such changes may result from reorientation of the molecules. If a paramagnetic center is not excited by the inversion pulse, changes its resonance frequency, and is then excited by the echo subsequence, it does not need to relax to contribute to the echo signal. To avoid this, the inversion pulse should have an excitation bandwidth that is larger than possible frequency changes by spectral diffusion. Alternatively, one can use a saturating pulse that is longer than the maximum delay time T_{max}. Such a pulse excites all spins that are accessible by spectral diffusion within the time scale of the experiment. In this saturation recovery experiment, the echo amplitude is zero at $T = 0$ and increases as $1-\exp(-T/T_1)$.

On the other hand, spectral diffusion may be the process of interest, as it is directly related to the dynamics of the paramagnetic centers. Spectral diffusion can be separated from longitudinal relaxation by first measuring T_1 using the saturation recovery technique, and then measuring the decay of the stimulated echo with time T (Fig. 13), which is much more sensitive to spectral diffusion. As the two-pulse echo, the stimulated echo experiment starts with a $\pi/2$ pulse that generates transverse magnetization and a subsequent delay τ during which the magnetization acquires phase $\phi = \Omega_S\tau$. However, at this

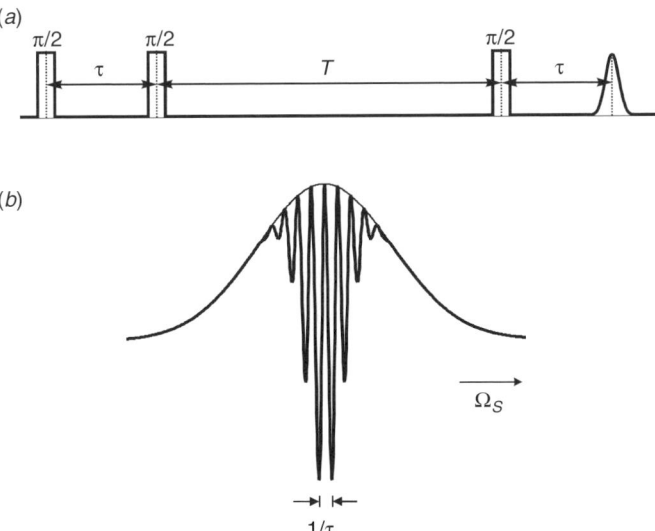

Fig. 13. Stimulated echo experiment. (*a*) Pulse sequence. (*b*) Polarization grating created by the first two $\pi/2$ pulses with interpulse delay τ in a Gaussian ESR line (simulation).

point a $\pi/2$ pulse is applied instead of the π pulse of the two-pulse echo sequence. The $\pi/2$ pulse converts transverse magnetization with zero phase ($+x$) to negative longitudinal magnetization ($-z$), it does not influence magnetization with phase $+y$ ($\phi = 90°$) or $-y$ ($\phi = 270°$), and it converts magnetization with phase $-x$ ($\phi = 180°$) to positive longitudinal magnetization ($+z$). As the magnetization before this pulse is equally distributed over the xy plane, only part of it is transferred to longitudinal magnetization. The remaining transverse magnetization decays much faster and does not contribute to the stimulated echo. If necessary, it can be eliminated by phase cycling of the pulses.[8] The longitudinal magnetization after the second $\pi/2$ pulse is described by $\cos(\Omega_S \tau)$. By considering the limited excitation bandwidth of the pulses, this corresponds to a polarization grating as shown in Fig. 12b. During the following variable delay of duration T, the grating decays with time constant T_1 due to longitudinal relaxation. In addition, changes in the resonance frequency of spin packets lead to exchange of polarization along the Ω_S axis, that is, to a smearing of the grating. In the limit of much faster spectral diffusion compared to longitudinal relaxation, the grating is transformed to a broad unstructured hole in the ESR line that resembles the excitation profile of the $\pi/2$ pulses.

The final $\pi/2$ pulse transforms the longitudinal magnetization (polarization) to transverse magnetization. The subsequently detected signal can be considered as an FID of the polarization pattern. While the FID of a broad unstructured hole decays within the dead time after the pulse and cannot be observed, the FID of the polarization grating has the form of the Fourier transform of this grating. Since an oscillation with period $1/\tau$ in angular frequency domain transforms to a delta peak at time τ in time domain, this FID appears as an echo at time τ after the last $\pi/2$ pulse. As a function of delay T, the amplitude of this echo decays with $\exp(-T/T_1)$, but is additionally attenuated by spectral diffusion. The contribution by spectral diffusion can be easily recognized even if T_1 is not known a priori, since the decay by spectral diffusion is faster for finer gratings, for longer interpulse delays τ.

Additional pulsed ESR experiments have been used, which are beyond the scope of this introductory chapter. An overview of these experiments, as well as on the theoretical background of pulsed ESR, can be found in Ref. 8.

ACKNOWLEDGMENTS

G. Jeschke gratefully acknowledges financial support by a Dozentenstipendium of Fonds der Chemischen Industrie. Research in the laboratory of S. Schlick is currently supported by grants from the Polymer Program of the National Science Foundation, the University Research Program of the Ford Motor Company, and the Fuel Cell Activity Center of General Motors.

REFERENCES

1. Carrington, A.; McLachlan, A.D. *Introduction to Magnetic Resonance, with Applications to Chemistry and Chemical Physics*, Harper & Row: New York, 1967.
2. Alger, R.S. *Electron Paramagnetic Resonance: Techniques and Applications*, Wiley-Interscience: New York, 1968.

REFERENCES

3. Abragam, A.; Bleaney, B. *Electron Paramagnetic Resonance of Transition Ions*, Clarendon: Oxford, UK, 1970.
4. Poole, C.P., Jr. *Electron Spin Resonance: A Comprehensive Treatise on Experimental Techniques*, 2nd ed., John Wiley & Sons, Inc.: New York, 1983.
5. Pilbrow, J.R. *Transition Ion Electron Paramagnetic Resonance*, Clarendon: Oxford, UK, 1990.
6. *Modern Pulsed and Continuous-Wave Electron Spin Resonance*, Kevan, L., Bowman, M.K., Eds.; John Wiley & Sons, Inc.: New York, 1990.
7. Weil, J.A.; Bolton, J.R.; Wertz, J.E. *Electron Paramagnetic Resonance: Elementary Theory and Practical Applications*; John Wiley & Sons, Inc.: New York, 1994.
8. Schweiger, A.; Jeschke, G. *Principles of Pulse Electron Paramagnetic Resonance*, Clarendon: Oxford, UK, 2001.
9. Wasserman, A.M. In *Specialist Periodical Reports — Electron Spin Resonance*; Gilbert, B.C., Davies M.J., Murphy D.M., Eds.; Royal Society of Chemistry: Cambridge, 1996; Vol. 15, pp. 115–152.
10. Goldfarb, D. In *Specialist Periodical Reports — Electron Spin Resonance*; Gilbert, B.C., Davies, M.J., Murphy, D.M., Eds.; Royal Society of Chemistry: Cambridge, 1996; Vol. 15, pp. 186–243.
11. Smirnov, A. In *Specialist Periodical Reports — Electron Spin Resonance*; Gilbert, B.C., Davies, M.J., Murphy, D.M., McLauchlan, K.A., Eds.; Royal Society of Chemistry: Cambridge, 2002; Vol.18, pp. 109–136.
12. Rånby, B.; Rabek, J.F. *ESR Spectroscopy in Polymer Research*, Springer-Verlag: Berlin, 1977.
13. *The Effects of Radiation on High-Technology Polymers.* Reichmanis, E., O'Donnell, J.H., Eds.; ACS: Washington, DC, 1989.
14. Hill, D.J.T.; Le, T.T.; O'Donnell, J.H.; Perera, M.C.S.; Pomery, P.J. In *Irradiation of Polymeric Materials: Processes, Mechanisms, and Application;* Reichmanis, E., Frank, C.W., O'Donnell, J.H., Eds.; ACS: Washington, DC, 1993.
15. Carswell, T.G.; Garrett, R.W.; Hill, D.J.T.; O'Donnell, J.H.; Pomery, P.J.; Winzor, C.L. In *Polymer Spectroscopy*; Fawcett, A.H., Ed.; Wiley: Chichester, UK, 1996; Chapt. 10, pp. 253–274.
16. *Macromolecule-Metal Complexes*; Ciradelli, F., Tsuchida, W., Wöhrle, D., Eds.; Springer: Berlin, 1996.
17. *Spin Labeling, Theory and Applications;* Berliner, L.J., Ed.; Academic Press: New York, 1976.
18. *Spin Labeling II, Theory and Applications;* Berliner, L.J., Ed.; Academic Press: New York, 1979.
19. *Biological Magnetic Resonance. Spin Labeling;* Berliner, L.J., Reuben, J., Eds.; Plenum: New York, 1989; Vol 8.
20. Motyakin, M.V.; Schlick, S. In *Instrumental Methods in Electron Magnetic Resonance, Biological Magnetic Resonance, Vol. 21;* Bender, C.J., Berliner, L.J., Eds.; Kluwer Academic/Plenum Publishing Corporation: New York, 2004; pp. 349–384.
21. *Molecular Motions in Polymers by E.S.R.*, Boyer, R.F., Keineth, S.E., Eds. Symposium Series Vol. 1; MMI Press: Harwood, Chur, 1980.
22. Cameron, G.G.; Davidson, I.G. In *Polymer Spectroscopy*; Fawcett, A.H., Ed.; John Wiley & Sons, Inc.: Chichester, UK, 1996; Chapt. 9, pp. 231–252.
23. Gerson, F.; Huber, W. *Electron Spin Resonance Spectroscopy of Organic Radicals*, Wiley-VCH: Weinheim, 2003.
24. Molin, Yu.N.; Salikhov, K.M.; Zamaraev, K.I. *Spin Exchange*, Springer: Berlin, 1980.

25. Spiess, H.W. *J. Polym. Sci. A* **2004**, *42*, 5031.
26. Owenius, R.; Engstrom, M.; Lindgren, M.; Huber, M. *J. Phys. Chem. A* **2001**, *105*, 10967.
27. Altenbach, C.; Greenhalgh, D.A.; Khorana, H.G., Hubbell, W.L. *Proc. Natl. Acad. Sci. USA* **1994**, *91*, 1667.
28. Nordio, P.L. In *Spin Labeling: Theory and Applications;* Berliner, L.J., Ed.; Academic Press: New York, 1976; Chapt. 2, pp. 5–52.
29. Bednarek, J.; Schlick, S. *J. Am. Chem. Soc.* **1991**, *113*, 3303.
30. Leporini, D.; Schädler, V.; Wiesner, U.; Spiess, H.W.; Jeschke, G. *J. Chem. Phys.* **2003**, *119*, 11829.

2

DOUBLE RESONANCE ESR METHODS

GUNNAR JESCHKE

Max Planck Institute for Polymer Research, Mainz, Germany

Contents

1. Introduction 26
2. Spin–Spin Couplings 27
   2.1. Dipole–Dipole Coupling 27
   2.2. Exchange Coupling 30
   2.3. Fermi Contact Interaction 30
   2.4. Spin Density in p and d Orbitals 31
3. Electron–Electron Double Resonance 31
   3.1. General Considerations 31
   3.2. Experimental Techniques 33
   3.3. Analogy to Form Factor and Structure Factor in Scattering 35
   3.4. Direct Computation of Distance Distributions 38
4. Electron–Nuclear Double Resonance 39
   4.1. General Considerations 39
   4.2. Experimental Techniques 40
   4.3. High-Field ENDOR 42
5. Electron Spin Echo Envelope Modulation 44
   5.1. Principles 44
   5.2. Experimental Techniques 44
   5.3. Data Analysis 47
6. Conclusions 49
Acknowledgments 50
References 50

Advanced ESR Methods in Polymer Research, edited by Shulamith Schlick.
Copyright © 2006 John Wiley & Sons, Inc.

1. INTRODUCTION

Electron spin resonance spectra provide information on the type of paramagnetic center: radical, transition metal ion, or crystal defect. If g-values and hyperfine splittings (liquid state) or g and hyperfine tensors (solid state) can be extracted, additional information is obtained on the molecular structure in the immediate vicinity of the atom(s) on which the spin is centered.[1] For strongly coupled paramagnetic centers, as, for example, in molecular magnets, such spectra may also contain information on dipole–dipole and exchange coupling between the centers. Finally, for spin probes with well-known ESR parameters, such as nitroxides, line shape analysis of continuous wave (CW) ESR spectra yields information on the rotational dynamics of the probe.[2] Electron spin resonance spectroscopy thus directly probes the vicinity of a paramagnetic center on the length scale of a few angstrom. Information on that length scale may, however, not be complete. Hyperfine couplings to nuclei in neighboring molecules are usually unresolved even if these molecules are in direct contact with the spin-bearing molecule.

For many applications in polymer science, intermolecular interactions and structural information on a somewhat longer range, up to a few nanometers, is of considerable interest. Length scales between 0.5 and 8 nm correspond to electron–nuclear or electron–electron couplings between 100 kHz and a few megahertz (MHz). Given that typical lifetimes of electron and nuclear spin states are longer than a few microseconds (µs), interactions of such a magnitude can, in principle, be measured. However, they are not resolved in ESR spectra, as there are too many of these interactions and, in solids, the lines are broadened due to anisotropy of the g-value and of the larger hyperfine couplings of nearby nuclei.

The long-range information is contained in weak couplings between distant spins. Such couplings are discussed in Section 2. They can be extracted by *separation of interactions*, that is, by techniques that detect a certain type of small interaction in the presence of larger interactions. The most important class of such techniques are double-resonance experiments. By electron–electron double resonance (ELDOR) it is possible to separate weak couplings between two electron spins from all other interactions. The accessible frequency range from 15 MHz down to 100 kHz corresponds to a distance range between 1.5 and 8 nm. Principles, experimental techniques, and data analysis for such ELDOR techniques are described in Section 3. By electron–nuclear double resonance (ENDOR) weak couplings between an electron spin and a nuclear spin can be measured (Section 4). The accessible frequency range is approximately the same. As such hyperfine couplings often have a Fermi contact contribution that is not easily related to spin–spin distances, it may be more difficult to extract precise structural information than it is for electron–electron couplings. However, in many cases even semiquantitative information is helpful. The Fermi contact contribution can usually be neglected for intermolecular hyperfine couplings. The hyperfine couplings are then purely dipolar, so that ENDOR directly provides distance information for supramolecular structures.

ENDOR techniques work rather poorly if the hyperfine interaction and the nuclear Zeeman interaction are of the same order of magnitude. In this situation, electron and nuclear spin states are mixed and formally forbidden transitions, in which both the electron and nuclear spin flip, become partially allowed. Oscillations with the frequency of nuclear transitions then show up in simple electron spin echo experiments. Although such electron spin echo envelope modulation (ESEEM) experiments are not strictly double-resonance techniques, they are treated in this chapter (Section 5) because of their close relation and complementarity to ENDOR. The ESEEM experiments allow for extensive manipulations of the nuclear spins and thus for a more detailed separation of interactions.[3] From the multitude of such experiments, we select here combination-peak ESEEM and hyperfine sublevel correlation spectroscopy (HYSCORE), which can separate the anisotropic dipole–dipole part of the hyperfine coupling from the isotropic Fermi contact interaction.

Double-resonance methods, such as ELDOR, can also be used to obtain information on the dynamics of paramagnetic species.[3] Such approaches are not considered in this chapter. Technical aspects and theory of CW ELDOR[4] and ENDOR[5] experiments will not be discussed, as pulsed techniques are nowadays more common, in particular for work on the highly viscous or solid systems that are typical for polymer research. Finally, this chapter is devoted exclusively to the description of the theoretical background and the concepts of double-resonance experiments. Applications are described in Chapter 7.

2. SPIN–SPIN COUPLINGS

2.1. Dipole–Dipole Coupling

The magnetic moments that are associated with electron and nuclear spins interact through space by the dipole–dipole coupling. This coupling is a pair interaction. Throughout this chapter we deal with experiments whose output data can be described as sums (ENDOR) or products (ELDOR, ESEEM) of pair contributions, which simplifies analysis tremendously. Furthermore, in all these experiments we can distinguish between an observer spin S and pumped spins I_i that are coupled to the observer spin. We may neglect the couplings of the pumped spins I_i among themselves. Therefore, we may restrict our general considerations to a spin pair of one observer spin S, which is always an electron spin, and one pumped spin I, which may be either an electron spin (ELDOR) or a nuclear spin (ENDOR, ESEEM).

All described experiments require that the electron Zeeman interaction of the electron spin S be much larger than all spin–spin couplings. Coupling terms containing S_x and S_y spin operators are thus negligible (nonsecular), as they act perpendicular to the quantization axis z. Furthermore, it is assumed that the g anisotropy is small or moderate, so that the quantization axis of the observer electron spin S coincides with the direction of the external magnetic field B_0. The Hamiltonian of the dipole–dipole (dd) interaction can then be written as in Eq. 1,

$$\hat{H}_{dd} = 2\pi v_{dd}[\hat{A} + \hat{B} + \hat{C}] \tag{1}$$

where the magnitude of the dipole–dipole interaction for two spins at a distance r from each other is quantified by the dipolar frequency.

$$|v_{dd}| = \frac{1}{r^3}\frac{\mu_0}{4\pi h}g\mu_B|g_I|\mu_I \tag{2}$$

The sign is positive if I is an electron spin and negative if I is a nuclear spin with positive nuclear g-value. In Eq. (2), μ_B is the Bohr magneton, g_I is the g-value of the pumped spin I, and μ_I is either the Bohr magneton (ELDOR) or the nuclear magneton (ENDOR, ESEEM). As all factors except for $1/r^3$ are fundamental constants, are known (nuclear g-values), or can be determined independently (electron g-values), the spin–spin distance can be computed directly from v_{dd}.

However, determination of v_{dd} from spectra is not trivial, as the terms below

$$\hat{A} = \hat{S}_z \hat{I}_z (1 - 3\cos^2 \theta) \tag{3a}$$

$$\hat{B} = -\frac{1}{2}(\hat{S}_x \hat{I}_x + \hat{S}_y \hat{I}_y)(1 - 3\cos^2 \theta) \tag{3b}$$

$$\hat{C} = -3\hat{S}_z \hat{I}_x \sin\theta \cos\theta \tag{3c}$$

depend on the angle θ between the spin–spin vector and the external magnetic field (Fig. 1), and the terms $\hat{B}$ and $\hat{C}$ may or may not influence the dipolar splittings. In the case of ELDOR, term $\hat{C}$ is always nonsecular and may be neglected. Only term $\hat{A}$ needs to be considered if the difference of the two microwave (mw) frequencies is much larger than v_{dd}. Under these experimental conditions, the difference between the resonance frequencies of the S and I spins in the absence of coupling must be

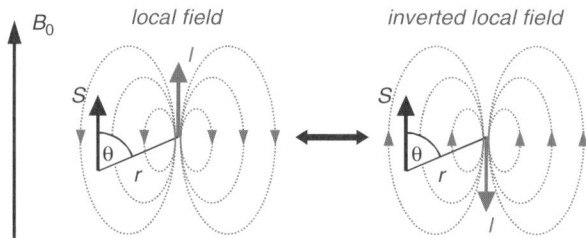

Fig. 1. Dipole–dipole coupling between two spins I and S. The local field imposed by the pumped spin I has a different sign for I being parallel (left) or antiparallel to the external field B_0. Hence, a flip of spin I shifts the resonance frequency of spin S.

much larger than the coupling, so that the term in Eq. 3b cannot mix the $|\alpha_S\beta_I\rangle$ with the $|\beta_S\alpha_I\rangle$ state of the two spins. Such mixing does occur if v_{dd} is comparable to the difference of resonance frequencies, and then term $\hat{B}$ needs to be included.

The ESEEM experiments only work if term $\hat{C}$ is significant. This term mixes the $|\alpha_S\alpha_I\rangle$ with the $|\alpha_S\beta_I\rangle$ state and the $|\beta_S\alpha_I\rangle$ with the $|\beta_S\beta_I\rangle$ state, that is states that differ in the magnetic quantum number of the nuclear spin. A mw pulse then excites not only the formally allowed transitions in which only the magnetic quantum number of the electron spin changes, but also formally forbidden transitions, in which both the electron and nuclear spin are flipped. Term $\hat{C}$ is significant (pseudo-secular) if the hyperfine coupling is of the same order of magnitude as the nuclear Zeeman interaction. At the usually applied magnetic fields of 0.35 T or 3500 G (X-band ESR) this applies to all nuclear isotopes for distances $r < 4$ Å. Electron–nuclear double resonance works best, and ENDOR spectra are most easily interpreted, if the pseudo-secular hyperfine coupling $\hat{C}$ can be neglected. This can be achieved by increasing the field and thus the nuclear Zeeman interaction (high-field ENDOR).

In the most simple case, where only term $\hat{A}$ needs to be considered, the dipolar spectrum is independent of the resonance frequencies of the two spins and is a simple Pake pattern (Fig. 2a). If term $\hat{B}$ also needs to be considered, both the shape of the pattern changes and the singularities shift (Fig. 2b). In this situation, numerical simulations may be required to extract v_{dd} and to determine the distance r.

Up to this point, we have assumed that the distance r and angle θ are sharply defined quantities. This is the case only if both spins are well localized on the length scale of r. While nuclear spins are always well localized on the relevant length scale of a few angstroms to a few nanometers, this is not necessarily the case for electron spins in π radicals or transition metal complexes. If an unpaired electron is significantly delocalized, the dipole–dipole coupling tensor has to be averaged over the spatial distribution of the electron spin. In general, the orientation dependence of the dipolar coupling terms is then no longer described by Eqs. 3a–c and the spectrum is no longer a Pake pattern, even if the pseudo-secular terms can be neglected. A detailed discussion of this situation is beyond the scope of this chapter.

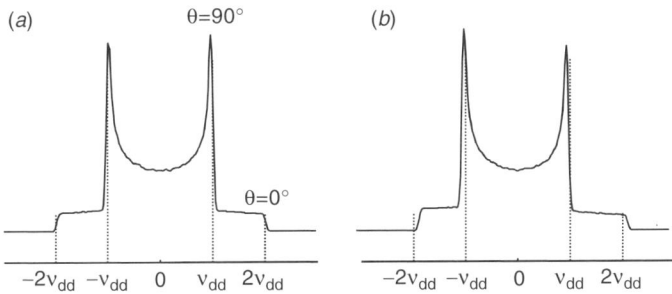

Fig. 2. Calculated dipolar powder spectra for two electron spins with resonance frequencies of 9.765 GHz (observer spin) and 9.700 GHz (pumped spins) and a spin–spin distance of (*a*) 2.5 nm and (*b*) 1.25 nm.

2.2. Exchange Coupling

If several unpaired electrons are localized in strongly overlapping orbitals or in orbitals that are very close to each other in space (mean dipole–dipole couplings larger than the electron Zeeman frequency), they are strongly coupled and are best treated as a single group spin $S > \frac{1}{2}$. Unpaired electrons that are localized in *weakly* overlapping orbitals are best treated as individual spins. As electrons are indistinguishable from each other, they can be exchanged even between weakly overlapping orbitals. This exchange leads to a small coupling of the two electrons.

Weak exchange (ex) coupling, which is typical for electrons that are separated by at least 5 Å, is an isotropic interaction that is described by the Hamiltonian in Eq. 4,

$$\hat{H}_{ex} = J(\hat{S}_x \hat{I}_x + \hat{S}_y \hat{I}_y + \hat{S}_z \hat{I}_z) \tag{4}$$

where J is the exchange coupling. Unfortunately, J is sometimes defined with opposite sign or may even be defined as one-half of the negative value of J in Eq. 4. Hence, comparison of J couplings from literature always requires some caution. In the definition used here, a positive sign of J corresponds to the case where the triplet state of the two electrons is higher in energy at zero field than the singlet state. This is antiferromagnetic coupling and corresponds to weak bonding overlap of the orbitals. The opposite case of $J < 0$ is ferromagnetic coupling, which corresponds to weak antibonding overlap.

Within the same class of compounds, weak exchange couplings decay exponentially with the distance between the two spins (centers of the spatial distribution). However, the decay constant and prefactor strongly depend on the type of bonding network and the conductivity of the medium between the two spins. Fully conjugated systems may have sizeable exchange couplings at distances as large as 3.6 nm.[6] If conjugation is broken by at least two single bonds between the spins and the medium is an isolator, exchange couplings are much smaller than the dipole–dipole coupling at distances > 1.5 nm. The modification of dipolar coupling patterns by J coupling has been discussed earlier in some detail.[7]

In solution, an exchange of electrons arises due to the short orbital overlap during radical–radical collisions. This effect leads to changes in the ESR line shape,[8] but is not important in the context of double-resonance experiments on soft matter or solids.

2.3. Fermi Contact Interaction

Unpaired electrons located in s orbitals have a finite probability to reside inside the atomic nucleus. Such contact between the electron and nuclear spin results in an isotropic coupling (Fermi contact interaction), which can be as large as 1480 MHz for the electron of a hydrogen atom or a few hundred megahertz (MHz) for electrons localized at a transition metal ion. Isotropic hyperfine couplings in organic radicals or in ligands of transition metal ions do not usually exceed 100 MHz. Intermolecular isotropic hyperfine couplings between 1 and 2 MHz have been observed for hydroxylic protons hydrogen bonded to nitroxides[9] and for fluorine nuclei of polyvinylidenefluoride doped with a nitroxide.[10] In the latter case, the strong intermolecular coupling is caused by the high electronegativity of fluorine. In other cases, intermolecular hyperfine couplings are not expected to exceed 200 kHz.

2.4. Spin Density in p and d Orbitals

If some spin density is transferred to a p or d orbital on a certain nucleus, this spin density contributes to the dipole–dipole interaction between the two spins. Such a contribution to the dipole–dipole interaction may exceed the contribution from through-space interaction between the nuclear spin and the center of electron spin density. The distance between the nucleus and the center of spin density cannot be computed directly from the anisotropic hyperfine coupling in this situation. Distance determination by ENDOR or ESEEM thus requires that either nuclei without significant spin density in p or d orbitals are used (protons, deuterons, alkali, and alkali earth metal ions) or that spin density transfer to the atom under consideration is negligible (intermolecular couplings).

3. ELECTRON–ELECTRON DOUBLE RESONANCE

3.1. General Considerations

Pulsed ELDOR experiments are designed to separate the coupling between two electron spins from other contributions to the spin Hamiltonian, namely, resonance offsets due to g-value dispersion, hyperfine couplings, and, for $S > \frac{1}{2}$, zero-field splitting (ZFS). This is achieved by refocusing all interactions, including the coupling between the two electron spins, in an echo experiment on the observer spin S. The spin–spin coupling is then reintroduced by a pump pulse that ideally exclusively excites spin I. Selection of the observer and pumped spin requires excitation pulses at two different frequencies, v_{obs} and v_{pump}, with separated excitation bands. The minimum excitation bandwidth of the pump pulse is the width of the dipolar Pake pattern, since both transitions of the pumped spin must be excited to fully reintroduce the coupling. It follows that distances of 2 nm or shorter require pulse lengths of 50 ns or shorter, so that the difference between v_{obs} and v_{pump} should be 30 MHz or larger. The resonance fields of the two spins must thus differ by > 10 G. If the total width of the ESR spectrum is narrower than that, single frequency techniques[11–13] have to be used for distance measurements. If the total width is larger, a larger frequency difference $v_{obs} - v_{pump}$ is better, as it diminishes the residual overlap of the excitation bands and the influence of the pseudo-secular term of the dipole–dipole coupling in Eq. 3b. Distances down to 1.8 nm can be measured quite well with observer pulses of 32-ns length, a pump pulse of 12-ns length, and a frequency difference of 65 MHz.

Selection of observed and pumped spins is possible even if both paramagnetic centers of the pair are chemically identical, because the resonance frequency depends on orientation and, for centers with strongly hyperfine coupled nuclei, on nuclear magnetic quantum number m_I. Both orientation and nuclear magnetic quantum number are constant during an experiment. For example, at X-band frequencies, a pump pulse applied to the global maximum of the nitroxide ESR spectrum excites radicals with $m_I(^{14}N) = 0$ at all orientations with respect to the magnetic field, but only a small fraction of orientations for radicals with $m_I(^{14}N) = +1$ and -1. An observer pulse applied near the low-field edge excites radicals with $m_I(^{14}N) = +1$ whose molecular z axis is nearly parallel with the external field.

The lower end of the distance range for reliable measurements by pulse ELDOR techniques is set by three complications. First, if the dipolar splitting exceeds the excitation bandwidth of the pump pulse, the pump pulse typically excites only one of the transitions of the dipolar doublet. In this situation, the coherence is not exchanged between the two transitions of the observer spin, but is rather converted to zero-quantum or double-quantum coherence involving both spins. The magnetization is thus lost and does not contribute to the signal. Second, a larger dipolar splitting increases the probability that observer spins are excited by the pump pulse. Third, if the dipolar splitting is of the order of the difference of the resonance frequencies in the absence of coupling, the pseudo-secular term of the coupling (Eq. 3b) induces frequency and phase shifts. The three complications are not independent of each other. For these reasons, measurements of distances and, in particular, distance distributions of nitroxides < 1.75 nm should be interpreted with caution. As a rule, pulsed ELDOR distance measurements are very precise (± 0.05 nm) when dipolar broadening in the CW ESR spectra is absent or weak (< 0.5 mT), and less precise if the CW ESR spectrum exhibits obvious broadening.

The higher end of the distance range is determined by the maximum time t_{max} for which dipolar evolution of the observer spin can be measured. Generally, this depends on the transverse relaxation time T_2 of the observer spin, so that measurements are best performed at low temperatures, where T_2 reaches its maximum. In soft matter at X-band frequencies, $T = 50$ K is often a good choice. Due to contributions from instantaneous diffusion, the apparent T_2 also depends on concentration.[3] For measuring very long distances, concentrations as low as 100 µM may be optimum. In any case, concentrations > 500 µM are detrimental. If S and I spins correspond to chemically different species with different T_2, observer spin S should be the more slowly relaxing spin. This advice may, however, conflict with the requirement to maximize the fraction of pumped spins, which is fulfilled by pumping at the global maximum of the total ESR spectrum of both species. If such a conflict exists, it is advisable to test both settings of pump and observer frequency.

If concentrations are between 50 and 500 µM and relaxation times are not diminished by proximity of other, very fast relaxing paramagnetic centers, maximum dipolar evolution times of at least 1 µs can usually be achieved. This allows for measurement of distances up to ≈ 4.5 nm. In most systems, $t_{max} = 2.5$ µs is feasible, corresponding to a distance range of up to 6 nm. In fairly rigid protonated matrices without methyl groups, such as o-terphenyl, $t_{max} = 6$ µs can be achieved, so that distances of 8 nm can be measured. To estimate the width of a distance distribution centered at 8 nm, $t_{max} = 20$ μs is required, which necessitates the use of deuterated matrices.[14] Measurements with $t_{max} = 20$ μs allow for determinations of reliable shapes of distances distributions up to 5 nm with a resolution of 0.1 nm.[15]

3.2. Experimental Techniques

Any pulsed ELDOR experiment starts with a π/2 pulse that generates transverse magnetization (coherence) on one or both transitions of the observer spin (Fig. 3). To understand the basic principle, it is sufficient to consider the fate of coherence on one of the two transitions, for example, the one belonging to the α state of the pumped spin I (Fig. 4a). In general, this transition has an offset Δv from the observer mw frequency. In a time t after the π/2 pulse, it thus acquires a phase Δvt. At this time, a π pulse is applied on the pump frequency. This pulse flips the pumped spin I from its α to its β state (Fig. 1), thus transferring the coherence to the other transition of spin S (Fig. 4b) and changing the resonance frequency of spin S by $-d$. Here, d is the spin–spin coupling including dipole–dipole *and* exchange contributions. In the remaining time $\tau-t$ before the second pulse at the observer frequency, the coherence acquires a phase $(\Delta v-d)(\tau-t)$. The total phase before this pulse is thus $\Delta v\tau-d(\tau-t)$. The π pulse on the observer frequency inverts the phase of the coherence, but does not change the frequency. During time τ before echo formation, the coherence thus acquires a phase $-(\Delta v-d)\tau$. The total phase ϕ_{echo} at the time of echo formation is

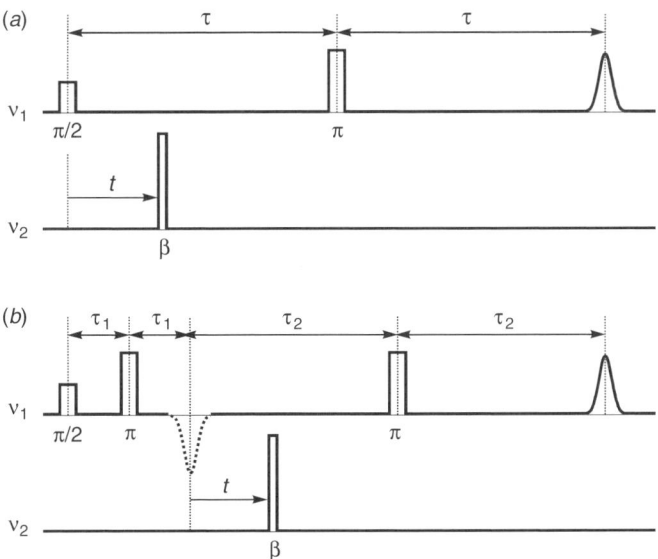

Fig. 3. Pulsed ELDOR (DEER) sequences. (*a*) Basic PELDOR experiment. Interpulse delay τ is fixed, interpulse delay *t* is varied, and echo intensity is recorded. (*b*) Four-pulse DEER. In the constant-time version, interpulse delays τ_1 and τ_2 are fixed, interpulse delay *t* is varied, and the intensity of the second echo (solid line) is recorded. For the variable-time version, see the text.

therefore dt. Echo intensity is proportional to $\cos(\phi_{\text{echo}}) = \cos(dt)$. Consideration of the other observer spin transition gives the result $\phi_{\text{echo}} = -dt$. As $\cos(dt) = \cos(-dt)$, both cases correspond to the same echo modulation. The spin–spin coupling d can thus be measured by recording the echo intensity as a function of interpulse delay t and Fourier transformation of the data.

Due to the orientation dependence of d and a possible distribution of distances r, d is usually distributed. Oscillations for different d values interfere destructively at long times t. In particular, there is always a large number of I spins of remote paramagnetic centers with small intermolecular couplings that lead to a decay of the signal toward zero at infinite times. For a homogeneous spatial distribution of the remote radicals, this decay is exponential.[16] Full information can only be obtained by measuring the echo decay from the very beginning at $t = 0$, where all oscillations interfere constructively.

With the basic pulsed ELDOR (PELDOR) sequence shown in Fig. 3a, such a measurement at $t = 0$ corresponds to overlapping pump and observer pulses. This leads to signal distortions unless the two frequencies are applied to two well-isolated modes of a bimodal resonator.[16] Furthermore, pulses at the two frequencies have to be amplified in two separate high-power amplifiers. The requirement for such specialized hardware and the restriction to a fixed frequency difference imposed by the bimodal resonator is overcome by using the four-pulse double electron-electron resonance (DEER) experiment (Fig. 4b).

In this experiment, the phase acquired by the observer spin coherence during the first interpulse delay of length τ_1 is exactly compensated by the phase acquired during the second interpulse delay of length τ_1 after the π inversion pulse. At time $2\tau_1$, the coherence thus has zero phase. This time can thus be identified with dipolar evolution time $t = 0$ and corresponds to the time immediately after the $\pi/2$ pulse in the basic PELDOR experiment. Spin evolution for $t > 0$ is analogous to basic PELDOR with time τ_2 replacing time τ. For $\tau_1 \geq 200$ ns, interference between pump and observer pulses can be safely neglected at $t = 0$.

The pulse sequences shown in Fig. 4a and b feature fixed interpulse delays τ or τ_2 that must be slightly longer than the maximum dipolar evolution time t_{max}. Such constant-time experiments have the advantage that the variation of the echo amplitude is not influenced by relaxation. Loss of magnetization of the observer spins by transverse relaxation is the same at all times t. The distribution of dipolar frequencies d can thus be determined free from relaxational broadening, which is an absolute requirement for measurements of distances > 3 nm in most systems. On the other hand, constant-time experiments have the disadvantage that the entire data set is measured with a relaxational loss of magnetization that corresponds to t_{max}. In principle, data points at $t < t_{\text{max}}$ can be measured with smaller loss. As relaxation is exponential, this can lead to a significant sensitivity gain. Part of this gain is lost again by the necessity to measure a reference data set that contains only the relaxational decay and dividing two data sets.[15] Nevertheless, such a variable-time DEER experiment still provides better sensitivity in most cases. The reference data set is measured by varying τ_2 in the sequence shown in Fig. 4b using fixed τ_1 and fixed $t = 0$. The recoupled data set is measured with the same sequence, by

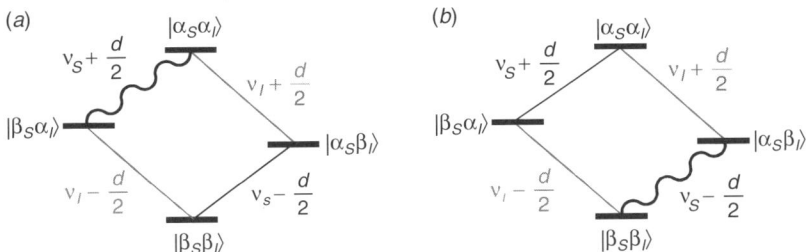

Fig. 4. Energy level scheme, transition frequencies, and coherence transfer in pulsed ELDOR experiments. (*a*) Situation before the pump pulse. (*b*) Situation after the pump pulse.

varying t and τ_2 simultaneously with the same time increment. In other words, in the reference experiment, the delay between the first observer π pulse and the pump pulse is constant, while in the recoupled experiment the delay between the pump pulse and the second observer π pulse is constant.

In all cases, the primary data set is a variation $V(t)$ of echo intensity with time. Normalization to the echo intensity at time zero gives the dipolar evolution function $D(t) = V(t)/V(0)$. The dipolar evolution function contains the information on the distribution $P(r)$ of spin-to-spin distances.

3.3. Analogy to Form Factor and Structure Factor in Scattering

Structures of polymer materials on nanometer length scales are mostly determined by small angle scattering techniques,[17] such as small angle X-ray scattering (SAXS) and small angle neutron scattering (SANS). For distances up to at least 5 nm, ESR spectroscopy can complement results from scattering, since spin labeling opens up a different and very versatile approach to contrast variation. Less sample is needed than for SANS and more precise distance measurements are possible. Combined scattering and ESR studies require a common framework for the description of structures.

Scattering experiments are based on different scattering cross-sections of atoms in the particles of interest and in the surrounding medium. The variation of scattering intensity with the wave vector q [scattering curve $I(q)$] is the product of an instrumental constant and of two q-dependent contributions related to the structure of the investigated material, the form and the structure factors. The form factor depends on the shape of the particle, while the structure factor contains information on any regularity of the spatial distribution of the particles. Symmetry of the particles may be nicely seen in the scattering curves that correspond to reciprocal space, but in many cases data are more easily interpreted in real space.[18] Real-space data are obtained by an indirect Fourier transformation technique. Indirect Fourier transformation is an ill-posed problem just as the computation of distance distributions from pulsed ELDOR data that is described below.

The "pair distance distribution function" $p(r)$ obtained in small-angle scattering from the form factor $P(q)$ is analogous to the distance distribution $P(r)$ obtained in pulsed ELDOR from the dipolar evolution function $D(t)$. In particular, for a *homogeneous* particle, $p(r)$ from scattering and $P(r)$ from ELDOR are the same. Complementary information can be obtained by spin-labeling only specific sites, such as the surface of the particle or the ends of a polymer chain.

Scattering and pulsed ELDOR differ profoundly in formation of the signal. Scattered waves totally cancel by destructive interference if there are no regularities in the structure. For particles whose positions are uncorrelated (particles homogeneously distributed in a matrix) the structure factor is a constant, $S(q) = 1$. In contrast, homogeneously distributed spins in pulsed ELDOR contribute a background function $B(t)$,

$$B(t) = exp\left(-\frac{2\pi\mu_0 g_s g_I \mu^2_B}{9\sqrt{3}\hbar} \lambda c_I t\right) \quad (5)$$

where c_I is the total concentration of I spins and λ is a modulation depth parameter that depends on the spectrum of spin I, on the excitation position within this spectrum, and on instrumental parameters (pulse width, pulse power, pulse shape, shape, and width of the resonator mode). On a given spectrometer and for a given type of samples, such as nitroxide spin labels, λ can be determined by calibration with a sample of known concentration c_I.

Such a signal contribution from a homogeneous distribution may be advantageous, as it allows for precise determination of particle concentrations and for the detection of subtle deviations from a homogeneous distribution by a deviation of $B(t)$ from an exponential decay. In scattering, such deviations contribute only a very broad background to $I(q)$ that is easily missed. The situation in pulsed ELDOR is illustrated in Fig. 5 for the enrichment or depletion of ionic spin probes close to a charged surface. By using the analytical solution of the Poisson–Boltzmann equation for a charged planar plate in an electrolyte solution,[19] we have calculated the concentration profiles of monovalent ions in solution with the opposite (Fig. 5a) or the same (Fig. 5c) charge as the platelet. Assuming that the observer spins S are attached to the surface of the platelet and the pumped spins are located at the ions in solution, we obtain the dipolar evolution functions $B(t)$ shown as solid lines in Fig. 5b and d, respectively. For enrichment of unlike charges near the surface, $B(t)$ has a positive deviation from an exponential fit (dashed line in Fig. 5b) at short times t. This positive deviation generally occurs if short distances are overrepresented compared to a homogeneous distribution. For depletion of like charges near the surface, $B(t)$ has a negative deviation from an exponential fit (dashed line in Fig. 5d) at short times t. The negative deviation generally occurs if short distances are underrepresented.[20]

In scattering data one can sometimes identify a regime, that is, a range of q values, where scattering intensity I scales as q^{-x}, where x is the dimensionality of the particle. For example, rodlike (one–dimensional' 1D) particles exhibit q^{-1} scaling, while flat (two–dimensional' 2D) particles exhibit q^{-2} scaling.[18] Such scaling is recognized by a linear dependence in plots of $\log(I)$ versus $\log(q)$. Analogously, the

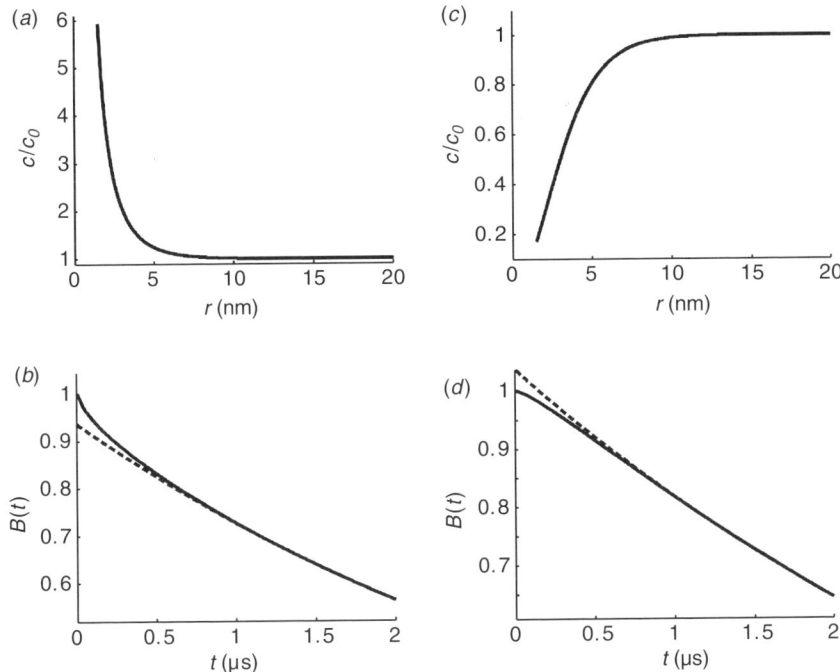

Fig. 5. Distribution of local concentrations c/c_0 and corresponding background functions $B(t)$ for monovalent charged ionic spin probes in a dispersion of charged planar platelets in water (simulation based on the Poisson–Boltzmann equation). A concentration of 1.67 mM of a monovalent salt corresponding to a Debye screening length of 7.5 nm was assumed. (*a*) Local concentration as a function of distance r from the platelet surface for counterions with unlike charge (enrichment near the surface). (*b*) Dipolar evolution function for counterions with unlike charge (solid line) and fit of the data between t = 1 and 2 μs by an exponential decay (dashed line). (*c*) Local concentration as a function of distance r from the platelet surface for counterions with like charge (depletion near the surface). (*d*) Dipolar evolution function for counterions with like charge (solid line) and fit of the data between t = 1 and 2 μs by an exponential decay (dashed line).

dipolar evolution function for a homogeneous distribution of spin labels in x dimensions is given in Eq. 6.,

$$D_x(t) = exp\left(-\alpha\, t^{x/3}\right) \qquad (6)$$

where $\alpha = k\lambda\rho$ is proportional to the density ρ of the I spins.[21,22] The dependence of the proportionality constant k on x can be inferred from Ref. 22. Plots of $\log(-\log(D(t)))$ versus $\log(t)$ exhibit a linear dependence with slope of $x/3$. The distribution of spin-carrying divalent counterions in semidilute frozen polyelectrolyte solutions could be fitted by a superposition of components with $x = 1$ and $x = 3$.[23] The dependence of the relative amount of the two components on polyelectrolyte concentration suggested that the first component corresponds to counterions in the vicinity of an isolated and locally stretched chain, while the second component corresponds to counterions in regions where chains overlap.

3.4. Direct Computation of Distance Distributions

The primary result of a pulsed ELDOR measurement is a distribution of dipolar couplings d. This information is contained either in the primary time-domain data (variation of echo intensity as a function of dipolar evolution time t) or in the dipolar spectrum obtained from these data by Fourier transformation. Time- and frequency-domain data contain exactly the same information, since Fourier transformation is a linear operation. However, some features are easier to recognize in time domain (e.g., quality of least-squares fitting of the data) and others in frequency domain (e.g., orientation selection by missing parts of the Pake pattern).

The wanted information is usually the distribution of distances $P(r)$. In other words, the distribution of orientations $P(\theta)$ has to be factored out of the data.[24,25] This is possible if exchange coupling J is negligible and $P(\theta)$ is known. Usually, a uniform distribution of spin–spin vector orientation with respect to magnetic field, $P(\theta) = \sin(\theta)$, is a fairly good approximation at X band, as we have found even for rigid biradicals. Some deviations can be seen in frequency domain, but with proper data processing, as explained below, they do not cause artifacts in the distance distribution $P(r)$.

The mathematical problem of data analysis is thus to convert $P(d)$ to $P(r)$. This problem is ill-posed, that is, a mild distortion of $P(d)$ by noise can cause a strong distortion of $P(r)$. The reason is that the signal from a certain distance r can be almost perfectly modeled by a superposition of signals from distances $r + \Delta r_1$ and $r - \Delta r_2$. In contrast to the true solution, this superposition may also fit part of the noise and may thus be preferred in a least-squares solution of the problem. As a result, one obtains an illusory set of narrow peaks in the distance range where the true distribution exhibits just one broad peak (for illustrative examples, see Ref. 26).

Two strategies can stabilize the solution. First, the algorithm should allow only for nonnegative values of $P(r)$. This is reasonable, as a distance distribution cannot have negative contributions. Second, if the primary data are fitted equally well by a broad distribution and a complicated pattern of narrow peaks, we know that the broad distribution is more likely the correct solution. This can be considered in data analysis by requiring that the solution $P(r)$ be both reasonably smooth and a good fit of the original data. In Tikhonov regularization,[27,28] the compromise between smoothness and fit quality is quantified by a regularization parameter α. By selecting a larger α, smoothness is given higher priority, while a smaller α provides a better fit. If the solution is computed for different values of α and a smoothness criterion is plotted against the deviation between fit and data, one often obtains an L-shaped curve. For very small α, smoothness improves drastically with increasing α, while the root-mean-square deviation changes only slightly (undersmoothing, vertical stroke of the L). For very large α, smoothness does not change anymore, while the deviation increases drastically with increasing α (oversmoothing, horizontal stroke of the L). The optimum regularization parameter corresponds to the corner of the L.[29,30]

4. ELECTRON–NUCLEAR DOUBLE RESONANCE

4.1. General Considerations

It is advisable not to rely uncritically on such an optimum regularization parameter, as the L curve is also influenced by noise. Instead, one should use existing information about the investigated material to estimate a minimum realistic width of peaks in the distance distribution.

The interaction between electron and nuclear spins causes line splittings in both ESR and nuclear magnetic resonance (NMR) spectra. However, since usually many nuclear spins I_k are coupled to a single electron spin S, the NMR spectra are easier to interpret and are better resolved. The number of ESR transitions, N_{ESR}, increases multiplicatively with the number of nuclear spins,

$$N_{\mathrm{ESR}} = \prod_{k=1}^{m} (2I_k + 1) \tag{7}$$

and the frequency of each transition is influenced by *all* the hyperfine couplings A_k. In contrast, the number of NMR transitions, N_{NMR}, increases only additively with the number of nuclear spins,

$$N_{\mathrm{NMR}} = 2(2S+1) \sum_{k=1}^{m} I_k \tag{8}$$

and the frequency of each transition is influenced by only one hyperfine coupling. This is illustrated by the ENDOR spectrum in Fig. 6 that was computed for the same spin system as the ESR spectrum in Fig. 5 of Chapter 1.

A further resolution advantage arises in the ENDOR spectra since the line width is limited by the longitudinal relaxation time T_{1e} of the electron spins or the transverse relaxation time T_{2n} of nuclear spins, rather than by the transverse relaxation time T_{2e} of the electron spins. Since in solids or soft matter $T_{1e}, T_{2n} > T_{2e}$, ENDOR lines are usually narrower than ESR lines.

In addition to being more simple and better resolved than an ESR spectrum, an ENDOR spectrum also contains additional information. Consider first-order expressions

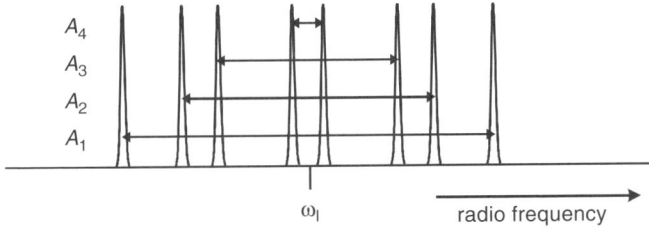

Fig. 6. Isotropic ENDOR spectrum for a system consisting of four nuclear spins with $I_k = \frac{1}{2}$ coupled to a single electron spin with $S = \frac{1}{2}$.

for the frequencies of ESR and ENDOR transitions of a spin system consisting of one electron spin $S = 1/2$ and one nuclear spin I. The ESR frequencies are given in Eq. 9,

$$v_{EPR} = v_S + m_I A \tag{9}$$

where v_S is the electron Zeeman frequency, $m_I = -I, -I+1, ..., I$ is the magnetic quantum number of the nucleus, and A is the hyperfine coupling. They thus do not contain any information on the nature of the coupled nucleus. The ENDOR frequency of the transition between states $|m_I\rangle$ and $|m_I+1\rangle$ is given by Eq. 10,

$$v_{ENDOR} = |v_I + m_s A + (2m_I + 1)v_Q| \tag{10}$$

where v_Q is the nuclear quadrupole splitting. For $I = \frac{1}{2}$ the ENDOR frequencies thus contain information on the nuclear Zeeman frequency, and for $I > \frac{1}{2}$ additional information on the nuclear quadrupole coupling. The information is best obtained at high fields, where the nuclear Zeeman interaction dominates. The multiplet of ENDOR lines is then centered at v_I, and the nuclear isotope can easily be assigned using a table of g_n values. By analyzing the dependence of ENDOR spectra as a function of the ESR observer field, it is possible to determine the full hyperfine and nuclear quadrupole tensors and their orientations with respect to the g-tensor.[31,32]

In complex materials or for structures with a low degree of order, such a detailed analysis is impossible. Information on the structure can then be obtained by quantifying proximity of nuclei of a certain element to the electron spin. For protons, distance information can be derived from the dipolar part of the hyperfine coupling. This approach may fail for nuclei with $I > \frac{1}{2}$ as, according to Eq. 10, ENDOR frequencies are influenced by nuclear quadrupole couplings. Proximity of nuclei can thus be characterized most precisely for nuclei with $I = \frac{1}{2}$ or for nuclei, such as deuterium and ^{6}Li, which have very small nuclear quadrupole couplings. In general, ENDOR techniques work best for nuclei with high or at least moderate gyromagnetic ratio, such as ^{1}H, ^{19}F, ^{31}P, and ^{13}C. For ^{19}F, small s spin densities already lead to a sizeable Fermi contact interaction. This simplifies semiquantitative detection of proximity of ^{19}F to a spin probe,[33] but makes it nearly impossible to determine a distance of closest proximity. The ENDOR distance measurements to nuclei in neighboring molecules work particularly well for phosphorus,[34] as ^{31}P has a natural abundance of 100% and is neither ubiquitous nor extremely rare in materials.

4.2. Experimental Techniques

In an ENDOR experiment, magnetization from electron spins has to be transferred to nuclear spin transitions. Resonant radio frequency (rf) irradiation can then change this magnetization and the change can be detected again on an electron spin transition. By plotting the ESR signal as a function of the irradiated rf, one obtains the spectrum of the nuclear spin transitions (ENDOR spectrum).

In solid materials, such experiments are best done with pulse techniques.[3] In Davies ENDOR (Fig. 7),[35] the mw pulses excite only a narrow frequency range within the ESR line. The first pulse with flip angle π inverts the magnetization of the resonant spins in a narrow frequency band and thus burns a hole into the ESR line (point II in Fig. 7). A resonant rf pulse partially shifts this hole into side holes, as it changes the nuclear spin state and thus the resonance frequency of the electron spin transition. The frequency shift is equal to the hyperfine coupling A. As the rf pulse drives $|\alpha_I\rangle \to |\beta_I\rangle$, as well as $|\beta_I\rangle \to |\alpha_I\rangle$ transitions, part of the hole is shifted to the left and another part to the right. A third part is unchanged, as the rf pulse excites only one of the two lines of the hyperfine doublet. The two final mw pulses detect the depth of the center hole as a spin echo signal. An off–resonance rf pulse corresponds to maximum hole depth at point II (rf pulse induces no change), while an on-resonant pulse corresponds to reduced hole depth (point III).

The Davies ENDOR experiment requires that the frequency shift A is larger than the hole width. It is thus best suited for large hyperfine couplings. For small couplings, the experiment needs to be performed with very long mw pulses to create a very narrow hole. This leads to low sensitivity, as only a small fraction of the ESR line contributes to the ENDOR signal.

This problem is solved by the Mims ENDOR experiment (Fig. 8).[36] Here the first two mw pulses create an oscillatory polarization grating, that is, a comb of holes over a broader range in the ESR line (point I in Fig. 8). The period of this grating $1/\tau$ can be adapted to the expected magnitude of the hyperfine couplings by changing interpulse delay τ. The rf pulse again leads to a shift of parts of the hole pattern by $\pm A$. The shifted gratings and the center grating interfere destructively, leading to a reduction in the amplitude of the stimulated echo. If the shift corresponds to phase inversion, $A = 1/(2\tau)$, the grating and the stimulated echo vanish completely. In contrast,

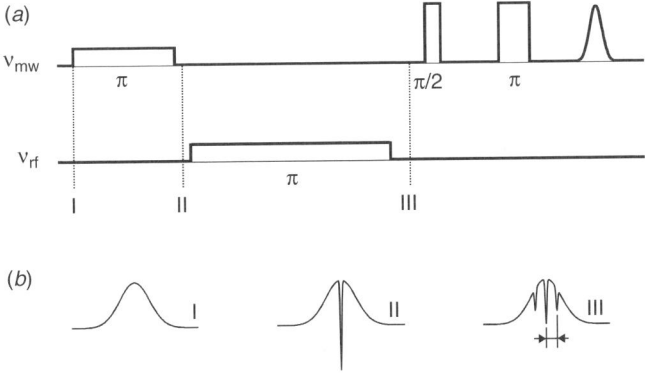

Fig. 7. Davies ENDOR experiment. (*a*) Pulse sequence. The frequency of the rf pulse is varied and integrated echo intensity is recorded. (*b*) Changes in the ESR line. At point I before the first pulse, the whole line comprises equilibrium polarization. The first π pulse burns a hole into the line (point **II**). A resonant rf pulse shifts one-half of the hole to two side holes at frequency difference $\pm A$ (point **III**).

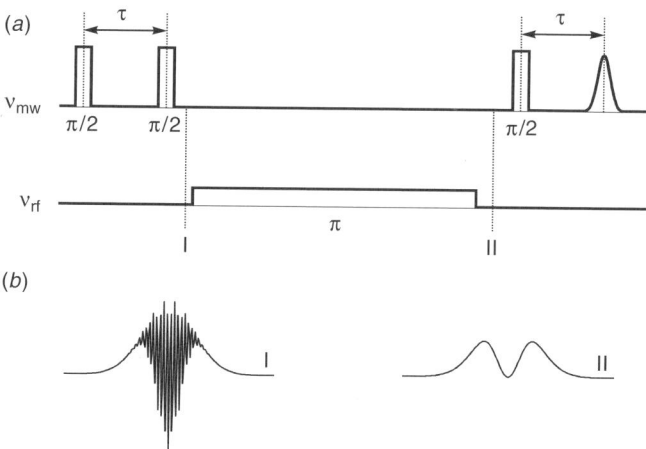

Fig. 8. Mims ENDOR experiment. (*a*) Pulse sequence. The frequency of the rf pulse is varied and integrated intensity of the stimulated echo is recorded. (*b*) Changes in the ESR line. The two mw π/2 pulses create a grating in the ESR line (point **I**). A resonant rf pulse shifts one-quarter of the grating to the right (frequency difference +*A*) and another one-quarter to the left (frequency difference –*A*). These shifted gratings may interfere destructively with the remaining grating in the center, so that just a broad hole remains (point **II**). In this case, no echo is formed.

if $A = 1/\tau$, the shifted patterns and the central pattern interfere constructively and there is almost no ENDOR effect on the stimulated echo (blind spot).

In general, blind spots occur at $A = n/\tau$, that is, the first one at $A = 0$. Accordingly, measurement of small couplings requires a long interpulse delay τ. On the other hand, echo intensity decreases due to transverse relaxation with increasing τ. The optimum interpulse delay for measuring very small couplings is $\tau \approx T_2$, where T_2 is the transverse relaxation time of the electron spins.[34] For moderate hyperfine couplings, the blind spot behavior can be overcome by averaging Mims ENDOR experiments for different τ values. Such an approach is superior in sensitivity with respect to Davies ENDOR for hyperfine couplings, smaller than $\approx$ 5 MHz.[37] Larger hyperfine couplings are best measured by Davies ENDOR.

4.3. High-Field ENDOR

Overlap of signals from different chemical elements is an inherent weakness of ENDOR spectroscopy at conventional ESR frequencies of $\approx$ 9.6 GHz (X band) as can be seen in Fig. 9. To overcome this problem, pulse ENDOR techniques have been developed that separate ENDOR signals according to the nuclear Zeeman frequency or the magnitude of the hyperfine coupling.[3] Provided the hardware is available, high-field ENDOR provides a more simple alternative. At an ESR frequency of 95

GHz (W band), nuclear Zeeman frequencies for most elements exceed the hyperfine couplings, and so do differences of nuclear Zeeman frequencies between ^{1}H and ^{31}P, between ^{31}P and ^{27}Al, and between ^{27}Al and ^{14}N. At 95 GHz, there may still be overlap of signals in groups of nuclei with similar gyromagnetic ratio. Such groups are, for example, ^{1}H and ^{19}F or ^{13}C, ^{23}Na, and ^{27}Al. ENDOR at even higher frequencies may sometimes be required to separate signals from these elements and is currently developed in several groups around the world.

High-field ENDOR also improves orientation selection, which allows for determination of the relative orientation of the hyperfine tensor with respect to the g tensor and also leads to resolution enhancement.[3,31,32] For observation at the extreme positions (edges) of the ESR spectrum, only molecules with the g_x or g_z principal axes along the magnetic field contribute, and single-crystal like spectra are observed.

At conventional fields, ENDOR spectra are usually independent of the sign of the hyperfine coupling. At high fields, one often observes a pronounced difference between ENDOR intensities at frequencies above and below the nuclear Zeeman frequency.[38] For high-spin systems ($S > \frac{1}{2}$) at temperatures close to or below the Zeeman temperature $T_Z = g\mu_B B_0 / k_B$, such a difference arises from different polarization of ESR transitions $|m_S\rangle \leftrightarrow |m_S + 1\rangle$ with different m_S. This is in turn due to the fact that the Zeeman energy is comparable to or exceeds the thermal energy $k_B T$. For $S = \frac{1}{2}$ systems, the intensity asymmetry stems from a build-up of nuclear polarization during consecutive pulse experiments. Such build-up occurs when longitudinal nuclear relaxation rates and electron-nuclear cross-relaxation rates are smaller than the repetition rate of the experiment. This condition is more often encountered at high fields, where the relaxation rates tend to be lower.

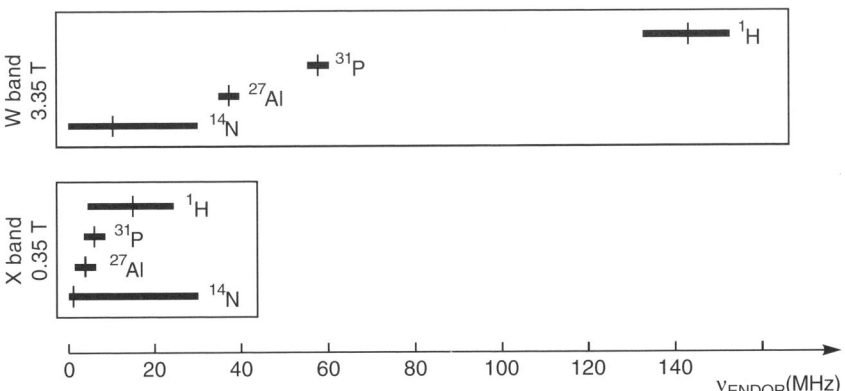

Fig. 9. Nuclear Zeeman frequencies (vertical markers) and ENDOR frequency ranges (horizontal bars) for selected isotopes at magnetic fields of 0.35 T (conventional ENDOR) and 3.35 T (high-field ENDOR).

5. ELECTRON SPIN ECHO ENVELOPE MODULATION

In some cases, ENDOR experiments are not the best choice for obtaining information on nuclei coupled to the electron spin. When the hyperfine coupling and the nuclear Zeeman frequency are of the same order of magnitude, transition probabilities for certain ENDOR frequencies become very small. Furthermore, for small nuclear Zeeman frequencies, all ENDOR frequencies may be < 2 MHz, where ENDOR sensitivity is relatively poor. Counting the number of nuclei of a given type that are coupled to an electron spin also appears to be difficult with ENDOR techniques. Another incentive for pursuing an alternative to ENDOR comes from the wealth of information that can be obtained by time-domain experiments, in particular, two-dimensional spectroscopy.[39] Although time-domain ENDOR is possible,[3,40] it generally suffers from small excitation bandwidth of the rf pulses and signal losses during the large effective deadtime in the nuclear frequency dimension. Time-domain ENDOR experiments therefore tend to have low sensitivity. When nuclear Zeeman frequencies and hyperfine couplings are of the same order of magnitude, and ENDOR is therefore problematic, ESEEM techniques are a viable alternative.

5.1. Principles

For an anisotropic hyperfine interaction, the local field imposed by the electron spin at the site of the nuclear spin is not in general parallel to the z axis defined by the external magnetic field. If the hyperfine interaction is much smaller than the electron Zeeman interaction, but comparable to the nuclear Zeeman interaction $\nu_I I_z$, the secular term $A\, S_z I_z$ and the pseudo-secular term $B\, S_z I_x$ are relevant. The pseudo-secular B term differs from zero except along the principal axes of the hyperfine tensor and leads to a deviation of the local field from the z direction. To see this, consider a system of an electron spin $S = \frac{1}{2}$ coupled to one nuclear spin $I = \frac{1}{2}$. The z component of the total field is given by $\nu_I + A/2$ for the $|\alpha_S\rangle$ state of the electron spin and by $\nu_I - A/2$ for the $|\beta_S\rangle$ state, and due to the nonzero B term, the local fields at the nuclear spin are not parallel in these two states (Fig. 10a). A mw pulse that excites the electron spin thus partially converts nuclear magnetization $S^\alpha I_z$ that is along the local field in the $|\alpha_S\rangle$ state to nuclear magnetization $S^\beta I_x$ that is perpendicular to the new local field (Fig. 10b). In other words, there is state mixing. The quantum numbers of both the electron and nuclear spin may change during a transition, when the usual selection rule $\Delta m_S = \pm 1$ does not strictly apply. Such forbidden transitions, in which both the electron and nuclear spin are excited, allow for measurements of nuclear frequencies by applying exclusively mw pulses.

5.2. Experimental Techniques

Consider a two-pulse echo experiment consisting of an mw $\pi/2$ pulse for excitation of transverse magnetization; an interpulse delay τ in which this magnetization defocuses due to a dispersion of resonance frequencies; a π pulse that inverts the phase of the magnetization; and another delay τ in which the magnetization refocuses. After this pulse sequence, a spin echo is observed.[3] Now, we examine coherence on the allowed transition $|\beta_S\alpha_I\rangle \leftrightarrow |\alpha_S\alpha_I\rangle$ excited by the first pulse (Fig. 10c). During the subsequent π pulse, the coherence branches as indicated in Fig. 10b. For part of the coherence only the phase is inverted, it remains on the same transition. Another part is transferred to the forbidden transition $|\beta_S\alpha_I\rangle \leftrightarrow |\alpha_S\beta_I\rangle$. The frequencies of these two transitions differ by the frequency v_α of the $|\alpha_S\alpha_I\rangle \leftrightarrow |\alpha_S\beta_I\rangle$ nuclear transition. If the echo intensity is measured as a function of interpulse delay τ (echo envelope), one thus observes two contributions. The coherence that

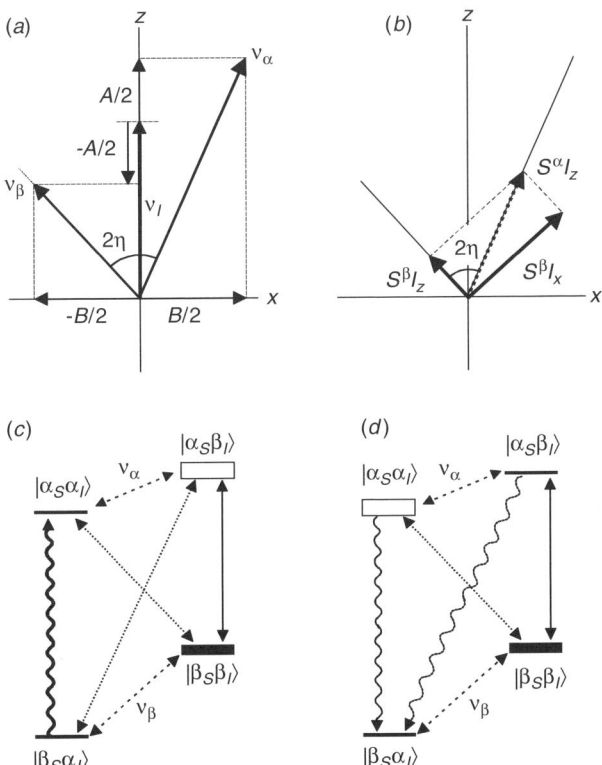

Fig. 10. Local fields at the nuclear spin and magnetization transfers for an anisotropic hyperfine coupling that is comparable in magnitude to the nuclear Zeeman interaction v_I. (a) The $|\alpha_S\rangle$ and $|\beta_S\rangle$ states of the electron spin differ in the sign of the components A $S_z I_z$ and B $S_z I_x$ of the hyperfine coupling. Hence, the quantization directions v_α and v_β of the nuclear spin differ by an angle 2η. (b) An mw pulse transfers longitudinal nuclear magnetization $S^\alpha I_z$ in the $|\alpha_S\rangle$ manifold of states (dotted arrow) to longitudinal magnetization $S^\beta I_z$ as well as transverse magnetization $S^\beta I_x\rangle$ in the $|\beta_S\rangle$ manifold of states. (c) Coherence on the allowed transition $|\beta_S\alpha_I\rangle \leftrightarrow |\alpha_S\alpha_I\rangle$ excited by an mw π/2 pulse. Allowed ESR transitions are shown as solid lines, forbidden ESR transitions as dotted lines, and nuclear transitions as dashed lines. (d) Coherence branching induced by an mw π pulse.

remains on the same transition provides a contribution to the echo that decays with the transverse relaxation time T_{2e} of the electron spins. The contribution from transferred coherence also decays, but in addition it oscillates with frequency ν_α. Hence, the echo envelope is modulated. The depth of this modulation depends on the probability of forbidden transitions.

A detailed examination of all coherence transfer pathways[3] reveals that the four frequencies seen in Eqs. 11–14

$$\nu_\alpha = \sqrt{(\nu_I + A/2)^2 + B^2/4} \tag{11}$$

$$\nu_\beta = \sqrt{(\nu_I - A/2)^2 + B^2/4} \tag{12}$$

$$\nu_+ = \nu_\alpha + \nu_\beta \tag{13}$$

$$\nu_- = |\nu_\alpha - \nu_\beta| \tag{14}$$

are observed and that the modulation depth is given by $k = \sin^2 2\eta$. The expressions for the frequencies and the factor $\sin 2\eta$ for a forbidden transfer are easily understood in the geometrical picture of Fig. 10a and b. The square in the modulation depth is due to the fact that any coherence transfer pathway that leads to echo envelope modulation contains either two forbidden transfers or one forbidden transfer and the observation of forbidden coherence.

Considering the resolution of the nuclear frequency spectrum, this two-pulse echo experiment is not optimal. The nuclear frequencies are here measured as differences of frequencies of the ESR transitions, so that the line widths correspond to those of ESR transitions. The nuclear transitions have longer transverse relaxation times T_{2n} and thus smaller line widths. In fact, if the second mw pulse is changed from a π pulse to a $\pi/2$ pulse, coherence is transferred to nuclear transitions instead of forbidden electron transitions. This coherence then evolves for a variable time T and thus acquires phase $\nu_\alpha T$ or $\nu_\beta T$. Nuclear coherence cannot be detected directly, but can be transferred back to allowed and forbidden electron coherence by another $\pi/2$ pulse. The sequence $(\pi/2)-\tau-(\pi/2)-T-(\pi/2)-\tau$ generates a stimulated echo, whose envelope as a function of T is modulated with the two nuclear frequencies ν_α and ν_β. The combination frequencies ν_+ and ν_- are not observed. The modulation depth is also $\sin^2 2\eta$. The lack of combination lines simplifies the spectrum and the narrower lines lead to better resolution. There is also, however, a disadvantage of this three-pulse ESEEM experiment. Depending on interpulse delay τ the experiment features blind spots.[3] Thus it needs to be repeated at several τ values.

For the measurement of small dipolar hyperfine couplings due to remote nuclei of a given element the blind spots can be easily avoided, because the blind spots depend on the nuclear frequencies that are very close to the known nuclear Zeeman frequency ν_I. Maximum modulation depth is obtained by adjusting τ to an antiblindspot $\tau = (2n+1)/(2\nu_I)$, where $n = 0,1,2,\ldots$

The lack of combination frequencies in three-pulse ESEEM may sometimes be detrimental. If the isotropic and dipolar hyperfine couplings are much smaller than v_I, the maximum of the sum combination frequency v_+ is given below.[41]

$$\max(v_+) = 2 \left| v_I + \frac{9}{16} \frac{v_{dd}^2}{|v_I|} \right| \quad (15)$$

Since v_I is known, one can determine v_{dd}, and hence the distance between the electron and nuclear spin, even in the presence of small, unknown isotropic hyperfine couplings. The second-order shift with respect to twice the nuclear Zeeman frequency is small. Hence, two-pulse ESEEM with its inferior resolution is not well suited for measuring this shift. The sum combination frequency can be introduced into stimulated-echo ESEEM by inserting an mw π pulse halfway through the evolution period of length T (sequence in Fig. 11 with $t_1 = t_2 = T/2$).

The same pulse sequence can be used for the 2D HYSCORE experiment,[42] which detects correlations between transitions of the same nuclear spin in the electron spin $|\alpha_S\rangle$ and $|\beta_S\rangle$ manifolds. Detection of such correlations helps to separate overlapping signals from different elements or to analyze spectra of nuclear spins with $I > \frac{1}{2}$. In the latter case, the simplification comes about since hyperfine couplings mainly lead to a frequency dispersion perpendicular to the diagonal while quadrupolar couplings mainly lead to a dispersion parallel to the diagonal. The higher order hyperfine shift parallel to the diagonal (Fig. 12) contains the same information as the sum combination peak in four–pulse ESEEM (Eq. 15) and can thus be used to separate the dipolar hyperfine coupling from the isotropic hyperfine coupling.

5.3. Data Analysis

The most basic use of ESEEM is the detection of proximity of nuclei of certain elements to an electron spin (local elemental analysis). This technique is based on the r^{-6} dependence of the ESEEM modulation depth k for weakly coupled nuclei:

$$k = \frac{9}{4} \left(\frac{\mu_0}{4\pi} \right)^2 \left(\frac{g\mu_B}{B_0} \right)^2 \frac{\sin^2 2\theta}{r^6} \quad (16)$$

ESEEM signals are thus detected only from nuclei that are at a radius of 6 Å or shorter from the electron spin.

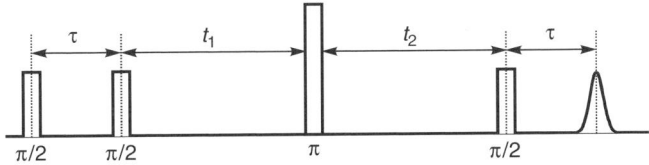

Fig. 11. Pulse sequence of the sum-combination (four-pulse ESEEM) and HYSCORE experiment. In the 1D four-pulse ESEEM experiment, $t_1 = t_2 = T/2$. In HYSCORE, t_1 and t_2 are incremented independently to provide a 2D data set.

While ENDOR lines correspond fairly well to powder patterns and exhibit clear line shape singularities, this is not the case for ESEEM lines. The reason can be seen in Eq. 16: modulation depth depends on orientation and vanishes at the $\theta = 90°$ singularity of a powder pattern as well as at the $\theta = 0°$ outer edge (cf. Fig. 2a). Furthermore, ESEEM spectra usually cannot be properly phased and have to be displayed as magnitude spectra rather than absorption spectra. The combination of these problems makes line shapes in 1D ESEEM spectra unreliable and hard to simulate. Line shape analysis in 1D ESEEM spectra is therefore strongly discouraged. One-dimensional ESEEM spectra are useful for well-defined coordination environments in transition metal complexes, in particular, if single crystals are available. This situation is, however, unusual in polymer applications.

If 1D ESEEM data are dominated by contributions from a single element, time-domain analysis can provide estimates for the distance of closest approach of nuclei of this element and of the average number of such nuclei. This technique relies on the distance dependence of the modulation depth (Eq. 16) and on the fact that the total ESEEM signal caused by several nuclei is the product of contributions from the single nuclei.[3] The modulation depth k_0 at time zero (Fig. 13) depends on both distance and number of the nuclei. The decay of the modulation is due to the frequency dispersion, which to first order depends only on the distance. By analyzing depth *and* decay of the modulation, the two parameters can thus be separated. A popular way of doing this is ratio analysis.[43] In this approach, the ESEEM data are reduced to the ratio R_{exp} between the upper and lower envelope of the echo decay (Fig. 13). For N nuclei at the same distance in the absence of orientational correlations,

$$\log(\log(R_{exp})) = \log(N) + \log(\log(R_I)) \tag{17}$$

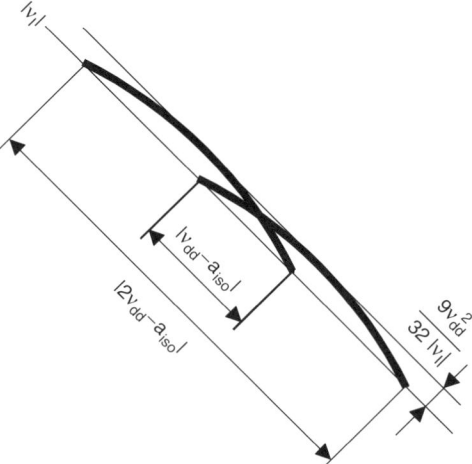

Fig. 12. Schematic HYSCORE powder pattern for an nuclear spin $I = 1/2$ coupled to an electron spin $S = 1/2$ with an axial hyperfine tensor. Intensity tends to zero at the edges and varies throughout the correlation ridges.

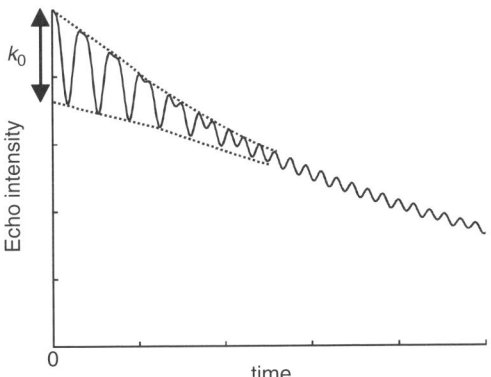

Fig. 13. Ratio analysis of the modulation depth in ESEEM data. The modulation depth k_0 at time t = 0 depends on the distance of the closest shell of nuclei and their number. The decay of the modulation, characterized by the time dependence of the ratio of the upper and lower envelopes (dotted lines), depends only on the distance of the closest shell of nuclei, but not their number.

where R_1 is the ratio expected for a single nucleus at that distance. Although such a model of one sphere of nuclei with the same distance of shortest approach is certainly idealized, ratio analysis can give useful hints for the structure of a material. Furthermore, trends of the modulation depth in a series of similar materials can be interpreted in terms of changing distances or coordination numbers. To use this approach in the presence of significant modulations from nuclei of two elements, one may record the data at a blind spot for one of the elements and suppress the unwanted contribution by digital filtering in spectral domain.[44]

The best way to verify the distance of closest approach and to detect contributions from several nuclei of the same element at different distances is an analysis of HYSCORE spectra. If the frequency shift according to Eq. 15 significantly exceeds the homogeneous line width $1/T_{2n}$ in the nuclear spectra, one can recognize curved ridges as shown in Fig. 13. In favorable cases, several of these ridges may be resolved. For deuterium, analysis of the HYSCORE patterns provides an estimate of the nuclear quadrupole coupling,[3,45] which can then be introduced as a fixed correction parameter in ratio analysis.

6. CONCLUSIONS

Double resonance and ESEEM experiments can reveal and quantify interactions that are unresolved in ESR spectra. As these interactions are often dominated by dipole–dipole contributions, such measurements can be used to determine distances between two electron spins (ELDOR) or between an electron spin and nuclear spins (ENDOR and ESEEM). While ELDOR measurements provide information on the distance range between 2 and 8 nm, ENDOR and ESEEM are sensitive in a range between 0.3 and 1 nm. In favorable cases, such as nitroxide spin probes and spin labels, the gap between the two ranges can be bridged by line shape analysis of CW ESR spectra (Chapters 1, and 3). Taken together, the whole toolbox of CW and pulsed ESR techniques can thus characterize the structure of materials with a low degree of order in the range between the shortest intermolecular contacts and typical extensions of polymer molecules or proteins. Success in this kind of structure characterization depends critically on selecting the right experiment or combination of experiments and on using proper techniques for data analysis.

ACKNOWLEDGMENTS

The author thanks Fond der Chemischen Industrie for a Dozentenstipendium. Helpful discussions with Dariush Hinderberger, Stefan Stoll, and Arthur Schweiger are gratefully acknowledged.

REFERENCES

1. Weil, J.A.; Bolton, J.R.; Wertz, J.E. *Electron Paramagnetic Resonance*, John Wiley & Sons, Inc.: New York, 1994.
2. Schneider, D.J.; Freed, J.H. In *Biological Magnetic Resonance*, Vol. 8, Berliner, J.H., Reuben J., Eds.; Plenum Press: New York, 1989, Chapt. 1.
3. Schweiger, A.; Jeschke, G. *Principles of Pulse Electron Paramagnetic Resonance*, Oxford University Press: Oxford, 2001.
4. Kevan, L.; Kispert, L. *Electron Spin Double Resonance Spectroscopy*, John Wiley & Sons, Inc.: New York, 1976.
5. Kurreck, H.; Kirste, B.; Lubitz, W. *Electron Nuclear Double Resonance Spectroscopy of Radicals in Solution*, VCH Publishers: New York, 1988.
6. Wautelet, P.; Bieber, A.; Turek, P.; LeMoigne, J.; André, J. *J. Mol. Cryst. Liq. Cryst.* **1997**, *305*, 55.
7. Jeschke, G. *Macromol. Rapid. Commun.* **2002**, *23*, 227.
8. Molin, Yu.N.; Salikhov, K.M.; Zamaraev, K.I. *Spin Exchange*. Springer: Berlin, 1980.
9. Kolodziejski, W.; Kecki, Z. *J. Magn. Reson.* **1977**, *28*, 63.
10. Jeschke, G.; Schweiger, A. *Chem. Phys. Lett.* **1995**, *246*, 431.
11. Borbat, P.P., Freed, J.H. In *Biological Magnetic Resonance,* Vol. 19; Berliner, L.J.; Eaton, S. S.; Eaton, G. R., Eds.; Kluwer: Amsterdam, The Netherlands, 2001, Chapt. 9, pp. 383–459.
12. Raitsimring, A.R. In *Biological Magnetic Resonance,* Vol. 19, Berliner, L.J., Eaton, S.S., Eaton, G.R., Eds.; Kluwer: Amsterdam, 2001, The Netherlands, Chapt. 10, pp. 461–491.
13. Jeschke, G.; Pannier, M.; Godt, A.; Spiess, H.W. *Chem. Phys. Lett.* **2000**, *331*, 243.
14. Jeschke, G.; Godt, A., in preparation.
15. Jeschke, G.; Bender, A.; Paulsen, H.; Zimmermann, H.; Godt, A. *J. Magn. Reson.* **2004**, *169*, 1.
16. Milov, A.D.; Maryasov, A.G.; Tsvetkov, Yu.D. *Appl. Magn. Reson.* **1998**, *15*, 107.

REFERENCES

17. Lindner, P.; Zemb, Th. *Neutrons, X–rays and Light: Scattering Methods Applied to Soft Condensed Matter.* North-Holland: Amsterdam, The Netherlands, 2002.
18. Glatter, O. In *Neutrons, X-rays and Light: Scattering Methods Applied to Soft Condensed Matter.* Lindner, P., Zemb, Th., Eds. North-Holland: Amsterdam, The Netherlands, 2002, Chapt. 4, pp. 73–102.
19. (a) Coalson, R.D.; Beck, T.L. *Numerical Methods for Solving Poisson and Poisson–Boltzmann Type Equations.* In *Encyclopedia of Computational Chemistry,* Vol. 3. (b) Schleyer, P.v.R.; Allinger, N.L.; Clark, T.; Gasteiter, J.; Kollman, P.A.; Schaefer, H.F.; Schreiner, H.F. Eds. Wiley: New York, 1998, pp. 2080–2100.
20. Milov A.D.; Tsvetkov, Yu.D. *Appl. Magn. Reson.* **2000**, *18*, 217.
21. Raitsimring, A.M.; Tregub, V.V. *Chem. Phys.* **1983**, *77*, 123.
22. Milov, A.D.; Tsvetkov, Y.D. *Appl. Magn. Reson.* **1997**, *12*, 495.
23. Hinderberger, D.; Spiess, H.W.; Jeschke, G. *J. Phys. Chem. B* **2004**, *108*, 3698.
24. Schäfer, H.; Mädler, B.; Volke, F. *J. Magn. Reson. A* **1995**, *116*, 145.
25. Jeschke, G.; Koch, A.; Jonas, U.; Godt, A. *J. Magn. Reson.* **2002**, *155*, 72.
26. Jeschke, G.; Panek, G.; Godt, A.; Bender, A.; Paulsen, H. *Appl. Magn. Reson.* **2004**, *26*, 223.
27. Tikhonov, A.N.; Arsenin, V.Y. *Solutions of Ill-Posed Problems*, John Wiley & Sons, Inc.: New York, 1977.
28. Weese, J. *Comput. Phys. Commun.* **1992**, *69*, 99.
29. Miller, K. *SIAM J. Math. Anal.* **1970**, *1*, 52.
30. Chiang, Y.-W.; Borbat, P.P.; Freed, J.H. *J. Magn. Reson.* **2005**, *172*, 279.
31. Rist, G.H.; Hyde, J.S. *J. Chem. Phys.* **1970**, *52*, 4633.
32. Hoffman, B.M.; Martinsen, J.; Venters, R.A. *J. Magn. Reson.* **1984**, *59*, 110.
33. Cramer, S.E.; Jeschke, G.; Spiess, H.W. *Macromol. Chem. Phys.* **2002**, *203*, 182.
34. Zänker, P.P.; Jeschke, G.; Goldfarb, D. *J. Chem. Phys.* **2004**, *122*, 024515.
35. Davies, E.R. *Phys. Lett. A* **1974**, *47*, 1.
36. Mims, W.B. *Proc. R. Soc. London A* **1965**, *283*, 452.
37. G. Jeschke, In *Biological Magnetic Resonance- ESR Spectroscopy in Membrane Biophysics*, Hemminga, M. Berliner, L.J., Eds.; Kluwer: New York, submitted.
38. Epel, B.; Manikandan, P.; Kroneck, P.M.H.; Goldfarb, D. *Appl. Magn. Reson.* **2001**, *21*, 287.
39. Ernst, R.R.; Bodenhausen, G.; Wokaun, A. *Principles of Nuclear Magnetic Resonance in One and Two Dimensions*, Clarendon Press: Oxford, 1987.
40. Höfer, P.; Grupp, A.; Mehring, M. *Phys. Rev. A* **1986**, *33*, 3519.
41. Reijerse, E.J.; Dikanov, S.A. *J. Chem. Phys.* **1991**, *95*, 836.
42. Höfer, P.; Grupp, A.; Nebenführ, G.; Mehring, M. *Chem. Phys. Lett.* **1986**, *132*, 279.
43. Ichikawa, T.; Kevan, L.; Bowman, M.K.; Dikanov, S.A.; Tsvetkov, Yu.D. *J. Chem. Phys.* 1979, *71*, 1167.
44. Hinderberger, D.; Spiess, H.W.; Jeschke, G. *J. Phys. Chem. B* **2004**, *108*, 3698.
45. Pöppl, A.; Böttcher, R. *Chem. Phys.* **1997**, *221*, 53.

3

CALCULATING SLOW-MOTION ESR SPECTRA OF SPIN-LABELED POLYMERS

KEITH A. EARLE
University at Albany, State University of New York, Albany, New York

DAVID E. BUDIL
Northeastern University, Boston, Massachusetts

Contents

1. Introduction   54
2. Physical Parameters of Nitroxide Labels   55
   2.1. Magnetic Tensors of the Nitroxide   55
   2.2. Diffusion Models   60
   2.3. Orientational Ordering in Spin-Labeled Polymers   63
   2.4. The Microscopic Order–Macroscopic Disorder Model   69
   2.5. Slowly Relaxing Local Structure Model   70
3. Other Stochastic Liouville Calculation Parameters   73
   3.1. Basis Set for the Stochastic Liouville Calculation   73
   3.2. Basis Set Symmetrization   74
   3.3. Basis Set Truncation Parameters   74
   3.4. Guidelines for Selecting Basis Set Truncation Parameters   76
   3.5. Basis Set Pruning   78
4. Nonlinear Least-Squares Analysis   79
Acknowledgments   81
Appendix: Program Availability   81
References   82

Advanced ESR Methods in Polymer Research, edited by Shulamith Schlick.
Copyright © 2006 John Wiley & Sons, Inc.

1. INTRODUCTION

Electron spin resonance (ESR) spectroscopy has been widely used to obtain information about the molecular dynamics of polymers. The method requires the introduction of a stable free-radical reporter group, such as a nitroxide, into the system. Nitroxide spin labels can be covalently attached to the polymer of interest, and can therefore serve as probes of the local backbone dynamics of the polymer, providing information on the local orientation, structure, dynamics, and environment.[1-3] A commonly used nitroxide is shown in Fig. 1. Depending on the ESR frequency, motion on time scales between 10^{-3} and 10^{-10} s may be investigated by this method, making it ideal to study the dynamics of macromolecules and macromolecular structures or assemblies.

The ESR spectrum of a nitroxide is sensitive to reorientational processes because of the inherent orientation dependence (anisotropy) of the magnetic interactions of the unpaired electron with the applied magnetic field, and with the magnetic nuclei on the label. The ESR spectrum is most sensitive when these interactions are modulated by rotation on a time scale τ_c that is comparable to the inverse of the frequency width, $\Delta\omega$, of the spectrum. When $\tau_c \Delta\omega \approx 1$, the motion is said to fall within the "slow-motional" regime for the given ESR frequency. At extremely short correlation times, the spectrum approaches the fast motion limit and one observes only the isotropic average of the magnetic interactions. At very long correlation times, a static distribution of all the possible probe orientations, the "rigid limit" spectrum, is observed.

Two major approaches have been employed to calculate the ESR spectrum of a paramagnetic species that is reorienting on the slow-motional time scale. One approach is the trajectory method,[4,5] which utilizes the time-dependent trajectories of axes that are fixed in the nitroxide frame to calculate the ESR spectrum directly. The trajectories may be generated either by simulating them using single-particle Brownian dynamics,[4] or by obtaining them directly from a molecular dynamics simulation.[5,6] An alternative approach is the stochastic Liouville equation (SLE),[7,8] which can be regarded as a generalized semiclassical master diffusion equation. In this description, the electronic and nuclear spins are treated quantum mechanically,

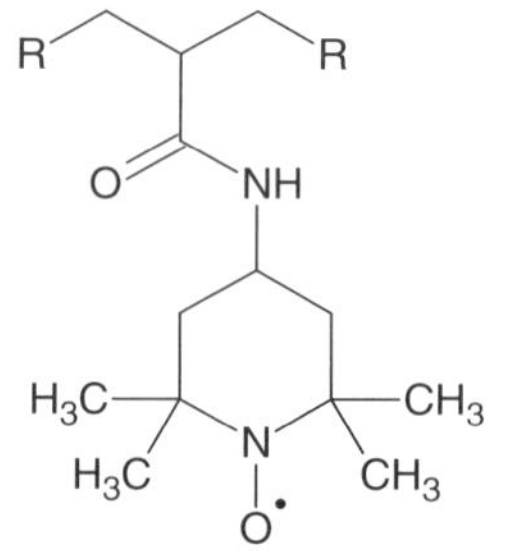

Fig. 1. A nitroxide that is commonly used in polymer spin labeling based on the commercially available 4-amino-2,2,6,6-tetramethyl piperidine-1-oxyl (TEMPAMINE) radical.

while the reorientational motion is treated classically, and parameterized in terms of rotational diffusion constants.

Of these two approaches, the SLE method enjoys much wider application to the analysis of spin-label spectra; while trajectory-based calculations remain time consuming, solution of the SLE has been rendered extremely efficient by the application of advanced sparse matrix algorithms.[9–11] On modern personal computers, a typical ESR spectrum at the conventional frequency of 9 GHz can be calculated by the SLE method in a fraction of a second. This efficiency is the major feature that enables the application of iterative analysis of the experimental line shape, as described below. Another important advantage of the SLE method is its versatility: it can be used to calculate spectra over the entire motional range of the probe, from the fast-motion limit to the rigid limit.

This chapter presents a detailed description of the analysis of slow-motional ESR spectra of spin-labeled polymers, using the SLE-based ESR line shape calculation developed by the Freed group at Cornell (EPRLL, as originally named by Schneider and Freed[10,11]). We will only present such details of the calculation as may be needed to understand how it may be most efficiently applied to ESR line shape analysis. Detailed expositions of the theory underlying the SLE equation and its implementation may be found elsewhere.[9,12–15]

We will start by surveying the parameters that are used to describe the magnetic interactions within the nitroxide label, as well as the parameters used to characterize its motion. Because of the significant number of parameters that may be required to describe the rotational dynamics, the best means of extracting information about probe structure, orientation, and dynamics is to fit a calculated ESR spectrum to the experimental one by least-squares minimization of the model parameters. The chapter describes some of the least-squares methods that have been successfully applied to this type of analysis and concludes with some illustrative examples. Specific applications to spin-labeled polymers and interpretation of the ESR parameters obtained by this analysis are described in greater detail in Chapter 6.

2. PHYSICAL PARAMETERS OF NITROXIDE LABELS

2.1. Magnetic Tensors of the Nitroxide

Analysis of a slow-motional ESR spectrum to obtain dynamic parameters requires prior knowledge of the magnetic interactions of the paramagnetic species (i.e., the parameters of the spin Hamiltonian). The two most important magnetic interactions of interest in nitroxide spin labels include (*1*) the hyperfine interaction between the electron and the nitroxide nitrogen, which may be either the ^{14}N ($I = 1$) or ^{15}N ($I = \frac{1}{2}$) nucleus; and (*2*) the *g*-factor anisotropy of the unpaired electron. Additional hyperfine interactions involving the ^{1}H nuclei on the nitroxide ring and vicinal methyl groups, which are typically not resolved in standard continuous wave (CW) ESR experiments, are therefore less important for obtaining dynamic information about the probe. However, these secondary hyperfine interactions give rise to an inhomogeneous line width that should be taken into account in the analysis of the spectra.

All orientation-dependent magnetic terms in the nitroxide may be represented as "tensors" (although not all of them are true tensors in the mathematical sense). These include the nitrogen hyperfine tensor, **A**, and the electronic g-tensor, **g**. Each tensor is represented in terms of three principal values and a set of three associated axis directions relative to the frame of the nitroxide. First, the axis orientations are discussed.

Theoretical[16] and experimental single-crystal studies[17] have shown that the axis directions of **g** (also called the *magnetic* axes x_m, y_m, and z_m) are oriented relative to the nitroxide frame as illustrated in Fig. 2. These axes are chosen so that the corresponding principal g-values obey the relation $g_x > g_y > g_z$. Thus, the z_m axis lies along the axis of the p_z orbital of the nitrogen, the x_m axis is perpendicular to z_m and lies approximately along the N–O bond direction, and the y_m axis is perpendicular to these.

The principal axis directions of the **A** tensor (x_A, y_A, z_A) are specified using the Euler angles that give their orientation relative to x_m, y_m, and z_m according to the standard convention.[18] These angles are referred to as the "magnetic tilt" angles, $\Omega_m = (\alpha_m, \beta_m, \gamma_m)$. Explicitly, they specify the following set of rotations, which are also illustrated in Fig. 3: (*1*) rotation about z_m by the angle α_m, (*2*) rotation about the new y axis by the angle β_m, and (*3*) rotation about the new z axis by the angle γ_m. This is equivalent to the following set of rotations in the original reference frame: (*1*) rotation about z_m by the angle γ_m, (*2*) rotation about y_m by the angle β_m, and (*3*) rotation about z_m by the angle α_m. In this convention, positive rotation angles specify a rotation that would advance a right-handed screw along the rotation axis.

In practice, the magnetic tilt angles are almost always taken to be zero (i.e., the **A** axes are assumed to coincide with the **g** axes); however, for nitroxides in

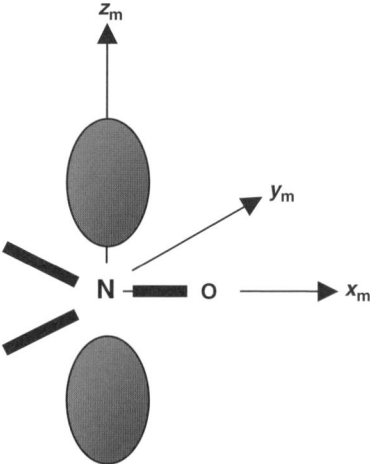

Fig. 2. Directions of the principal g (or magnetic) axes x_m, y_m, and z_m are oriented relative to the nitroxide molecular frame.

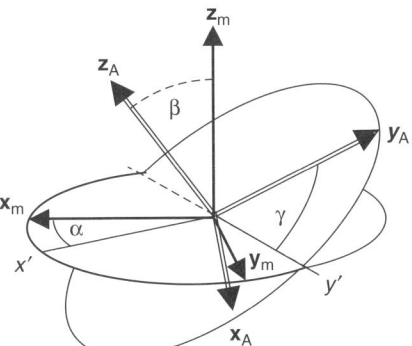

Fig. 3. Definition of the "magnetic tilt" angles, $\Omega_m = (\alpha_m, \beta_m, \gamma_m)$ that define the directions of the principal axes of the ^{14}N hyperfine tensor relative to the magnetic axes.

six-membered rings, single-crystal studies have revealed a small rotation in the x–y plane (represented by either α_m or γ_m) corresponding to "twisted boat" configurations of the ring.[17] Although such conformations are observed in single crystals, and interconversion between ring conformations should in principle contribute to the ESR line shape, conformational dynamics of the ring is typically neglected in line shape analysis of spin-labeled macromolecules.

In addition to **A** and **g**, the inhomogeneous line width tensor **W** may be specified. Here **W** is an ad hoc representation of orientation-dependent line broadening that arises from variations in the local electrostatic environment of the nitroxide. Inhomogeneous broadening of this type becomes particularly significant at higher ESR frequencies, where small variations in the **g** tensor are resolved. Moreover, this type of broadening is effectively anisotropic, since it increases in degree as the g-value gets further from the free-electron g-value. Thus, electrostatic inhomogeneity preferentially affects the g_x value of a nitroxide. The principal values of **W** represent the derivative peak-to-peak width of a Lorentzian line. Because **W** is generally used to represent g-factor inhomogeneity, the principal axes of **W** are taken to be the same as those of **g**.

The principal values of **g**, **A**, and **W** are most typically obtained from the rigid limit spectrum of the nitroxide. At high ESR frequencies (94 GHz and above), the features corresponding to the x, y, and z orientations of the nitroxide are well enough resolved that the principal values of **g** and **A** may be read directly from the spectrum, as illustrated in Fig. 4a. Specifically, at high field one may observe three groups of three features; the central feature of each group may be used to calculate the g-factor for the given orientation, and the spacing between the features gives the hyperfine interaction for that orientation. Although commercial instruments are available at frequencies up to 140 GHz, the most commonly used ESR frequency is 9 GHz. At this frequency, the features identifying the principal **g** and **A** values overlap significantly (cf. Fig. 4b), and it is often necessary to carry out a least-squares fitting to the experimental rigid limit line shape. Least-squares analysis is also the preferred method to determine the principal values of **W**.

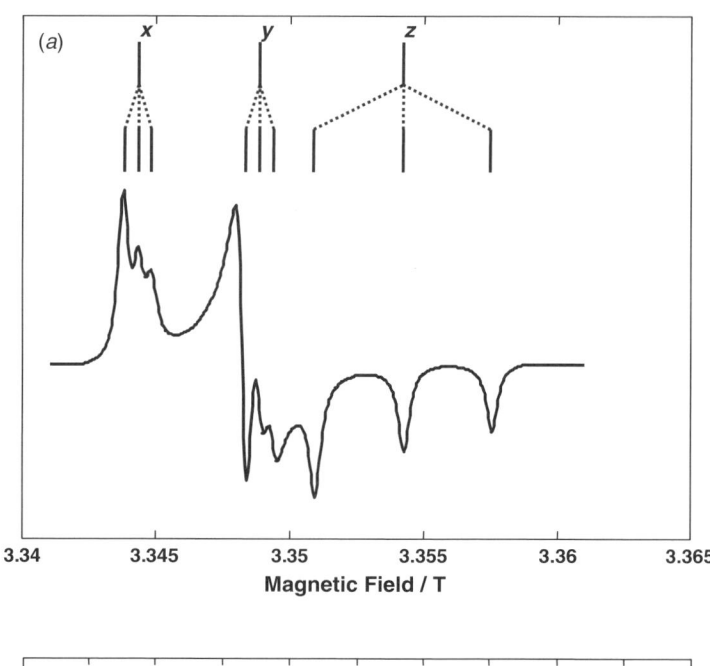

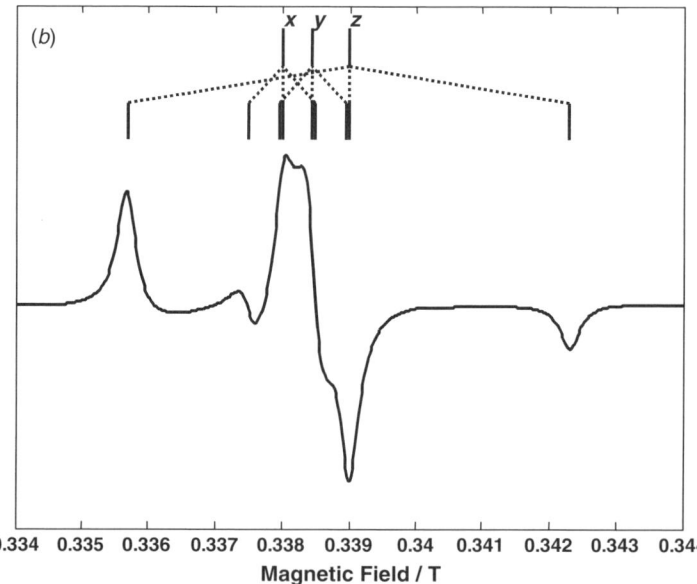

Fig. 4. Rigid limit spectra of a nitroxide at (*a*) 94 GHz and (*b*) 9 GHz. Lines indicate the location of "turning points" in the spectrum from which magnetic tensor parameters may be determined. The central lines of each group of three indicate the field positions from which the anisotropic g_x-, g_y-, and g_z- values are determined, and the spacing between the lines indicates the corresponding ^{14}N hyperfine splittings. Note the high degree of overlap in the 9-GHz spectrum.

In addition to the inhomogeneous line width tensor **W**, the EPRLL programs provide a way to account for inhomogeneous broadening that specifically arises from unresolved hyperfine interactions between the unpaired electron spin and surrounding nuclei (generally ^{1}H). This type of broadening is generally well approximated by a Gaussian line shape, and is specified as the derivative peak-to-peak width of a Gaussian line ($\Delta^{(0)}$). In liquid solution, the underlying hyperfine interactions may be treated as being isotropic; in oriented systems, such as liquid crystalline polymers or membranes, the broadening is orientation-dependent due to small anisotropies in the hyperfine interactions. Thus, the EPRLL programs provide a second Gaussian line width parameter $\Delta^{(2)}$ that accounts for the dependence of line width on the orientation angle Ψ of the membrane or liquid-crystal director (see Section 2.3) according to the definition $\Delta = \Delta^{(0)} + \Delta^{(2)}\sin^2\Psi$. It is important to note that this orientation dependence is defined relative to the director frame (defined below) and not to the magnetic axis system.

An alternative form for specifying tensor principal values that is sometimes used in the EPRLL-family programs is the "spherical" representation. In terms of the Cartesian components of a tensor **M**, namely, M_x, M_y, and M_z, its "spherical" components are defined as

$$M_1 = (M_x + M_y + M_z)/3$$

$$M_2 = M_z - (M_x + M_y)/2 \qquad (1)$$

$$M_3 = M_x - M_y$$

The components M_1, M_2, and M_3 are referred to as the isotropic, axial, and rhombic components of the tensor **M**, and they differ by constant factors from the respective conventional spherical tensor components, $M^{(0,0)}$, $M^{(2,0)}$, and $M^{(2,2)}$. The definition given here allows the components to be correlated directly with and/or estimated from the features of an experimental spectrum. This is illustrated in Fig. 5,

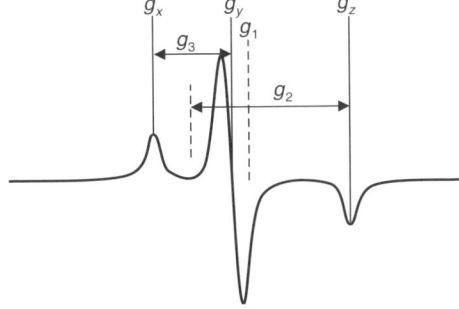

Fig. 5. Relationship between the pseudo-spherical tensor components and the turning points in the spectrum of a radical with g anisotropy but no nuclear hyperfine interactions. Note that the tensor components may be estimated directly from the features of the spectrum.

TABLE 1. Magnetic Parameters in the EPRLL Programs [a]

Symbol	Name	Description
g_x, g_y, g_z	**gx, gy, gz**	Principal values of the electronic **g** matrix:
g_1, g_2, g_3	**g1, g2, g3**	Cartesian x, y, and z components (unitless)
A_x, A_y, A_z	**ax, ay, az**	Principal values of the nuclear hyperfine
A_1, A_2, A_3	**a1, a2, a3**	tensor: Cartesian x, y, and z components (in gauss)
I	**in2**	Two times the total nuclear magnetic spin I.
$\alpha_m, \beta_m, \gamma_m$	**alpham, betam, gammam**	Euler angles $\alpha_M, \beta_M, \gamma_M$ specifying the tilt of the principal axes of the nuclear hyperfine interaction relative to those of the electronic Zeeman interaction (in degrees)
γ_n	**gamman**	Nuclear gyromagnetic ratio γ_N (in
$\Delta^{(0)}, \Delta^{(2)}$	**gib0, gib2**	Isotropic ($\Delta^{(0)}$) and orientation-dependent ($\Delta^{(2)}$) Gaussian inhomogeneous broadening $\Delta = \Delta^{(0)} + \Delta^{(2)}\sin^2\psi$ (Gaussian derivative peak-to-peak width in Gauss)
W_x, W_y, W_z	**wx, wy, wz**	Principal values of the orientation-dependent
W_1, W_2, W_3	**w1, w2, w3**	inhomogeneous line-broadening tensor, Cartesian x, y, and z components (Lorentzian derivative peak-to-peak width in gauss)

[a] (One-nucleus, one-electron spin $\frac{1}{2}$ system).

which for clarity shows the spectrum of an unpaired electron with g anisotropy, but without any resolved hyperfine splittings.

The spherical representation is included because it allows one to vary the tensor components in such a way that it remains axially symmetric, or maintains a constant trace. Such constraints can often be assumed based on other considerations (e.g., average hyperfine coupling from fast-motional spectra) before the least-squares fitting is undertaken. Thus, to maintain a constant isotropic value one should fix M_1 and vary M_2 and M_3; to maintain axial symmetry, M_3 should be set to zero, and M_1 and M_2 varied.

Table 1 summarizes the magnetic parameter names that are used in the EPRLL family programs to specify the spin Hamiltonian for a electron spin $S = \frac{1}{2}$ radical with a single significant nucleus, such as a nitroxide spin label.

2.2. Diffusion Models

This section summarizes some of the models for molecular motion that are implemented in currently available SLE programs for ESR line shape calculations. In the following description, parameters specifying rates or rate constants in units of reciprocal seconds (s^{-1}) will be expressed on a base 10 logarithmic scale. Thus, for example, a diffusional rate constant of $3.0 \times 10^7 \, s^{-1}$ would be represented by the parameter

PHYSICAL PARAMETERS OF NITROXIDE LABELS   61

value 7.48. The major reason for this convention is that most dynamic processes of the polymer or spin label are activated processes and therefore exhibit an approximately exponential dependence upon the temperature. A second advantage is that dynamic parameters expressed on this scale are of the same order of magnitude as the other quantities used as input to the SLE, which avoids some numerical problems that may arise in least-squares minimization when the search parameters are of significantly different magnitudes.

Before discussing the different parameters used in the diffusion operator SLE formalism, it is necessary to define a number of different coordinate systems, depicted in Fig. 6a and b. The first frame of interest is the *director frame* (x_D, y_D, z_D), which is fixed relative to the structure of the polymer to which the label is attached. The z_D axis is used to define the energy potential that imposes orientational order on the

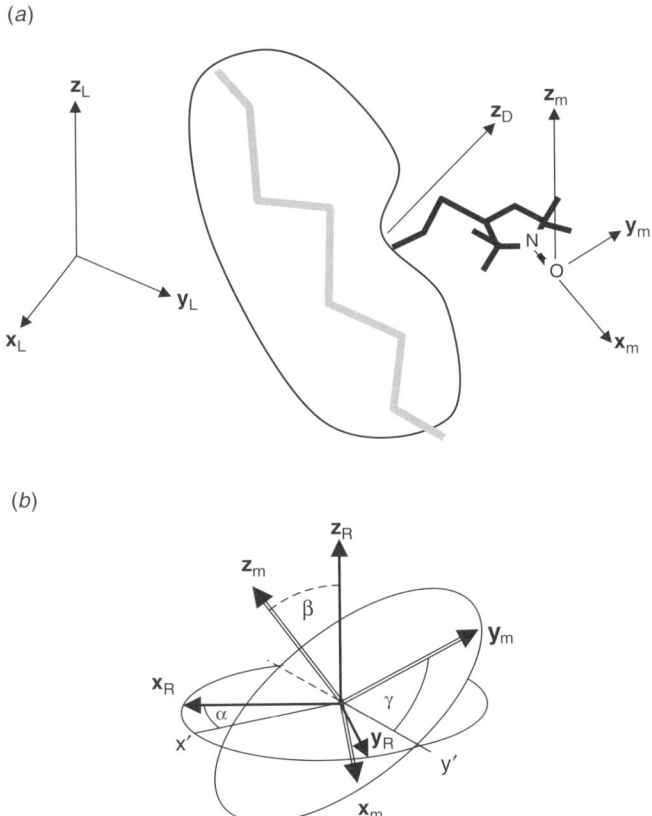

Fig. 6. (a) Different coordinate systems (laboratory: L, director: D, and magnetic: m) used to define motion parameters for a nitroxide spin label. (b) Diffusion rotation angles used to define the magnetic axes relative to the diffusion axes. Note that the reference system for these angles is the diffusion frame, whereas the reference system is the magnetic (g) frame for the magnetic tilt angles (cf. Fig. 3).

probe molecule as described above. The *magnetic frame* (x_m, y_m, z_m) already defined above is fixed relative to the structure of the nitroxide label. Another important frame is the principal axis system of the rotational diffusion tensor, or *rotational diffusion frame* (x_R, y_R, z_R), which is fixed relative to the magnetic frame. The orientations of the principal diffusion axes are specified by the *diffusion tilt angles* $\Omega_D = (\alpha_D, \beta_D, \gamma_D)$, which are the Euler angles of the magnetic axes in the rotational diffusion frame as defined above. These Euler angles are defined in exactly the same way as those shown in Fig. 3 and the accompanying discussion. However, it is important to note that the base frame for the magnetic tilt angles is the magnetic (g-tensor) axis system, whereas the base frame for the diffusion tilt angles is the diffusion axis system (cf. Fig. 6b with Fig. 3), even though both sets of angles involve the magnetic frame.

The standard model for diffusive motion in polymers is Brownian diffusion, which occurs as a series of infinitesimal reorientational steps. This model is most appropriate for intermediate-to-large sized spin probes and spin-labeled macromolecules, where the macromolecule is much larger than any solvent molecules. Because of this broad applicability, the Brownian diffusion model is the most widely used. This type of rotational diffusion is completely analogous to the one-dimensional random walk used to describe translational diffusion in standard physical chemistry texts, with the difference that the steps are described in terms of a small rotational step $\delta\theta$ that can occur in either the positive or negative direction. In three dimensions, rotations about each of three principal axes of the nitroxide must be taken into account. A diffusion constant may be defined for each of these rotations motions, in a way that is completely analogous to the definition of translational diffusion constant for the one-dimensional random walk.

The orientations of the principal diffusion axes x_R, y_R, z_R depend critically on the hydrodynamic properties and geometry of the spin label as well as its attachment to the polymer backbone. The rotational diffusion constants R_x, R_y, and R_z associated with each axis are in general different from each other; however, nitroxides that are covalently tethered to a polymer backbone generally exhibit a principal rotation axis, around which the probe rotation is significantly faster than for any other axis. This principal direction is assigned to be the z_R axis and its orientation is typically determined by the tether bonds of the nitroxide to the polymer backbone.

In cases where the overall rotation of a polymer molecule may be neglected because of its size, the combined motion of the spin label and local polymer chain segment may be approximated using a single rotational diffusion tensor.[19,20] This approximation assumes that the local polymer chain segment has an isotropic rotational diffusion coefficient R_S and the spin label rotates with diffusion coefficient R_I around the tether by which it is attached to the polymer. In this case, the rotational diffusion tensor will have axial symmetry, with $R_\perp = R_S$ and $R_\parallel = R_S + R_I$.

The EPRLL programs also allow for non-Brownian rotational diffusion, which implies a discrete, step motion of the spin probe. Two limiting models are available: (*1*) jump diffusion, and (*2*) approximate free diffusion. In currently available implementations of the SLE line shape calculation, non-Brownian models may not be used with an orienting potential, and only with the assumption of an axially symmetric diffusion tensor. For these reasons, and since Brownian motion is usually an excellent approximation for large molecular systems, such models are only rarely applied in the analysis of spin-labeled polymers.

In addition to the rotational diffusion constants $R_\| = R_z$ and $R_\perp = R_x = R_y$, non-Brownian diffusion around each axis is specified using the *model parameters*, $P_\| = R_\|\tau_\|$ and $P_\perp = R_\perp\tau_\perp$, where $\tau_\|$ and $\tau_\perp$ are the non-Brownian *residence times*. The physical interpretation of the residence time for a given axis depends on the specific model used. In jump diffusion, the molecule remains stationary for an average time τ after which it jumps to a new orientation, specified by an angle of rotation about the specified axis. The root-mean-square reorientation angle (in rad) is given by the formula $\theta_{avg} = \langle\theta^2\rangle = 6R_i\tau_i$, where i specifies the axis perpendicular or parallel to z_R. In the approximate free diffusion model, the molecule rotates freely about axis i with a rotational rate R_i, but instantaneously reorients at an average interval τ, after which it continues its free rotation. Like all of the other dynamic parameters in the EPRLL family programs, the non-Brownian residence time products are specified on the $\log_{10}$ scale.

In addition to the rotational diffusion models mentioned above, it is possible to account approximately for the effects of isotropic Heisenberg spin exchange between probe molecules with a rate specified by ω_{SS} (in units of rad s^{-1}). This parameter may become important in microscopically heterogeneous polymers, where aggregation of the spin probes is possible, and it has been applied to account for spin–spin interactions of spin labels adsorbed to the surfaces of zeolites.[21,22] In its original implementation, ω_{SS} was used to simulate the line broadening of the fast-motional spectrum that occurs at high-spin probe concentrations, reflecting the encounter rate between spin-bearing molecules. However, the most recent program modification allows exchange to take place between spin labels of arbitrary orientation.[14]

2.3. Orientational Ordering in Spin-Labeled Polymers

The EPRLL programs include a mechanism to describe the tendency of a spin label or spin probe to become partially ordered within its local environment, a situation that is encountered frequently in spin-labeled polymers. Probe ordering is characterized through the use of an orienting potential that governs the tendency of the spin probe to align relative to the polymer molecule. It may also be regarded as imposing constraints upon the spin-label motion, so that the probe samples only a restricted range of orientations as it moves.

The orienting potential was originally intended for use with materials, such as liquid crystals and membranes, where the molecules align with a macroscopic direction called the *director*. In polymer systems, the director is usually defined microscopically, reflecting the fact that local structural features, such as the polymer backbone or microscopic crystalline or liquid-crystalline domains of the polymer, influence the label alignment.

Figure 7 illustrates one way in which the local director may be defined in spin-labeled polymer systems. As the spin label moves relative to the polymer, its diffusion axes trace trajectories in a reference frame that is fixed relative to the polymer. The orientation trajectory may be regarded as a path traced by the vector on the

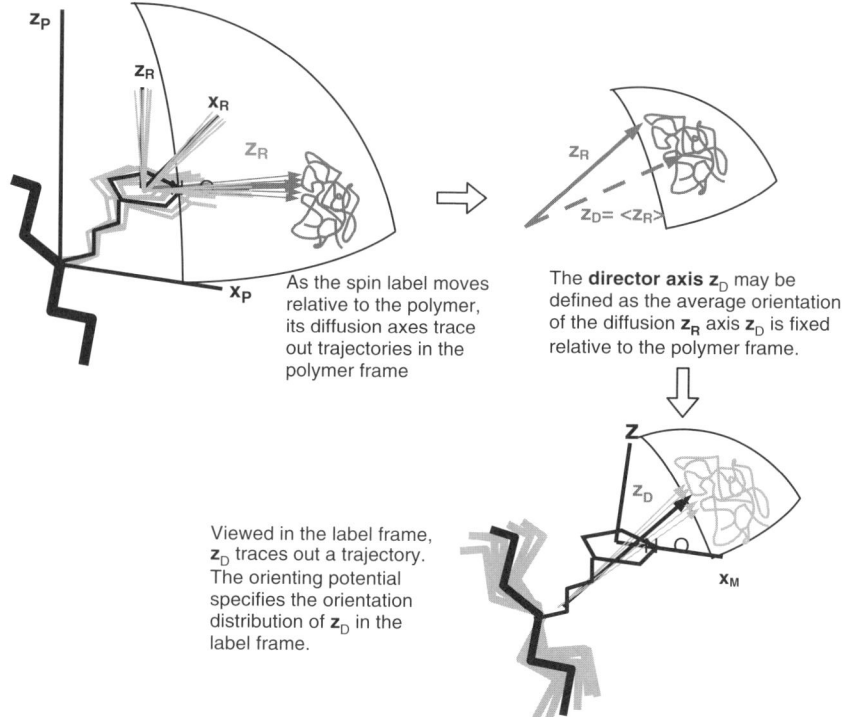

Fig. 7. Definition of the local director axis in spin-labeled polymer systems.

surface of a sphere. The director axis z_D may be defined as the average orientation of the diffusion z_R axis over the course of its motion. The z_D vector is thus fixed relative to the polymer frame. However, viewed in the spin-label frame, z_D traces a trajectory. The orienting potential specifies the orientation distribution of z_D in the rotational diffusion frame.

The orienting potential is expressed as a function of the polar angles (θ,ϕ) of the director in the rotational diffusion axis frame. It is most conveniently included in the SLE equation by expanding it in a series of spherical harmonic functions $\mathcal{D}_K^L$ as follows:

$$U(\Omega) = -\sum_{L,K} c_{LK} \mathcal{D}_K^L(\theta,\phi) = -\sum_{L,K} c_{LK} \left[D_{0K}^L(\theta,\phi) + D_{0-K}^L(\theta,\phi) \right] \qquad (2)$$

Each function $\mathcal{D}_K^L$ is a symmetric combination of generalized spherical harmonic functions $D_{0K}^L(\Omega)$, where Ω is the set of Euler angles that specifies the orientation of the director with respect to the label frame. The symmetrical combination of spherical harmonics with $+K$ and $-K$ indices ensures that each of the $D_{0K}^L(\Omega)$ functions (and thus the potential U) is real valued.

The EPRLL-family programs include additional restrictions on the terms in Eq. 2. The summation is restricted to even values of L and K that are ≤ 4. Thus, there are only five terms in the potential, represented by the coefficients c_{20}, c_{22}, c_{40}, c_{42}, and c_{44}. These restrictions are equivalent to two assumptions about the ordering potential: first, that the orienting potential has at least twofold symmetry, and second, that the principal ordering axes of the nitroxide are identical to the principal diffusion axes. The observables obtained by least-squares analysis of the ESR spectrum are the coefficients c_{LK}, which are expressed as energy in units of kT.

The observable effect of an orienting potential upon the ESR spectrum is the anisotropic distribution of orientations produced by the potential. The equilibrium distribution of orientations obeys the Boltzmann relation,

$$P(\theta,\phi) = \frac{\exp[-U(\Omega)/kT]}{\int_\Omega \exp[-U(\Omega)/kT]d\Omega} \quad (3)$$

where P is the probability of finding the director axis z_D with polar angles in the diffusion frame between (θ,ϕ) and $(\theta + d\theta, \phi + d\phi)$.

One may obtain some idea about the shapes of the orienting potentials produced by each term in Eq. (2) by recognizing their equivalence with the more familiar shapes of atomic orbitals, which are also based on spherical harmonics. For example, the $L = 2$, $K = 0$ function is similar to the d_z^2 orbital, with a positive lobe along the $+z$ direction and a negative lobe in the xy plane, perpendicular to z. Likewise, the $L = 2$, $K = 2$ corresponds to the $d_{x^2-y^2}$ orbital. Each of the possible terms in the orienting potential function is given with its explicit functional form and the related atomic orbital designation in Table 2. The corresponding population distributions $P(\theta,\phi)$ for each function are displayed in Fig. 8. Two columns are shown for the two possible signs of the coefficient for each function in order to emphasize the regions for which the function is positive or negative. Figure 8 makes apparent that the orienting potential and related population distribution function refer to the distribution of the director in the label diffusion frame. For example, the $L = 2$, $K = 0$ function has a positive lobe along the $\mathbf{z}_R$ axis and a negative lobe or ring in the $\mathbf{x}_R$-$\mathbf{y}_R$ plane. Thus, a potential

TABLE 2. Terms in the Energy Potential Used to Describe Spin Label Orientation

Term	Functional Form	Analogous Atomic Orbital
$\mathcal{D}_0^2(\theta)$	$\frac{1}{2}(3\cos^2\theta - 1)$	d_{z^2}
$\mathcal{D}_2^2(\theta,\phi)$	$\sqrt{\frac{3}{2}}\sin^2\theta\cos2\phi$	$d_{x^2-y^2}$
$\mathcal{D}_0^4(\theta)$	$\frac{1}{8}[(35\cos^2\theta - 30)\cos^2\theta + 3]$	g_{z^4}
$\mathcal{D}_2^4(\theta,\phi)$	$\sqrt{\frac{5}{8}}\sin^2\theta(7\cos^2\theta - 1)\cos2\phi$	$g_{z^2(x^2-y^2)}$
$\mathcal{D}_4^4(\theta,\phi)$	$\sqrt{\frac{35}{8}}\sin^4\theta\cos4\phi$	$g_{x^4-y^4}$

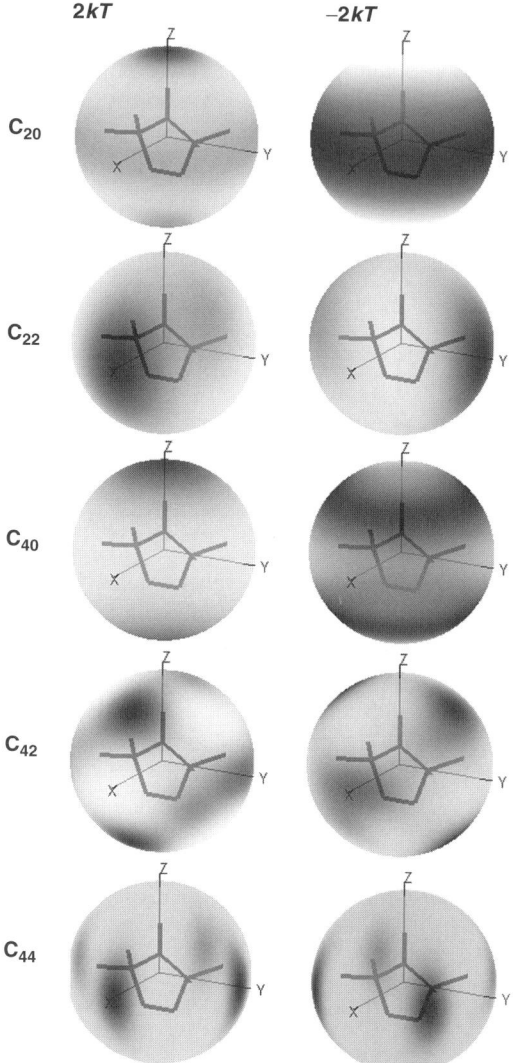

Fig. 8. Population distributions $P(\theta,\phi)$ calculated for each of the functions used in the orienting potential expansion given in Table 2.

with a positive c_{20} coefficient will have a minimum for $\theta = 0$ (i.e., along the z_R axis) by virtue of the negative sign in Eq. 2. In turn, this will produce a maximum in the orientation population distribution at this orientation according to the Boltzmann relation in Eq. 3, as shown in the top plot on the left-hand side of Fig. 8. In contrast, a negative c_{20} will produce a minimum in the potential at $\theta = 90°$, leading to a maximum in the director orientation distribution in the x_R-y_R plane, as shown in the top plot on the right-hand of Fig. 8. By analogy, the $L = 2$, $K = 2$ function has positive lobes along the $\mathbf{x}_R$ axis and negative lobes along the $\mathbf{y}_R$ axis, so that a positive c_{22} coefficient produces director orientations distributed along $\mathbf{x}_R$ and a negative c_{22} produces orientations along $\mathbf{y}_R$ (cf. plots second from the top in Fig. 8). Similar arguments apply to the higher order functions shown in the figure.

In practice, satisfactory fits to experimental ESR spectra at 9 GHz can usually be obtained by varying only the c_{20} and c_{22} coefficients. However, higher frequencies afford greater orientational resolution, so that higher order terms in the potential are often required to achieve satisfactory fits to high-frequency spectra. Even within the limitations placed on the orienting potential function expansion, the distributions in Fig. 8 make clear that rather complex orientation distributions can be reproduced by an appropriate combination of potential coefficients. Figure 9 illustrates two of the more commonly encountered distribution shapes that reflect potentials with significant contributions from at least two of the functions given in Table 2.

The first commonly encountered distribution is an "elongated spot" shown along the top of Fig. 9, which arises from combinations of the c_{20} and c_{22} coefficients. This type of distribution is observed when the probe experiences different degrees of orientation around different axes, as depicted schematically in the cartoon in Fig. 10a. Assuming that z is the ordering axis as shown in Fig. 9, the direction of the elongation

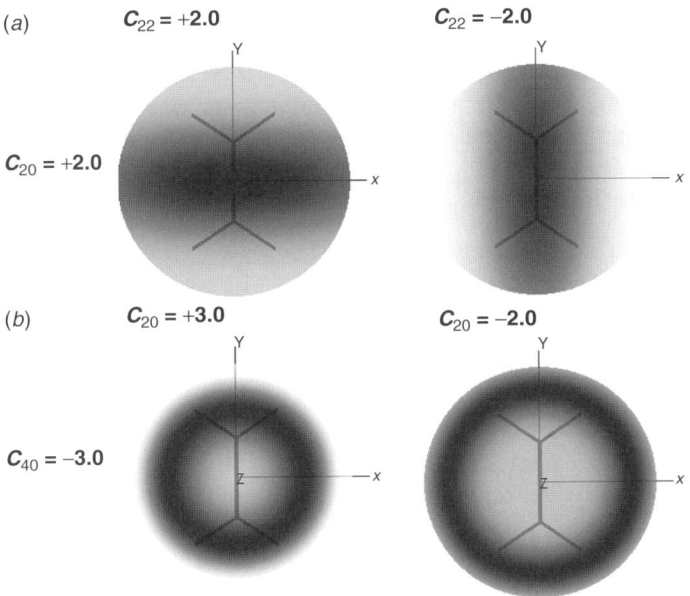

Fig. 9. Two commonly encountered shapes of director axis distributions relative to the nitroxide frame, given by potentials with significant contributions from at least two of the functions given in Table 2. (*a*) Elongated spot given by different combinations of the c_{20} and c_{22} coefficients; note that the orientation of the long axis of the spot depends on the sign of c_{22}. (*b*) Cone distributions given by different combinations of the c_{20} and c_{40} coefficients, with $c_{40} < 0$. The angle of the cone is determined by the ratio c_{40} / c_{20}.

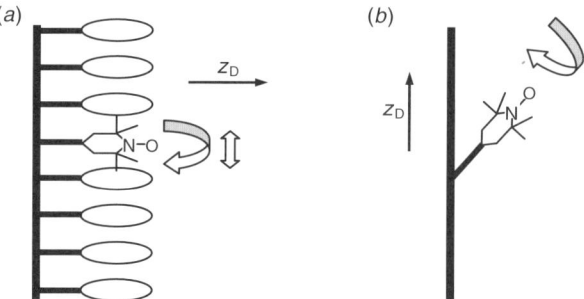

Fig. 10. Cartoons depicting situations in which a spin-labeled polymer may yield the types of local director distribution shapes shown in Fig. 9. (a) Director perpendicular to polymer chain direction, determined by interactions between spin-label and liquid-crystalline side chains. Different ordering around two axes leads to the elongated distributions shown at the top of Fig. 9b Director parallel to main-chain direction, with spin label held at a fixed angle with respect to the chain, leading to the conical distributions shown at the bottom of Fig. 9.

is determined by the sign of c_{22}: for positive c_{22}, the spot is elongated in the xz plane, and for negative c_{22}, it is elongated in the yz plane. Appropriate combinations of c_{20} and c_{22} can be used to produce the same distribution with x or y as the ordering axis. Barnes and Freed have given a useful set of relations that can be used to accomplish these transformations.[23]

A second commonly encountered distribution is the cone distribution illustrated along the bottom of Fig. 9, and in Fig. 10b. This type of distribution is produced by a combination of the c_{20} and c_{40} coefficients, with $c_{40} < 0$. The half-angle of the cone is determined by the ratio of c_{20} and c_{40} as follows:

$$\cos^2\theta_{\mathrm{cone}} = \left(\frac{3}{7} - \frac{6}{35} \frac{c_{20}}{c_{40}} \right) \tag{4}$$

Although c_{40} must be negative to obtain a conical distribution, c_{20} is positive for $\theta_{\mathrm{cone}} < 49.1°$, and negative for $\theta_{\mathrm{cone}} > 49.1°$. Any half-angle between 0 and 90° can be modeled in this way, which enables one to model most situations of practical interest in spin-labeled polymers.

In the presence of an orienting potential, another model for spin-label diffusion that is available from the EPRLL family programs is the "anisotropic viscosity" model, originally described by Polnaszek et al.[24] The anisotropic viscosity model is primarily used for macroscopically ordered fluids such as membranes and liquid crystals and its applicability to spin-labeled polymer systems is limited to cases where the polymer may be aligned macroscopically to a relatively high degree of order. The dynamic model that is most commonly applied to polymers in the presence of a local orienting potential is some form of the microscopic order–macroscopic disorder (MOMD) model described in Section 2.4.

2.4. The Microscopic Order–Macroscopic Disorder Model

In most cases, the local directors in a polymer sample are not aligned with each other, and therefore cannot be aligned with a macroscopic direction. That is, the polymer sample consists of locally ordered domains that are randomly oriented with respect to the laboratory frame as illustrated schematically in Fig. 11. The model describing this situation is known as the MOMD model. As originally described for spin-labeled biological membranes and proteins,[25] the model assumes that the spin probe undergoes microscopic molecular ordering with respect to a local director; however, the local directors in the sample are randomly oriented and rigidly fixed with respect to the laboratory frame. This situation is realized in many types of polymers with crystalline or liquid-crystalline phases, and may also apply to spin labels where the local dynamics is constrained by the tether bond or by interaction with polymer side chains.

The approach to calculating a MOMD spectrum using the EPRLL programs is straightforward: The spectrum must be integrated over the distribution of local director orientations Ψ. Figure 12 shows the component spectra corresponding to selected tilt angles that are integrated to give MOMD spectra at both 9 and 94 GHz. In the calculation program, a MOMD spectrum is specified simply by setting the number of tilt angles orientations, n_{ort}. It is important to allow enough tilt angles to ensure proper sampling of director orientations and convergence of the MOMD spectrum. At 9-GHz frequencies, between 10 and 30 orientations are generally needed, with a

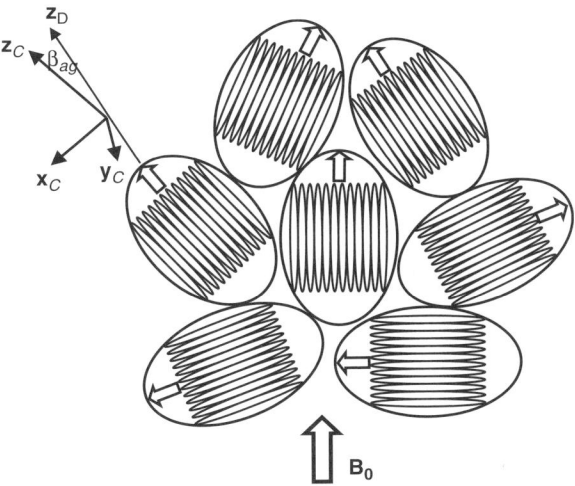

Fig. 11. A representation of a MOMD sample showing local ordering of microdomains with directors z_D (heavy arrows), but macroscopic disorder due to the random orientations of the microdomains in a three-dimensional disordered 'mosaic'. The applied magnetic field B_0 points along the laboratory z axis. The upper left shows the axis system defined for the case of a slowly relaxing local structure. Motion of the microdomains is defined relative to the "cage" system x_C, y_C, z_C, and β_{ag} is the tilt angle between the principal cage diffusion axis z_C and the local director z_D.

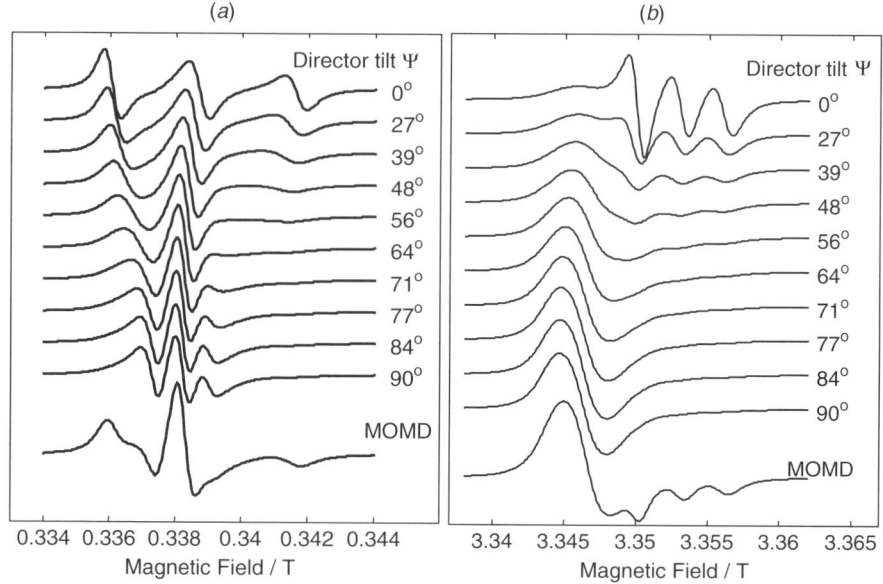

Fig. 12. Components of a MOMD ESR spectrum of a nitroxide spin label, showing the individual spectra from domains with different tilt angles, and integrated MOMD spectrum at the bottom, for 9 GHz (left-hand side) and 94 GHz (right-hand side).

greater number of orientations required when the line width for a single orientation is small. More orientations are also required at higher frequencies, where the orientation resolution is greater.

The EPRLL family programs also admit a "partial MOMD" model described by Barnes and Freed.[23] This model includes microscopic ordering with respect to a local director just as in the MOMD model; however, the directors are assumed to be partially ordered with respect to a macroscopic ordering axis. The tilt angle Ψ in this case refers to the angle between the spectrometer field and the macroscopic ordering direction. The macroscopic ordering is described by a potential that is analogous to the microscopic spin-label ordering potential; however, it only includes the $L = 2$, $K = 0$ term with coefficient b_{20}. Although the partial MOMD model was originally conceived to treat partially ordered membrane dispersions, it should be applicable to such systems as spin-labeled polymer fibers that are partially aligned with a macroscopic direction.

The dynamic parameters for the EPRLL program are summarized and defined in Table 3.

2.5. Slowly Relaxing Local Structure Model

A more general dynamic model permits motion of the potential-defining environment on the ESR time scale, as might occur in a slowly tumbling spin-labeled polymer. In such a system, the ordering potential is considered to define the constrained

TABLE 3. Dynamic Parameters of EPRLL Programs

Symbol	Name	Description
$\alpha_D, \beta_D, \gamma_D$	**alphad, betad, gammad**	Diffusion tilt angles in degrees
$c_{20}, c_{22}, c_{40},$ $c_{42}, c_{44},$	**c20, c22, c40, c42, c44**	Orienting potential coefficients in units of kT
b_{20}	**b20**	Director orienting potential coefficient for $L=2$, $K=0$ in units of kT
ω_{ss}	**oss**	Log_{10} of Heisenberg spin exchange rate in $\sec^{-1}$
$P_\perp, P_\parallel$	**pmxy, pml**	Model parameter for non-Brownian motion around z axis
Ψ	**psi**	Director tilt angle in degrees
R_x, R_y, R_z R, N, N_{xy}	**rx, ry, rz** **rbar, n, nxy**	Cartesian (and spherical) components of the rotational diffusion tensor ($\log_{10}$ of diffusion constant in s^{-1})
$\hat{R}_\parallel, \hat{R}_\perp$	**djf, djfprp**	Anisotropic viscosity rotational diffusion rate constants parallel and perpendicular to director

environment of the tethered spin probe. This constrained environment slowly reorients with respect to the laboratory frame as the polymer tumbles in solution, providing the coupling to the over-all, or global, diffusion.

The slowly relaxing local structure (SRLS) model describes this type of composite dynamics using only a few modes of diffusion. The main justification for this simplification is the limited resolution of the observed ESR line shapes. In the SRLS model, the spin probe is assumed to be reorienting in a local environment that is relaxing on a longer time scale. In applications to macromolecular systems, the faster motion describes the internal dynamics, while the slower motions account for the global rotation of the macromolecule.

It is important to note that the SRLS model contains the MOMD model as a limiting case, namely, the rigid limit for the slower global motion. Another significant limiting case occurs when the internal dynamics is in the fast-motion limit, called fast internal motion (FIM) model. In the FIM model, the internal dynamics lead to a partial averaging of the magnetic tensors, which is quantified by an effective-order parameter. The rotational diffusion tensor then describes the global motion.[26]

The general SRLS model requires a number of additional parameters to describe the global diffusion modes.[27–29] Three rotational diffusion constants, R_x^0, R_y^0, and R_z^0 are used to characterize the reorientation of the SRLS cage. They are expressed on the

$\log_{10}$ scale, and may be expressed as the pseudo-spherical components $\overline{R}^0$, N^0, and N^0_{xy} by analogy with the probe diffusion parameters described above. Finally, the general SRLS model includes a tilt angle β_{ag} between the principal cage diffusion axis $\mathbf{z}_C$ and the local director $\mathbf{z}_D$. The SRLS dynamic parameters are summarized in Table 4.

Often, a simplified SRLS model is sufficient to describe experimental data. Most commonly, only one additional dynamical parameter, $R^C_\perp$, is needed to describe the cage diffusion. Also, the mean-field cage potential can often be approximated with cylindrical symmetry, so that only the $L = 2$ parameters need be retained in the cage orienting potential.

Although the SRLS model has not yet been extensively applied to spin-labeled polymer systems, there are a number of studies on model and biological systems that illustrate its potential utility in this application. SRLS is needed in cases where both the local and global dynamics induce significant averaging of the spectrum. For example, time scales typical of global macromolecular modes of motion generally fall within the slow-motion regime near ESR frequencies of 9 GHz; thus, it is important to account for both the global and local dynamics at this frequency. In contrast, the global modes of motion are often in the rigid limit at high ESR frequencies. Thus it is possible to determine local dynamic parameters using the MOMD model at high frequency, then fix these parameters in a constrained SRLS analysis of 9-GHz spectra to characterize the global dynamics. The work of Liang and Freed, who derived the MOMD limit from the more general SRLS model,[26] validated this general approach.

The SRLS mechanism has been invoked to explain the segmental rotational dynamics of spin-labeled polystyrene in toluene solution.[20] This study provides an example of a system in which the dynamic cage is defined by constrained diffusion of the spin label attached to a polymer. The SRLS model is appropriate to analyze these spectra, since the local ordering environment is coupled to the global tumbling modes of the polymer. Moreover, the high (250–GHz) frequency ESR spectra of this

TABLE 4. SRLS Parameters of ESR Line Shape Programs

Symbol	Name	Description
$a_{20}, a_{22}, a_{40},$ a_{42}, a_{44}	**a20, a22, a40,** **a42, a44**	Coefficients of coupling potential between SRLS cage and spin probe in units of kT
b_{10}, b_{20}	**b10, b20**	Coefficients of orienting potential for SRLS cage in units of kT
$\underline{R}_x^\circ, R_y^\circ, R_z^\circ$ $\overline{R}^\circ, N^\circ, N_{xy}^\circ$	**r0x, r0y, r0z** **r0bar, n0, n0xy**	Cartesian (and spherical) components of the rotational diffusion tensor for the SRLS cage ($\log_{10}$ of diffusion constant in s^{-1})
β_{ag}	**bag**	Cage tilt angle between the SRLS cage principal diffusion axis and the SRLS cage director

system are not sensitive to the global tumbling modes of the polymer, as they are too slow on the 250-GHz ESR time scale.

3. OTHER STOCHASTIC LIOUVILLE CALCULATION PARAMETERS

This section describes a number of input parameters used to control the stochastic Liouville calculation itself, the most notable of which are the parameters that specify the basis set used to carry out the calculation.

3.1. Basis Set for the Stochastic Liouville Calculation

The solution of the SLE may be expressed as the following matrix equation:

$$I(\Delta\omega) = \langle\langle v|[(-i\mathbf{L}+\Gamma)+\mathbf{I}\Delta\omega]^{-1}|v\rangle\rangle \tag{5}$$

in which I is the intensity of the spectrum, ω is the frequency, $\mathbf{L}$ is the Liouville superoperator that governs the quantum mechanical evolution of the electron and nuclear spins, Γ is the stochastic superoperator that governs the spin probe motion, $\mathbf{I}$ is the unit matrix, and v is a vector that projects out the observable magnetization and equilibrium orientation distribution of the probe. Although we will not go into the details of this calculation, it is necessary to understand a few important features of Eq. 5 in order to apply it to ESR line shape analysis.

The most important feature of the matrices and vectors in Eq. 5 is that they are constructed in a vector space, or *basis set*, that consist of the direct product of generalized spherical harmonic functions representing the rotational degrees of freedom of the probe, and spin functions that represent its spin degrees of freedom. Proper specification of the basis set is therefore an important requirement for accurate calculation of the slow-motional spectrum.

Each basis function in the EPRLL program is specified by five quantum numbers: L, K, M, p_I, q_I, with L, K, and M specifying the generalized spherical harmonic function, and the transition indices p_I and q_I specifying the spin functions. Note here that p_I and q_I refer only to nuclear spin states. In the general formulation of the SLE, it would also be necessary to include the indices p_S and q_S for the electronic spin states; however, the EPRLL programs make use of the high-field approximation, which implicitly restricts the calculation to the $p_S = 1$, $q_S = 0$ subspace.[13] The basis set indices and the physical quantities they represent are summarized in Table 5.

We briefly mention that, for the SRLS model described above, up to three additional basis indices are needed to represent motion of the solvent cage, namely, L^C, K^C, and M^C. These indices are specified by truncation parameters and have the same general physical meaning (in the context of cage diffusion) as the L, K, and M indices in Table 5. The different approaches to optimizing basis sets described in Sections 3.2 and 3.3 also apply to the cage indices; however, for the sake of simplicity, the SRLS model in these sections will not be considered.

TABLE 5. Basis Set Indices and Their Physical Interpretations

Index	Physical Interpretation	Range
L	Quantum number for total rotational angular momentum	(0 to L_{emax})
K	Quantum number for projection of rotational angular momentum on laboratory Z axis	($-L$ to L)
M	Quantum number for projection of rotational angular momentum on molecular Z axis	($-L$ to L)
p_I	Nuclear spin transition index: net change in Z projection of nuclear magnetic moment	($-2I$ to $2I$)
q_I	Nuclear spin transition index: total number of nuclear spin quanta involved in a transition	($|p_I|-2I$ to $2I-|p_I|$)

3.2. Basis Set Symmetrization

The full basis set of basis functions used in the SLE calculation includes both positive and negative K and M indices. However, the entire range of indices is required only in the general case; in many cases of interest, it is possible to reduce the number of K and M values in the basis set. This reduction is accomplished by two so-called "symmetrization" transformations called the K and M symmetrizations. To illustrate what this means, the K-symmetrized basis set is constructed by taking symmetric and antisymmetric combinations of basis functions with the same absolute value of K. That is, each pair of functions $|L, K, M, p^I, q^I\rangle$ and $|L, -K, M, p^I, q^I\rangle$ in the original basis are used to form two new functions proportional to $(|L, K, M, p^I, q^I\rangle \pm |L, K, M, p^I, q^I\rangle)$. The new functions are specified by their (non-negative) K index and a symmetry index j^K, which is $+1$ for the symmetric and -1 for the antisymmetric combination. The M symmetrization is similar, although this transformation also includes specific combinations of spin transition indices.[13,14]

Although the K and M indices are always non-negative in the symmetrized basis, the EPRLL programs accept negative values for the K and M truncation indices, using the convention that negative K or M values specify j^K or $j^M = -1$. Many typical SLE calculations do not require basis functions with j^K or $j^M = -1$. The special circumstances requiring use of negative K and M indices are discussed below.

3.3. Basis Set Truncation Parameters

For any of the basic EPRLL-family programs, it is necessary to specify the range of basis set indices that will be used to construct the SLE matrix equation. This is done via the five *truncation parameters*, L_{emx}, L_{omx}, K_{mx}, M_{mx}, and p^I_{mx}, which specify the maximum values that may be assumed by each quantum number in the basis. Separate maxima are specified for even and odd L values, and no truncation parameter is needed for the q_1 index. In cases where negative K and M indices are required, two additional truncation parameters are also used: K_{mn} and M_{mn}.

In general, higher truncation values are needed to represent slower motions. This may be understood by the following analogy. A rigid limit ESR spectrum is calculated by averaging the spectrum over all possible orientations of the paramagnetic species relative to the spectrometer field. It is therefore necessary to sample the orientations with sufficient resolution to achieve a smooth line shape; an insufficient number of orientations produces "ripples" in the line shape. By analogy, higher L, K, and M indices correspond to sampling of a larger number of orientations. Thus, if the truncation indices used for a given slow-motional calculation are not large enough, "ripples" appear in the calculated slow-motional spectrum, as shown in Fig. 13.

Since the matrix size, and thus the processor time, for a given calculation increase geometrically with the truncation indices, calculations can quickly become quite time-consuming as the basis set is increased. Thus, it is desirable to optimize the basis by finding the smallest set necessary for acceptable convergence of the calculated spectrum, particularly for MOMD spectra. When the truncation parameters are all set to their respective acceptable minima, the corresponding list of basis elements is referred to as the *minimum truncation set* (MTS), and the truncation parameters are called the MTS parameters.

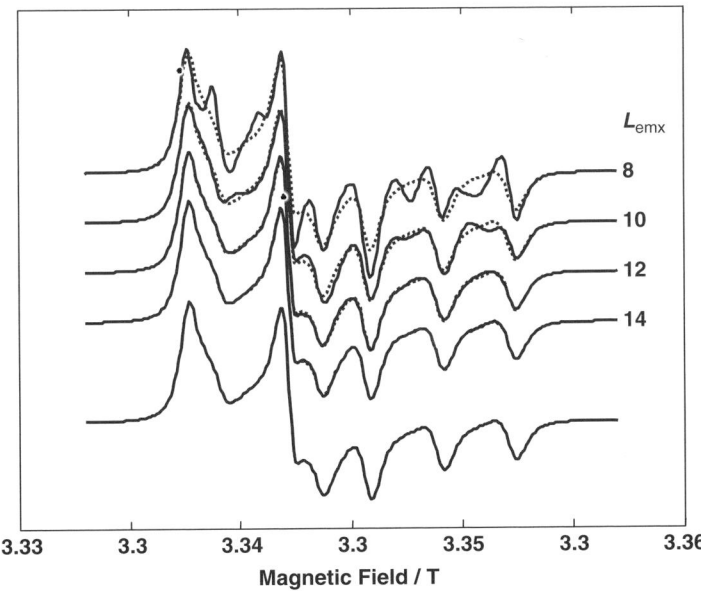

Fig. 13. Effects of using an incomplete basis set on the calculated slow-motional 94 GHz ESR spectrum of a nitroxide. The converged spectrum calculated with $L_{emx} = 22$ is shown at the bottom, and by the dashed lines above. Spectra calculated with indicated maximum L-index (L_{emx}) values are shown by dashed lines and compared with the converged spectrum. Note the appearance of significant oscillations in the spectrum as L_{emx} decreases.

3.4. Guidelines for Selecting Basis Set Truncation Parameters

This section gives some rules of thumb for selecting MTS parameters to optimize the speed of a given slow-motional line shape calculation. They are intended as an approximate guide rather than a definitive set of rules, since the MTS needed for each problem may be slightly different. In general, the reader is recommended to use the basis set "pruning" procedure described below if optimal calculation efficiency is desired. With the availability of fast modern processors, it is often unnecessary to optimize the basis set to the greatest possible extent. Particularly for applications in least-squares fitting, it is desirable to balance computational efficiency with the overhead of having to reoptimize the basis set when the model parameters change significantly.

As mentioned in the last section, the L and K truncation parameters must in general be increased as the motion slows. If the calculation involves an ordering potential with nonzero director tilt angles (including MOMD spectra), the M truncation parameter must also be increased. Separate maxima are specified for even and odd L values. The reason for this is that the number of odd L values required depends on the symmetry of the system: in general, no odd L values are needed at all in the case of axial symmetry, and higher odd values are required as systems deviate further from axial symmetry.

The basic strategy recommended for obtaining appropriate MTS parameters is therefore to start by determining L_{emx} according to the ESR frequency and anticipated minimum $R_\perp$ diffusion parameter. One may then set the remaining parameters (L_{omx}, K_{mx}, and M_{mx}) relative to the L_{emx} value. Figure 12 shows the L_{emx} value obtained over a range of rotational rates for three different ESR frequencies using the swept-fields conjugate gradient method introduced by Vasavada et al.[30] and described below in the section on basis set pruning.

The curves shown in Fig. 14 were calculated from empirical expressions that can be used to estimate the L_{emx} needed for a given calculation from the ESR frequency and $R_\perp$ without performing a swept-fields calculation. Although an axial diffusion tensor was assumed in deriving these expressions, the method may also be applied when a rhombic ($R_x \neq R_y$) diffusion tensor is anticipated, using the lesser of the quantities R_x, R_y.

The expressions used to calculate the curves shown in Fig. 14 are obtained as follows. First, the value of $\log_{10} R_\perp$ corresponding to the onset of the fast-motion regime is approximated as $\log_{10} R_{\perp,max} = 8.40 + 0.00745\nu$, where ν is the frequency in gigahertz (GHz). Above this value, $L_{emx} = 2$ is sufficient to calculate the spectrum. For lower values of $R_\perp$, the plotted curves are calculated for $\nu = 9$, 94, and 140 using Eq. 6,

$$L_{emx} = A(\nu)\left(\log_{10}R_{\perp,min} - \log_{10}R_\perp\right)^2 + 2 \tag{6}$$

where A is a frequency-dependent constant given by $3.07 + 0.00679\nu$ (where ν again is in units of GHz). The parameter L_{emx} should be rounded to the next highest even integer above the value returned by this equation.

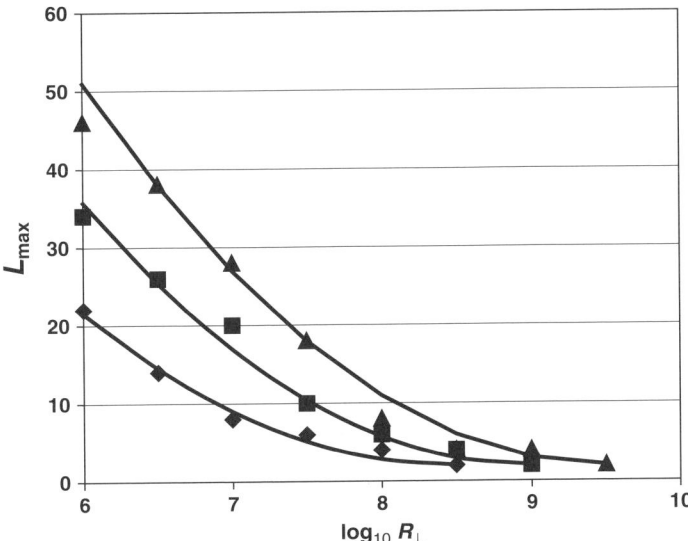

Fig. 14. The L_{emx} values required for convergence of the slow-motional ESR spectrum of a nitroxide as a function of $\log_{10}R_\perp$ at frequencies of (♦) 9 GHz (□) 94 GHz, and (▲) 140 GHz. The L_{emx} values were obtained using the swept-fields calculation of Vasavada et al.[34] with a tolerance of 0.003. Solid lines show empirical functions used to estimate L_{emx} as described in the text.

Once L_{emx} is determined, the appropriate values of L_{omx} and K_{mx} may be set relative to it. For most nitroxide spectra, $L_{omx} = 0.7\ L_{emx}$ (rounded to the next highest odd integer) is sufficient, although this factor can be reduced somewhat at frequencies at or < 9 GHz, where the spectrum is dominated by the nearly axial ^{14}N hyperfine anisotropy. The ratio between K_{mx} and L_{emx} depends on the rotational anisotropy parameter N. For an isotropic rotational diffusion tensor, it is generally sufficient to take $K_{mx}=0.6\ L_{emx}$; when $N \geq 10$ ($\log_{10} N \geq 1$) one may generally use $K_{mx}=0.3\ L_{emx}$; and when $N \leq 0.1$ ($\log_{10} N < -1$) it is necessary to use $K_{mx} = 0.9\ L_{emx}$, in each case rounding to the next highest even integer.

In the absence of an orienting potential, or for zero director tilt in the presence of a potential, M_{mx} may be taken equal to the p^I_{max} value. For MOMD calculations, it is recommended simply to set M_{mx} equal to K_{mx}.

The K_{mn} index should be set to zero unless either the α_D or the γ_D diffusion angle is nonzero. For relatively small tilt angles, only small negative values of K_{mn} are required; however, for angles approaching 45°, K_{mn} should be chosen equal to $-K_{mx}$. In this case, $|K_{mn}|$ and $|K_{mx}|$ may be reduced by a factor of ~0.7 relative to the optimal K_{mx} in the absence of tilt. If the tilt angles are to be varied in a least-squares procedure, however, it is best not to reduce the magnitude of K_{mx} and K_{mn}.

The M_{mn} parameter should generally be set to zero unless there is a significant nuclear Zeeman interaction present. Even in this case, only small negative values of M_{mn} will be needed; however, since the optimal index should depend on the relative magnitudes of the nuclear Zeeman and nuclear hyperfine interactions, the most reliable way to set M_{mn} is to use the basis set pruning method described below.

The nuclear transition index p^I varies from $-p^I_{mx}$ to $+p^I_{mx}$. It is recommended always to set p^I_{mx} equal to twice the total nuclear spin ($2I$). Although the larger values of p^I are not needed for the very fastest motions, calculations for nitroxides in this motional regime are already rapid enough that eliminating unneeded p^I values results in very little savings in computation time.

3.5. Basis Set Pruning

Practical experience has shown that significant reduction of the basis set below that specified by the MTS is possible. That is, not every member of a MTS is needed for a given calculation. However, the pattern of omissions is rather irregular and cannot easily be formulated as a simple set of rules. The alternate strategy implemented in the EPRLL family of programs is simply to keep a list of all the important basis indices using a strategy called "pruning", which has been described in detail.[30]

To prune a basis set, one determines "weighting factors" for each vector in an oversized basis set by finding the maximum of the projection of the basis vector on the solution vector evaluated at a selection of fields across the spectrum. This is done by rearranging Eq. 5 and solving the matrix equation $(i\Delta\omega\mathbf{I} + \Gamma - i\mathbf{L})|u\rangle = |v\rangle$ using a given value of $\Delta\omega$ to obtain a solution vector $|u\rangle$. This calculation is repeated for a number of $\Delta\omega$ values that span the spectrum (typically ≈ 20 at 9 GHz and 40 at 94 GHz). This is referred to as the *swept-field* calculation in the original literature about the EPRLL program.[10,11] For MOMD calculations, it is necessary to repeat the swept-field calculation at each of the director tilt angles used in the calculation (n_{ort}).

Once the swept-field calculation is accomplished, the maximum projection of each basis vector on the set of $|u\rangle$ vectors is tabulated, normalized, and stored in a basis set file as a set of *significance* or *weighting factors* for each element of the basis vector. The vectors in the starting basis set may be selected into a "pruned" set by specifying a *pruning tolerance*. Only basis vectors with weighting factors larger than the pruning tolerance are retained in a pruned basis set.

It is possible to "over-prune" the basis (i.e., to specify too large a pruning tolerance) to the point it works only for a very narrow range of physical parameters. The symptoms of an insufficient basis set are often less apparent than in the case where the truncation parameters are too small. Although pruning tolerances of up to 0.03 have been used,[30] we recommend a more conservative value of 0.003 for general use. This tolerance represents a reasonable compromise between computational efficiency and flexibility. Even smaller tolerances may be required for the slowest motions at high ESR frequencies. It is advisable to check the calculation by reducing the pruning tolerance until the calculated line shape does not change significantly with the tolerance. This can be accomplished by starting with an oversized basis set, to avoid repeating an entire field-swept calculation for each relevant range of parameter values.

TABLE 6. Control Parameters Used in EPRLL Programs

Symbol	Name	Description
B_0	**b0**	Spectrometer field in gauss
	cgtol	Tolerance for conjugate gradient iterations
	range	Spectrum field range in gauss
ν	**freq**	Spectrometer frequency in GHz
ϕ	**phase**	Phase of spectrum in degrees: $0°$ = absorption, 90 = dispersion
	shiftr, shifti	Real and imaginary parts of diagonal shift term added to matrix in CG solution
	ideriv	Derivative mode (0 = absorption, 1 = 1st derivative)
p^I_{mx}	**ipnmx**	Maximum nuclear transition quantum number in basis set ($0 \leq$ ipnmx $\leq$ in2)
K_{mx}, K_{mn}	**kmx, kmn**	Maximum and minimum K quantum number in basis set (kmn $\leq 0 <$ kmx $<$ lemx)
L_{emx}, L_{omx}	**lemx, lomx**	Maximum even and odd L quantum number used in basis set ($0 <$ lomx $<$ lemx)
$m_\perp, m_\parallel$	**mprp, mpll**	Model flags for non-Brownian diffusion perpendicular and parallel to diffusion z (1=free, 2= jump diffusion)
M_{mx}, M_{mn}	**mmx, mmn**	Maximum and minimum M quantum number in basis set (mmn $\leq 0 <$ mmx $<$ lemx)
	ipdf	Diffusion model (0 = Brownian; 1 = non-Brownian, 2 = anisotropic viscosity)
n_{field}	**nfield**	Number of field positions in spectrum
n_{ort}	**nort**	Number of MOMD orientations (0 for no MOMD)
n_{step}	**nstep**	Maximum number of conjugate gradient steps allowed in tridiagonal matrix calculation

The control parameters used in the EPRLL programs are summarized and defined in Table 6.

4. NONLINEAR LEAST-SQUARES ANALYSIS

The physical parameters in the models described above are most usually obtained from experimental slow-motional ESR spectra using nonlinear least-squares (NLLS) analysis, in which a subset of m parameters is iteratively refined to minimize the quantity

$$\chi^2 = \frac{1}{n-m} \sum_{i=1}^{n} (y_i - f(B_i, \mathbf{x}))^2 \qquad (7)$$

where (B_i, y_i) are the experimental field and intensity values, f is the line shape function, and $\mathbf{x}$ is a vector of search parameters. Assuming that a global minimum in χ^2 is located at $\mathbf{x} = \mathbf{x}^*$ the parameter values in $\mathbf{x}^*$ are taken to be the "solution", that is, the values given by the experimental spectrum.

A wide variety of methods are available to accomplish the minimization described above.[31] In all but the simplest cases, programs designed to carry out NLLS using the SLE line shape calculation are computationally quite demanding, so it becomes important to reduce the number of iterations during the fitting procedure as much as possible, especially when fitting for many parameters. This can be achieved by using a suitable fitting algorithm, as well as by gaining an understanding of the behavior of the model itself and the correlations among the fitting parameters. We conclude by briefly discussing the two most common approaches, which include (1) variants of the basic Gauss–Newton method, including the Levenberg–Marquardt algorithm,[15] and (2) the Nelder–Mead, or "downhill simplex", search.[32,33]

The most widely distributed program for least-squares analysis is the NLSL program distributed by the Freed group at Cornell, which is based on the Levenberg–Marquardt method.[15] This method is designed to provide quadratic convergence from an arbitrary starting point to a minimum in χ^2. In principle, this should be significantly faster than downhill search methods such as the Nelder–Mead algorithm.[31] One potential drawback of Gauss–Newton methods is that they require estimation of the Hessian, or curvature, matrix, that is, an m by m matrix of second derivatives $H_{ij} = \partial^2(\chi^2)/\partial x_i \partial x_j$. This matrix is typically estimated by calculating the first derivatives of χ^2 using the forward-difference approximation, which requires $m + 1$ line shape calculations per iteration. The rapid convergence of Gauss–Newton methods generally compensates for this computational burden, and they have the additional advantage that parameter uncertainties may be estimated from the curvature matrix.[15]

However, a more serious difficulty with such methods is that accurate derivatives require much closer convergence of the SLE calculation to the "true" spectrum. Basis sets and tolerance parameters that lead to a satisfactory spectrum can still produce unacceptable oscillations in the derivatives of that spectrum with respect to the calculation parameters, significantly slowing convergence of the minimization. Our own side-by-side comparison of the Levenberg–Marquardt and Nelder–Mead methods suggests that, although the downhill simplex algorithm requires more iterations to locate the minimum, the minimization is often accomplished with a comparable number of line shape calculations. Given that the Gauss–Newton method is more computationally demanding for each line shape calculation, the Nelder–Mead search may be a competitive alternative in many cases.

With the widespread use of commercially available computational software packages, it has become possible to adapt ESR line shape analysis to specific needs by simply interfacing it line shape calculation to the appropriate software. Several groups currently use the LabView[34] program based on the EPRLL calculation that was originated by the Hubbell group at UCLA.[35] Recently, a version of the EPRLL program that is callable from a wide variety of computational packages, such as Matlab,[36] MathCad,[37] and Mathematica,[38] has been released (see the section *Appendix: Program Availability* below). This capability should greatly facilitate the adaptation of slow-motional line shape calculations to include new methods for fitting multicomponent spectra,[39,40] obtaining global fits to multifrequency spectra,[26] or carrying out global fits including other types of physical measurements.[41] Such flexibility promises to significantly enhance the information obtainable from ESR spectra of spin-labeled polymers in the future.

ACKNOWLEDGMENTS

K.A. Earle thanks the ACERT center at Cornell University for the use of its facilities during the preparation of this manuscript. ACERT is supported by National Institutes of Health NCRR grant RR016292. D.E. Budil gratefully acknowledges support from the Institute for Complex Scientific Software in the College of Computer Science at Northeastern University.

APPENDIX: PROGRAM AVAILABILITY

The calculations described in this chapter refer specifically to the EPRLL programs distributed by the Center for Advanced ESR Technology (ACERT). The ACERT website has a number of useful programs available for FTP downloading in the directory ftp.ccmr.cornell.edu/pub/freed. This directory can be accessed by anonymous FTP to ftp.ccmr.cornell.edu. Numerous research groups all over the world have availed themselves of these programs. The programs source codes are stored in uncompressed format and may be obtained by standard text (ASCII) ftp transfer. Tar files are provided for compressed data transfer and contain all files in the specific directories and subdirectories except the executables to save space. The available software packages relevant to this chapter include the following:

- **PC:** Contains the original pc version of the CW spectrum simulation programs (Ref. 10).
- **PC.NEW:** A new addition that contains a version of the EPRLL programs suitable for running on a pc with Windows 95. Only the source files that have been changed from the EPRLL directory are included, plus the executables. See the README file on that directory.
- **EPRLL**: Contains the basic simulation program for CW spectrum calculation including the program EPRBL used to determine the truncated basis sets.
- **NLSL:** Contains the Windows-compatible least squares version of the above CW program using Marquardt–Levenberg minimization. See the subdirectory EXAMPLES for fitting examples to test (Ref. 15).

MATLAB: Contains the EPRLL line shape calculation engine (a dynamic load library, or DLL file) that is callable from MATLAB for use in customized least-squares optimization.

NLSL.SRLS: Performs fitting for multi-frequency ESR spectra using the Slowly Relaxing Local Structure (SRLS) model (Ref. 26).

Free use and distribution of these programs is permitted with suitable reference to the original publication (see above) in any published work resulting from the use of these programs or programs derived from them. The programs may be copied and distributed, so long as: (1) due credit is given by retaining the comment lines at the beginning of each source file in all copies; and (2) all copies must be distributed free of charge.

Every effort has been made to ensure that these programs are correct and thoroughly tested. However, the programs are distributed "AS IS", and all warranties, whether expressed or implied, as to correctness or fitness for any specific purpose are specifically disclaimed. In no event shall the authors be liable for any direct, consequential or incidental damages arising from the use of these programs. The READ.ME files are provided in each of the above directories giving further information on their contents and usage. Please check the file CORRECTIONS in the directories for important information and code updates.

Another popular and widely available package that may be used to model and fit ESR spectra of nitroxides in the slow-motional regime is EasySpin (S. Stoll and A. Schweiger, "EasySpin, a comprehensive software package for spectral simulation and analysis in EPR," *J. Magn. Reson.*, **178**: 42–55, 2006, available at http://www.easyspin.ethz.ch/). The programs also model spectra in the fast-motion and rigid limits, and are based in the Matlab computational package, affording significant graphical capabilities and ease of use. Compared to the EPRLL programs described in this chapter, EasySpin has the additional capability to approximate the slow-motion spectrum of a radical with multiple nuclei; however, the present version does not include a local ordering potential.

REFERENCES

1. *Spin Labeling I. Theory and Application,* Berliner, L.J., Ed.; Vol. 1. New York: Academic Press, 1976.
2. *Biological Magnetic Resonance,* Berliner, L.J., Reuben, J., Eds.; New York: Plenum Press, 1989; Vol. 8, 444 pp.
3. Likhtenshtein, G.I. *Biophysical Labeling Methods in Molecular Biology.* Cambridge: Cambridge University Press, 1993.
4. Robinson, B.; Slutsky, L.; Auteri, F. *J. Chem. Phys* **1992**, *96*, 2609.
5. Hakansson, P.; Westlund, P.; Lindahl, E.; Edholm, O. *Phys. Chem. Chem. Phys.* **2001**, *3*, 5311.
6. Steinhoff, H.-J.; Hubbell, W.L. *Biophys. J.* **1996**, *71*, 2201.

REFERENCES

7. Freed, J.H. In *Spin Labeling*, L.J. Berliner, Ed.; Academic Press: New York. 1976; p. 53.
8. Kubo, R.; Tomita, K. *J. Phys. Soc. (Jpn)* **1954**, *9*, 8.
9. Moro, G.; Freed, J.H. *J. Phys. Chem.* **1980**, *84*, 2837.
10. Schneider, D.J.; Freed, J.H. *Biol. Magn. Reson.* **1989**, *8*, 1.
11. Schneider, D.J.; Freed, J.H. *Adv. Chem. Phys.* **1989**, *73*, 387.
12. Moro, G.; Freed, J.H. *J. Chem. Phys.* **1981**, *74*, 3757.
13. Meirovitch, E.; Igner, D.; Igner, E.; Moro, G.; Freed, J.H. *J. Chem. Phys.* **1982**, *77*, 3915.
14. Lee, S.; Budil, D.E.; Freed, J.H. *J. Chem. Phys.* **1994**, *101*, 5529.
15. Budil, D.E.; Lee, S.; Saxena, S.; Freed, J.H. *J. Magn. Reson, Ser. A* **1996**, *120*, 155.
16. Ding, Z.; Gullà, A.F.; Budil, D.E. *J. Chem. Phys.* **2001**, *115*, 10685.
17. Brustolon, M.; Maniero, A.L.; Corvaja, C. *Mol. Phys.* **1984**, *51*, 1269.
18. Zare, R.N. *Angular Momentum: Understanding Spatial Aspects in Chemistry and Physics*; John Wiley & Sons, Inc., New York: 1988; 346 pp.
19. Marek, A.; Czernek, J.; Steinhart, M.; Labsky, J.; Stepanek, P.; Pilar, J. *J. Phys. Chem. B* **2004**, *108*, 9482.
20. Pilar, J.; Labsky, J.; Marek, A.; Budil, D.E.; Earle, K.A.; Freed, J.H. *Macromolecules* **2000**, *33*, 4438.
21. Liu, Z.; Ottaviani, M.F.; Abrams, L.; Lei, X.; Turro, N.J. *J. Phys. Chem. A* **2004**, *108*, 8040.
22. Ottaviani, M.F.; Lei, X.-G.; Liu, Z.; Turro, N.J. *J. Phys. Chem. B* **2001**, *105*, 7954.
23. Barnes, J.P.; Freed, J.H. *Biophys. J.* **1998**, *75*, 2532.
24. Polnaszek, C.F.; Bruno, G.V.; Freed, J.H. *J. Chem. Phys.* **1973**, *58*, 3185.
25. Meirovitch, E.; Nayeem, A.; Freed, J.H. *J. Phys. Chem.* **1984**, *88*, 3454.
26. Liang, Z.; Freed, J.H. *J. Phys. Chem. B*, **1999**, *103*, 6384.
27. Polimeno, A.; Freed, J.H. *J. Phys. Chem.* **1995**, *99*, 10995.
28. Sastry, V.S.S.; Polimeno, A.; Crepeau, R.H.; Freed, J.H. *J. Chem. Phys.* **1996**, *105*, 5753.
29. Sastry, V.S.S.; Polimeno, A.; Crepeau, R.H.; Freed, J.H. *J. Chem. Phys.* **1996**, *105*, 5773.
30. Vasavada, K.V.; Schneider, D.J.; Freed, J.H. *J. Chem. Phys.* **1987**, *86*, 647.
31. Press, W.H.; Teukolsky, S.A.; Vetterling, W.T.; Flannery, B.P. *Numerical Recipes in C: The Art of Scientific Computing*. New York: Cambridge University Press, 1992; 994 pp.
32. Duling, D.R. *J Magn Reson B* **1994**, *104*, 105.
33. Fajer, P.G. *Biophys. J.* **1994**, *66*, 2039.
34. Labview, National Instruments, Austin, TX.
35. Altenbach, C., personal communication.
36. Matlab, The Math Works, Inc., Natick, MA.
37. Mathcad, MathSoft Engineering and Education, Inc., Cambridge, MA.
38. Mathematica, Wolfram Research, Inc., Champaign, IL.
39. Baumann, B.A.; Liang, H.; Sale, K.; Hambly, B.D.; Fajer, P.J. *Biophys J.* **2004**, *86*, 3030.
40. Strancar, J.; Koklic, T.; Arsov, Z. *J Membr. Biol*, **2003**, *196*, 135.
41. Hustedt, E.J.; Cobb, C.E.; Beth, A.H.; Beechem, J.M. *Biophys J.* **1993**, *64*, 614.

4

ESR IMAGING

SHULAMITH SCHLICK
University of Detroit Mercy, Detroit, Michigan

Contents

1. Introduction 85
2. ESR Imaging: Hardware, Data Acquisition, and Software 87
   2.1. ESR Spectra in the Presence of Magnetic Field Gradients 87
   2.2. Hardware for ESR Imaging 88
   2.3. Intensity Profiling from 1D ESRI 90
   2.4. Line Shape Profiling from 2D Spectral-Spatial ESRI 92
3. ESRI Methods Applied to Polymeric Systems 93
   3.1. Motivation of Our Studies 95
Acknowledgments 96
References 97

1. INTRODUCTION

"…breathtaking opportunities disguised as insoluble problems."

—John W. Gardner[*]

The electron spin resonance ESR methods described in Chapters 1–3 can detect and determine the presence and intensity of species containing unpaired electron spins as average properties in *whole* samples. In some cases, however, the most

---

[*]J. W. Gardner (b 1912), Administrator, Secretary of Health, Education, and Welfare 1965–1968.

Advanced ESR Methods in Polymer Research, edited by Shulamith Schlick.
Copyright © 2006 John Wiley & Sons, Inc.

important information is the *distribution* of paramagnetic species along one (1D), two (2D), or three dimensions (3D) in the system studied. Examples include the spatial distribution of defects produced by irradiation, formation of radicals in polymers due to contact with oxygen at the sample edges, diffusion of paramagnetic tracers along a sample length, and, in biological samples, the distribution of nitroxides as probes during oxidative stress. This spatial information can be obtained by ESR imaging (ESRI).

Electron spin resonance spectroscopy can be transformed into an imaging method, ESRI, by measuring ESR spectra in the presence of magnetic field gradients. The gradients allow the encoding of spatial information in the ESR spectra and the separation of signals corresponding to different spatial elements. In this way, it is possible not only to verify the existence of the paramagnetic species in a sample, but also "to tell exactly where the signal came from".[1] The ESRI methodology is similar to that of nuclear magnetic resonance (NMR) imaging (NMRI, or magnetic resonance imaging, MRI).

The challenges in the application of the gradient approach to ESRI are numerous: First, higher gradients are needed compared to NMRI, usually 100–1000 times larger. Second, the ESR spectra are often complex, with multiple lines due to hyperfine interactions and *g*-value anisotropy; these signals complicate the imaging experiments, but in most cases do not add additional information. Third, most systems do not contain stable paramagnetic species on which imaging is based. The ESRI experiments are usually performed on paramagnetic transition metal ions, radicals produced by irradiation, or stable nitroxide radicals as dopants; in some experiments, triarylmethyl ('trityl') radicals are used as probes, because their ESR spectrum consists of a single narrow line, typically ≈ 50 mG in the absence of oxygen.

The feasibility of ESRI was first demonstrated in 1979 by Hoch and Day, who described the distribution of color centers in natural diamonds.[2] The instrumentation, software, and applications of ESRI have been described in a 1991 monograph[3] and were updated in recent reviews.[4–8] The early efforts described the type and stability of the gradients necessary for specific ESRI experiments and the software necessary for image reconstruction in spatial and spectral dimensions. These studies also investigated the feasibility of ESRI experiments in a variety of "phantom" samples, and discussed and estimated the spatial resolution. A phantom is a sample in which the distribution of paramagnetic centers is known; this sample is used to check the ESRI hardware and software.[9] The resolution in most ESRI experiments is of the order of 50–100 μm, but can vary widely, depending on the ESR line shapes and line widths. Because of the short relaxation of the electron spins, most ESRI experiments are performed in the continuous wave (CW) mode, unlike MRI, which is used in the pulsed mode. *In vivo* studies are usually performed at lower frequencies, typically < 2 GHz, which can accommodate large samples with high water content. Most experiments in materials science are performed at X band, ≈9 GHz. In the imaging experiments, gradients can be applied in the three spatial dimensions, and a spectral dimension can be added by the method of stepped gradients. The widening scope of ESRI studies was highlighted at the 2004 ESR Symposium in Denver,[10] with focus on spatially resolved degradation and software development for applications to polymers, and *in vivo* studies at 300 and 700 MHz for the detection of local oxygen profiles and the study of radical involvement in oxidative diseases.

Recent advances in the ESRI field, for example, the choice of pulsed versus CW experiments, combined ESR and NMR imaging, progress in resonators, and applications in various disciplines have been described in detail.[5-8] The focus of this chapter is on ESRI experiments and applications that have been used for, and are relevant to, polymeric systems. Some of the experimental details reflect the imaging system in our laboratory. Section 2 describes 1D spatial and 2D spectral-spatial ESRI experiments. A brief review of ESRI applications to solving important problems in polymeric systems is given in Section 3. This chapter concludes with an evaluation of ESRI methods and prospects for further applications.

2. ESR IMAGING: HARDWARE, DATA ACQUISITION, AND SOFTWARE

2.1. ESR Spectra in the Presence of Magnetic Field Gradients

In ESR imaging experiments, the microwave power is absorbed by the unpaired electrons located at point x, when the resonance condition, Eq. 1, is fulfilled.

$$\nu = (g\, \beta_e / h)\, (B_{\text{res}} + x\, G_x) \tag{1}$$

In Eq. 1, ν is the microwave frequency, g is the spectroscopic g-factor for the electron, β_e is the Bohr magneton, h is the Planck constant, and G_x is the linear magnetic field gradient (in Gcm^{-1}) at x. As in NMR imaging, the field gradients produce a correspondence between the spin location and the resonant magnetic field, B_{res}. If the sample consists of two point samples, for example, the distance between the samples along the gradient direction can be deduced if the field gradient is known. As an example, we present in Fig. 1 ESR spectra of a "phantom" consisting of two specks of the radical 2,2-diphenyl-1-picrylhydrazyl (DPPH) separated vertically by 1 cm inside a capillary placed in the ESR resonator. One signal only is detected for zero gradient, and two signals in the presence of the gradient, with a separation in gauss that increases with increasing gradient. The two signals have different heights because of different DPPH speck sizes.

The effect of the gradient on the ESR spectrum of a nitroxide radical is seen in Fig. 2, for a phantom consisting of two capillary tubes containing a solution of 2,2,6,6-tetramethyl-1-piperidinyloxyl (TEMPO) in benzene and separated by 3 mm. The top, spectrum (a), is the regular ESR spectrum (gradient-off). Spectrum (b) is the 1D image (gradient-on), using a gradient of 100 G cm^{-1} in the z direction, parallel to the static (Zeeman) magnetic field. The spatial distribution of the radicals is shown in (c), and is deduced from the two spectra, the 1D image and the gradient-off spectrum, and allows the calculation of the distance between the capillaries in the phantom.[11]

The spatial resolution is an important parameter in imaging, and can be defined in various ways, as discussed recently in Ref. 12; the resolution depends on the line width and line shape. Most commonly, the resolution is expressed as the ratio of the line width to the field gradient, $\Delta H/G$; this definition implies that two signals separated by one line width due to the field gradient can be resolved.

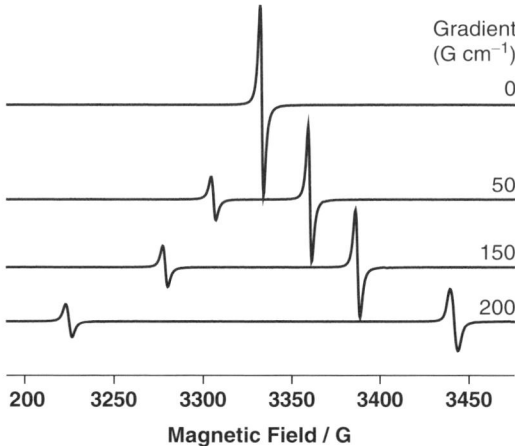

Fig. 1. The ESR spectra in the absence (top) and in the presence of the indicated magnetic field gradients. The sample consisted of two specks of DPPH at a distance of 1 cm in a capillary placed along the vertical (y) direction in the resonator. The different heights of the signals detected in the presence of gradients reflect the slightly different sizes of the DPPH specks.

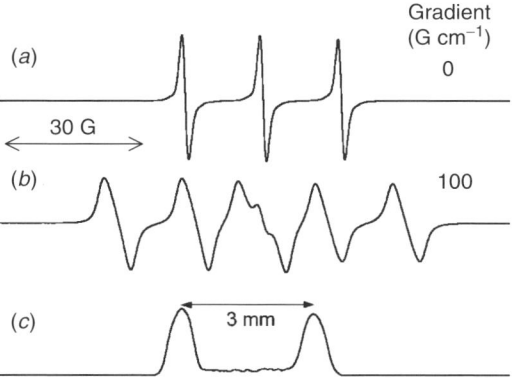

Fig. 2. One-dimensional ESRI at 298 K of a phantom consisting of two parallel capillary tubes separated by 3 mm and filled with a solution of TEMPO in benzene. (*a*) X-band ESR spectrum recorded in the absence, and (*b*) in the presence of the magnetic field gradient (100 G cm^{-1}). The radical distribution is shown in (*c*). (Redrawn from Ref. 9, with permission.)

2.2. Hardware for ESR Imaging

An ESR imaging system can be built with small modifications of commercial spectrometers: gradient coils fixed on the poles of the spectrometer magnet, regulated direct current (dc) power supplies, and required computer connections. In most systems, the software for image reconstruction in the spatial and spectral dimensions must be developed on site. ESR imagers built by Bruker Biospin Co. have recently become available for selected applications, mostly in biological applications of ESRI.[10]

Because the ESR spectrum is measured by scanning the Zeeman magnetic field, which is in the z direction, the gradients coils must supply gradients $\delta B_z/\delta x$ (along x), $\delta B_z/\delta y$ (along y), or $\delta B_z/\delta z$ (along z). For a cylindrical sample inserted in a rectangular resonator, the axis of the cylinder is along the y axis. Gradient coil arrangements that supply gradients along z and y directions are shown in Fig. 3. The Helmholtz coils in Fig. 3a consist of two coils with the current in opposite directions as indicated, and supply a gradient in the z direction. The gradient obtained in the center of symmetry depends on the number of turns in the coils, the diameter R, the distance between the coils D, and the current I. The "figure eight" coil arrangement for the cylindrical coils in Fig. 3b produces a gradient along the vertical, y, direction of the sample; this is often the direction of choice for the gradient direction in 1D experiments, because it is along the longest dimension that can be visualized in ESRI experiments. The same figure eight coils can be wired to produce a gradient in the z direction. The coils are cooled by water and protected by a temperature sensor that interrupts the current when the temperature is above a set limit.

The two sets of coils, each consisting of a figure eight coil, are fixed on the poles of the spectrometer magnet. Depending on the electrical wiring of the two sets, the coils in our laboratory supply a maximum linear field gradient of ≈ 320 G cm^{-1} in the direction parallel to the external magnetic field (z axis), or ≈ 250 G cm^{-1} in the vertical direction (along the long axis, y, of the microwave resonator), with a constant control current of 20 Ampères applied to each power supply. The coils are positioned so that the zero point of the gradient field coincides with the center of the resonator.

To calibrate the gradient and check its linearity, a sample consisting of two specks of DPPH at a distance of 10 mm is used. A straight line obtained by plotting the

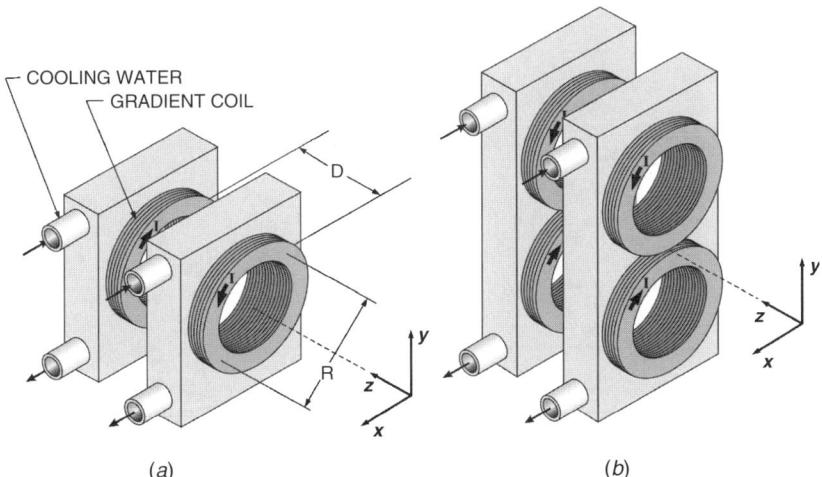

Fig. 3. Gradient coils for the z direction, $\delta B_z/\delta z$ (a) and the y direction, $\delta B_z/\delta y$ (b). Here R is the diameter of the coils and D is the distance between the coils.

separation in gauss (as shown in Fig. 1 for selected gradients) as a function of the current indicates the linearity of the gradient, as required. This calibration is repeated regularly.

2.3. Intensity Profiling from 1D ESRI

In the general case, the sample contains a distribution of paramagnetic centers and the ESR spectrum in the presence of the magnetic field gradient is a superposition of signals from paramagnetic centers located at different positions. If the distribution is along a given direction, the intensity profile can be obtained from 1D ESRI experiments, with the magnetic field gradient along the direction of the distribution. The two ESR spectra needed to deduce the profile are shown in Fig. 4: spectrum measured in the absence of magnetic field gradient, $F_0(B)$, and the 1D image, $F(B)$, measured in the presence of the gradient.

Mathematically, $F(B)$ is a convolution of $F_0(B)$ with the distribution function of the paramagnetic centers, Eq. 2,

$$F(B) = \int_{-\infty}^{\infty} F_0(B - B^*)C(B^*)dB^* \qquad (2)$$

where $B^* = B_0 - xG$, $B_0 = h\nu/g\beta_e$, and $C(B^*)$ is the intensity distribution (profile) of the paramagnetic centers along the gradient direction. It is essential to note that the convolution expressed in Eq. 2 is correct only if the ESR line shape *has no spatial dependence*. This requirement has dictated the conditions for data acquisition in the 1D ESRI study of degradation processes described in Chapter 9.[13]

The profile can be obtained by deconvolution. Various optimization methods can then be applied. The process starts by assuming an initial distribution, which can be described by a set of parameters. Optimization methods use the convolution of this

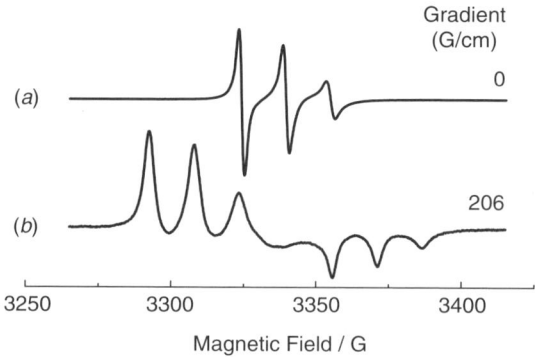

Fig. 4. The 1D ESRI at 340 K of a cylindrical sample (height ≈ 3 mm, diameter ≈ 4 mm) of heterophasic propylene–ethylene copolymer (HPEC) containing a nitroxide derived from a hindered amine stabilizer (HAS). X-band ESR spectrum recorded in the absence (*a*) and in the presence of a vertical magnetic field gradient of 206 G cm^{-1} (*b*).

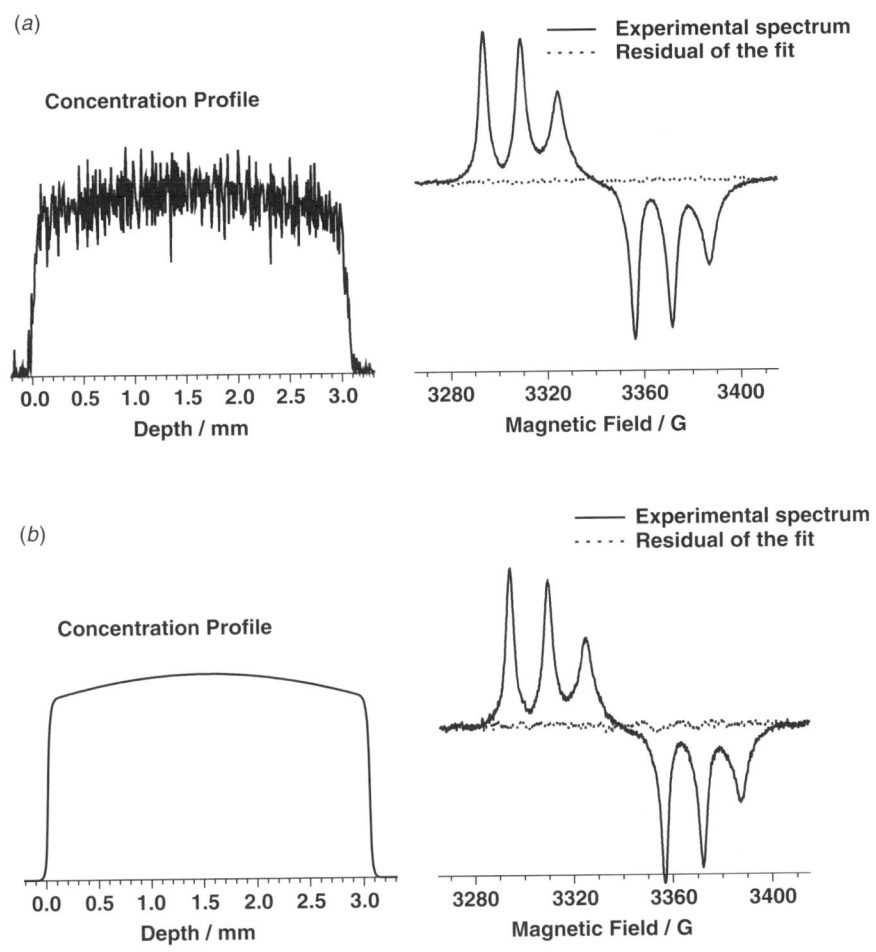

Fig. 5. Determination of concentration profiles in thermally treated HPEC containing a HAS-derived nitroxide. (a) Concentration profile obtained by deconvolution followed by Monte Carlo optimization (left), and 1D image and the residual to the fit (right). (b) Concentration profile obtained by simulation of the 1D image on the right with the Genetic algorithm (left), and the residual to the fit (right). The 1D image was obtained with a gradient of 200 G cm^{-1}.

initial distribution function with the experimental spectrum in the absence of gradient in order to calculate the 1D image. The deviation between the calculated and the experimental spectra is then minimized by an optimization procedure. In our initial ESRI studies, the concentration profiles of the radicals were deduced by Fourier transform followed by optimization with the Monte Carlo (MC) procedure.[11,13,14] The disadvantage of this method is the high-frequency noise present in the optimized profiles. In more recent publications, the intensity profile was fitted by analytical functions and convoluted with the ESR spectrum measured in the absence of the field gradient in order to simulate the 1D image. The best fit was obtained by variation of the type and parameters

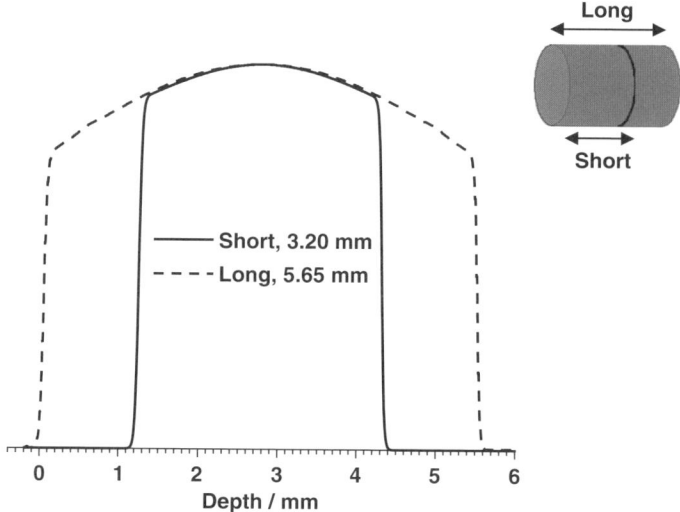

Fig. 6. Sensitivity profile of the resonator in the y direction for short (3.20 mm) and long (5.65 mm) samples.

of the analytical functions chosen (e.g. Gauss or Boltzmann functions) in order to obtain good agreement with the 1D image, and selected by visual inspection.[15,16]

Lately, the genetic algorithm for minimization of the difference between simulated and experimental 1D images was implemented; this procedure allowed the best fit to be chosen automatically.[17,18] A typical genetic algorithm (GA) is patterned after the Darwinian principle of reproduction and survival: creation of the initial population, calculation of the fit to experimental data, selection of the couples, crossover (reproduction), and mutation. The approach and terminology are adopted from biology and resemble fundamental steps in evolution. Profiles obtained by Monte Carlo optimization and the GA algorithm are compared in Fig. 5.

The concentration profile obtained by the various methods must be corrected for the sensitivity of the resonator in the direction of the gradient, as seen in Fig. 6 for two lengths of cylindrical samples. This correction is usually necessary when the dimension of the sample is >2 mm.

2.4. Line Shape Profiling from 2D Spectral-Spatial ESRI

These ESRI experiments provide the ESR spectrum as a function of a spatial coordinate. The data collection consists of projections that examine an object viewed in the H (spectral) and L (spatial) coordinates, as shown in Fig. 7. The angle α is between the L coordinate and the direction of a given projection: spatial information only is obtained when $\alpha = 0°$, and spectral information only when $\alpha = 90°$. The maximum attainable α value in a given experiment, α_{max}, is given in Eq. 3,

$$\tan \alpha_{max} = (L/\Delta H)\, G_{max} \qquad (3)$$

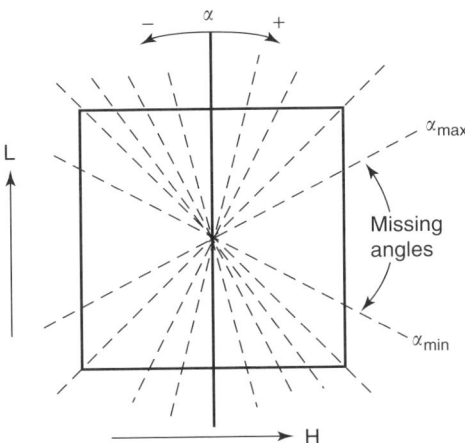

Fig. 7. Two-dimensional spectral–spatial ESRI images by a stepped gradient, and the cone of missing angles.

where L is the sample length, ΔH is the spectral width, and G_{max} is the maximum gradient.

Each 2D image is reconstructed from a complete set of projections, collected as a function of the magnetic field gradient, using a convoluted back-projection algorithm.[19,20] The number of points for each projection (512–1024) is kept constant. The maximum sweep width is $SW_{max} = \sqrt{2}\Delta H/\cos\alpha_{max}$. For a width $\Delta H \approx 65$ G, which is typical for the slow-motional spectral component of a nitroxide radical present in irradiated polymers, a sample length of 4 mm, and a maximum field gradient of 250 G cm^{-1} along the vertical axis, we obtain $\alpha_{max} \approx 60°$ and $SW_{max} = 169$ G. A complete set of data for one image consists of 64–256 projections, taken for gradients corresponding to equally spaced increments of α in the range 0–180°; of these projections, typically 41 or 43 (out of 64) are experimentally accessible, and the rest are projections at missing angles (for α in the intervals 60 to 120°, as seen in Fig. 7. The projections at the missing angles are often assumed to be the same as those at the maximum experimentally accessible angle α_{max}, or determined by the projection slice algorithm (PSA) with several iterations.[13–18]

3. ESRI METHODS APPLIED TO POLYMERIC SYSTEMS

The ability to perform ESRI is restricted to a small number of groups worldwide. While most efforts are directed to the study of biological applications,[21–25] a small number of studies on polymeric systems have appeared: Information on the spatial distribution of paramagnetic molecules deduced from ESRI experiments has been used successfully for measurement of the macroscopic translational diffusion. Diffusion coefficients, D, of paramagnetic diffusants can be deduced from an analysis of the time dependence of the concentration profiles along a selected axis of the sample. The determination of diffusion coefficients for spin probes in liquid crystals and model membranes, and the effect of polymer polydispersity, have been described in a series of papers by Freed and co-workers.[26] In our laboratory, the diffusion coefficients of paramagnetic guests in ion-containing polymers, polymer solutions, crosslinked polymers swollen by solvents and self-assembled polymeric surfactants have been determined by 1D ESRI.[27–32] These papers represent an effort to move beyond phantoms, and to extract quantitative information from ESRI experiments. Moreover, in some cases these experiments allow the comparison of macroscopic diffusion coefficients in the presence of a concentration gradient (measured by ESRI) with the microscopic diffusion coefficients (measured by pulsed field-gradient NMR).

Some of these studies have resulted in measurements of diffusion coefficients that can be deduced only by ESRI. An example is the measurement of D at 300 K for nitroxide probes that differ in their hydrophobicity, and were doped in the various phases (micellar, hexagonal, lamellar, and reverse micellar) of the triblock copolymers poly(ethylene oxide)-b-poly(propylene oxide)-b-poly(ethylene oxide), $EO_mPO_nEO_m$, (commercial name Pluronics).[30] The self-assembling is due to the different hydrophobicity of the two blocks, PEO and PPO, in water as solvent. Ionic, neutral, and hydrophobic probes select specific sites in the self-assembled system, and these sites are reflected in the rate of transport of the probes: in the value of D. The cationic probe 4-(N,N,N-trimethyl)ammonium-2,2,6,6-tetramethyl-piperidine-1-oxyl iodide (CAT1) and the hydrophobic probe 5DSE, the methyl ester of doxylstearic acid (5 indicates the carbon atom to which the doxyl group is attached) exhibited a contrasting transport behavior in aqueous solutions of Pluronic L64, $EO_{13}PO_{30}EO_{13}$: Although the molecular masses M are similar, 340 for CAT1 (213 for the cation), and 414 for 5DSE, the corresponding D values at each polymer content are very different. The ratio D_{CAT1}/D_{5DSE} is 35 in the micellar phase (polymer contents 20 and 35 % w/w), 11 in the hexagonal phase (polymer content 50 % w/w), and 3.1 in the mixed L_α (lamellar) and L_2 (reverse micellar) phase (polymer content 80 % w/w). The diffusion coefficient of a small neutral probe, perdeuterio-2,2',6,6'-tetramethyl-piperidone-N-oxide (PDTEMPONE, or PDT) that is located at the interface between water and the EO domains, decreases with increase in the polymer content, but the decrease is more prominent for polymer content in the range 10–30 % w/w. The range of D values measured in this study was 1.0×10^{-5} cm^{2}s^{-1} – 1.00×10^{-7} cm^{2}s^{-1}. These results have indicated that ESRI is the method of choice for the determination of the diffusion coefficients for guests present in low concentrations and located in various regions of self-assembled systems; this conclusion is relevant for drug delivery systems.

Degradation processes can be studied by ESRI in polymers containing hindered amine stabilizers (HAS); this approach was originally suggested by Ohno, who presented 2D spectral-spatial ESRI images of radicals in polypropylene (PP) containing two different stabilizers, but no detailed analysis.[33] The method is based on the formation of stable nitroxide radicals derived from HAS (termed HAS–NO) during ultraviolet (UV) irradiation or thermal treatment. In these systems, the original role of the nitroxides was to protect the polymer by scavenging reactive radicals produced by exposure to environmental factors. The ESRI based on HAS–NO represents an important step in the development of ESRI beyond phantoms, because the nitroxides are part of the system, and participate in, and reflect, on degradation processes. Quantitative studies based on this approach followed: Lucarini et al.[11,34,35] determined the distribution of the nitroxide radicals in UV irradiated polypropylene (PP) containing a hindered amine stabilizer (HAS) by 1D ESRI.

Ahn et al.[36] deduced the concentration profile of heat-induced radicals in a polyimide resin. *In vitro* degradation of poly(ortho esters) containing 30 mol% lactic acid has been studied by 1D and 2D spectral–spatial ESRI, based on pH-sensitive nitroxide spin probes.[37] Efforts to understand the failure of ultrahigh molecular weight polyethylene (UHMWPE) in orthopedic components has been studied by spectral–spatial ESRI in γ-irradiated UHMWPE.[38]

3.1. Motivation of Our Studies

Recent advances in the understanding of degradation processes are anchored on the finding that polymer degradation is often spatially heterogeneous. When the rate of oxygen diffusion is not sufficient to supply all the oxygen that can be consumed, only outside layers in contact with oxygen are degraded, whereas the sample interior is protected: this is the diffusion-limited oxidation (DLO) regime.[39–42] The DLO concept implies that in order to understand degradation and predict lifetimes of polymeric materials in different environments, it is necessary to develop profiling methods that determine the variation of the extent of degradation within the sample depth.

The presence of oxygen is crucial for oxidative degradation of polymers. The degradation depth, a, depends on the oxygen availability within sample depth, and is expressed in Eq. 4,[41,43]

$$a = \sqrt{\frac{D}{k_c}} \qquad (4)$$

where D is the diffusion constant of oxygen in the polymer, and k_c is the rate of oxygen consumption in degradation processes. The degradation profile for two values of a and a sample depth of 4 mm are shown in Fig. 8. For a degradation depth comparable with the sample depth, $a = 3.5$ mm in Fig. 8a, the degradation profile is essentially homogeneous; this is the case of low k_c, low oxygen consumption, and low degradation rates. For a degradation depth smaller than the sample depth, $a = 0.5$ mm in Fig. 8b, the degradation profile is heterogeneous; this is the DLO regime, with high k_c, high oxygen consumption, and high degradation rates.

The ability to visualize degradation profiles and variation of degradation processes within sample depth was the primary motivation for ESRI studies in our laboratories. We have developed ESRI methods for the study of spatially resolved degradation studies in heterophasic polymer systems containing HAS. The HAS-derived nitroxides provide the contrast needed for the ESRI experiments; probe the morphology of the system, in terms of glass transition characteristics and dynamics; and reflect the degradation process. Once ESRI data are collected and transformed into intensity profile (from 1D ESRI) and spectra as a function of sample depth (from 2D ESRI), the remaining challenge is to translate information extracted from ESRI

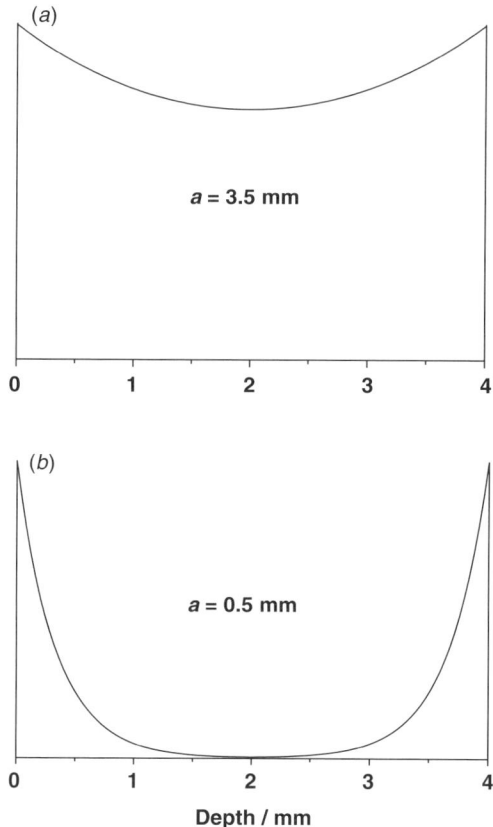

Fig. 8. Oxygen profile during degradation. (*a*) Nearly homogeneous profile when the degradation depth, *a*, is similar to sample depth. (*b*) Heterogeneous profile when the degradation depth, *a*, is much smaller than the sample depth.

into details on degradation kinetics and mechanism. The nitroxide radicals reflect not only the spatial extent of degradation, but also events that occur in different morphological domains: In recent studies of ABS polymers for example, 1D and 2D spectral–spatial ESRI images have enabled the visualization of selective damage, along the sample depth, in butadiene (B)-rich domains, compared to acrylonitrile/styrene (SAN)-rich domains. Details on some of these studies will be described in Chapter 9.

ACKNOWLEDGMENTS

Electron spin resonance and ESRI studies at UDM are supported by the Polymers and International Programs of NSF, by General Motors Fuel Cell Activities Program, and by University Research Grants from The Ford Motor Company. The contributions of UDM faculty and graduate students P. Eagle, S.-C. Kweon, Z. Gao, J.G. Bokria, B. Varghese, and W. Aris; postdocs and collaborators J. L. Gerlock, J. Pilar, A. Marek, K. Kruczala, T. Spalek, Z. Sojka, and M.V. Motyakin have been essential for the development of the ESRI methods in our laboratory.

REFERENCES

1. Lauterbur, P. *Nature (London)* **1973**, *242*, 190.
2. Hoch, M.J.R.; Day, A.R. *Solid State Commun.* **1979**, *30*, 211.
3. *EPR Imaging and in Vivo EPR*; Eaton, G.R., Eaton, S.S., Ohno, K., Eds.; CRC Press: Boca Raton, FL, 1991.
4. Eaton, S.S.; Eaton, G.R. In *Modern Pulsed and Continuous-Wave Electron Spin Resonance;* Kevan, L., Bowman, M.K., Eds.; Wiley-Interscience: New York, 1990; pp. 405–435.
5. Eaton, S.S.; Eaton, G.R. In *Specialist Periodical Reports — Electron Paramagnetic Resonance*; Gilbert, B.C., Davies, M.J., Murphy, D.M., Eds.; Royal Society of Chemistry: Cambridge, 1996; Vol. 15, pp. 169–185.
6. Eaton, S.S.; Eaton, G.R. In *Specialist Periodical Reports — Electron Paramagnetic Resonance*; Gilbert, B.C., Davies, M.J., Murphy, D.M., McLauchlan, K. A., Eds.; Royal Society of Chemistry: Cambridge, 2000; Vol. 17, pp. 109–129.
7. Lurie D.J. In *Specialist Periodical Reports — Electron Paramagnetic Resonance*; Gilbert, B.C., Davies, M.J., Murphy, D.M., McLauchlan, K. A., Eds.; Royal Society of Chemistry: Cambridge, 2002; Vol. 18, pp. 137–160.
8. Motyakin, M.V.; Schlick, S. In *Instrumental Methods in Electron Magnetic Resonance, Biological Magnetic Resonance,* Bender, C.J., Berliner, L.J., Eds.; Kluwer Academic/Plenum Publishing Corporation: New York, 2004; pp. 349–384.
9. Berliner, L.J.; Fujii, H. *Science* **1985**, *227*, 517.
10. EPR Imaging Workshop, 27th International EPR Symposium, Denver, CO, 1–5 August 2004.
11. Lucarini, M.; Pedulli, G.F.; Motyakin, M.V.; Schlick, S. *Prog. Polym. Sci.* **2003**, *28*, 331.
12. Van Kienlin, M.; Pohmann, R. In *Spatially Resolved Magnetic Resonance: Methods, Materials, Medicine, Biology, Rheology, Ecology, Hardware*; Blümler, P., Blümich, B., Botto, R., Fukushima, E., Eds.; Wiley-VCH: Weinheim, 1998; Chapt. 1, pp. 3–20.
13. (a) Motyakin, M.V.; Gerlock, J.L.; Schlick, S. *Macromolecules* **1999**, *32*, 5463. (b) Motyakin, M.V.; Schlick, S. *Macromolecules* **2001**, *34*, 2854.
14. Schlick, S.; Kruczala, K.; Motyakin, M.V.; Gerlock, J.L. *Polym. Degrad. Stab.* **2001**, *73/3*, 471.
15. Motyakin, M.V.; Schlick, S. *Polym. Degrad. Stab.* **2002**, *76*, 25.
16. Motyakin, M.V.; Schlick, S. *Macromolecules* **2002**, *35*, 3984.
17. Kruczala, K.; Bokria, J.G.; Schlick, S. *Macromolecules* **2003**, *36*, 1909.
18. Kruczala, K.; Aris, W.; Schlick, S. *Macromolecules* **2005**, *38*, 6979.
19. Kweon, S.-C. M.Sc. Dissertation, University of Detroit Mercy, 1993.
20. Kruczala, K.; Motyakin, M.V.; Schlick, S. *J. Phys. Chem. B* **2000**, *104*, 3387.

21. Halpern, H.J.; Yu, C.; Peric, M.; Barth, E.; Grdina, D.J.; Teicher, B. *Proc. Natl. Acad. Sci. USA,* **1994**, *91,* 13047.
22. Rosen, G.M.; Pou, S.; Halpern, H.J. *Methods Mol. Biol.* **1998**, *108,* 27.
23. Zweier, J.L.; Kuppusamy, P. In *Spatially Resolved Magnetic Resonance: Methods, Materials, Medicine, Biology, Rheology, Ecology, Hardware*; Blümler, P., Blümich, B., Botto, R., Fukushima, E., Eds.; Wiley-VCH: Weinheim, Germany, 1998; Chapt. 34, pp. 373–388.
24. Halpern, H.J.; Chandramouli, G.V.R.; Barth, E.D.; Williams, B.B.; Galtsev, V.E. *Curr. Top. Biophys.* **1999**, *23,* 5.
25. Mailer, C.; Robinson, B.H.; Halpern, H.J. *Magn. Reson. Med.* **2003**, *49,* 1175.
26. Xu, D.; Hall, E.; Ober, C.K.; Moscicki, J.K.; Freed, J.H. *J. Phys. Chem.* **1996,** *100,* 15856, and references cited therein.
27. Schlick, S.; Pilar, J.; Kweon, S.-C.; Vacik, J.; Gao, Z.; Labsky, J. *Macromolecules* **1995**, *28,* 5780.
28. Gao, Z.; Schlick, S. *J. Chem. Soc. Faraday Trans.* **1996**, *92,* 4239.
29. Schlick, S.; Eagle, P.; Kruczala, K.; Pilar, J. In *Spatially Resolved Magnetic Resonance: Methods, Materials, Medicine, Biology, Rheology, Ecology, Hardware*; Blümler, P., Blümich, B., Botto, R., Fukushima, E., Eds.; Wiley-VCH: Weinheim, Germany, 1998; Chapt. 17, pp. 221–234.
30. Degtyarev, E.N.; Schlick, S. *Langmuir* **1999**, *15,* 5040.
31. Pilar, J.; Labsky, J.; Marek, A.; Konak, C.; Schlick, S. *Macromolecules* **1999**, *32,* 8230.
32. Marek, A.; Labsky, J.; Konak, C.; Pilar, J.; Schlick, S. *Macromolecules* **2002**, *35,* 5517.
33. Ohno, K. In *EPR Imaging and In Vivo EPR*; Eaton S.S., Eaton, G.R., Ohno, K., Eds.; CRC Press: Boca Raton, FL, 1991; Chapt. 18, p. 181.
34. Lucarini, M.; Pedulli, G.F.; Borzatta, V.; Lelli, N. (a) *Res. Chem. Intermed.* **1996**, *22,* 581. (b) *Polym. Degrad. Stab.* **1996**, *53,* 9.
35. Lucarini, M.; Pedulli, G.F. *Angew. Makromol. Chem.* **1997**, *252,* 179.
36. Ahn, M.K.; Eaton, S.S.; Eaton, G.R.; Meador, M.A.B. *Macromolecues* **1997**, *30,* 8318.
37. Capancioni, S.; Abdellaoui, K.S.; Kloeti, W.; Herrmann, W.; Brosig, H.; Borchert, H.-H.; Heller, J.; Gurny, R. *Macromolecules* **2003**, *36,* 6135.
38. Timmons, G. Cited at the EPR Imaging Workshop, 27th International EPR Symposium, Denver, CO, 1–5 August 2004.
39. Gillen, K.T.; Clough, R.L. *Polymer* **1992**, *33,* 4359.
40. Billingham, N.C. In *Atmospheric Oxidation and Antioxidants,* G. Scott, Ed.; Elsevier, Amsterdam, 1993, Vol. II, Chapt. 4, pp. 219–277.
41. *Polymer Durability: Degradation, Stabilization and Lifetime Prediction*; Clough, R.G., Billingham, N.C., Gillen K.T., Eds.; Adv. Chem. Series 249, American Chemical Society: Washington, DC, 1996.
42. Wise, J.; Gillen, K.T.; Clough, R.L. *Radiat. Phys. Chem.* **1997**, *49,* 565.
43. De Bruiijn, J.C.M. In *Polymer Durability: Degradation, Stabilization and Lifetime Prediction*; Clough, R.G., Billingham, N.C., Gillen K.T., Eds.; Advanced Chemical Series 249, American Chemical Society: Washington, DC, 1996; Chapt. 36, pp. 599–620.

PART II

ESR APPLICATIONS

5

ESR STUDY OF RADICALS IN CONVENTIONAL RADICAL POLYMERIZATION USING RADICAL PRECURSORS PREPARED BY ATOM TRANSFER RADICAL POLYMERIZATION

ATSUSHI KAJIWARA
Nara University of Education, Takabatake-cho, Nara Japan

KRZYSZTOF MATYJASZEWSKI
Carnegie Mellon University, Pittsburgh, Pennsylvania

Contents
1. Introduction 102
2. Chain Length Dependence 105
   2.1. ESR Spectra of Propagating Radicals Formed in the Polymerization of *tert*-Butyl Methacrylate 105
   2.2. ESR Spectra of Model Radicals with Controlled Chain Lengths 106
3. Chain-Transfer Reactions 109
   3.1. ESR Spectra Observed in Polymerization of *tert*-Butyl Acrylate 109
   3.2. ESR Study of Model *tert*-Butyl Acrylate Radicals with Controlled Chain Lengths 111
   3.3. Polymeric Radicals and Chain-Length Dependent Changes in ESR Spectra 116
4. Penultimate Unit Effect 119
   4.1. Monomeric Model Radicals 120
   4.2. Dimeric Model Radicals 122
   4.3. Dimeric Radicals and the Penultimate Unit Effect 124
   4.4. Activation Rate Constants for Monomeric and Dimeric Alkyl Bromides and ESR Spectra 128
5. Conclusions 129

Advanced ESR Methods in Polymer Research, edited by Shulamith Schlick.
Copyright © 2006 John Wiley & Sons, Inc.

Acknowledgments 129
References 129

1. INTRODUCTION

This chapter describes the application of electron spin resonance (ESR) spectroscopy and controlled radical polymerization techniques to basic research on the chemistry of radical polymerizations. This combination can provide information on the chain length of propagating radicals, chain-transfer reactions to polymers, and penultimate unit effects in copolymerization, topics that have been difficult or impossible to study by direct detection of radicals.

Electron spin resonance is the most powerful method for investigation of radical species in radical polymerization. When well-resolved spectra can be observed, the spectra provide information not only on the structure, properties, and concentration of radicals, but also on the initiating and propagating (oligomeric and polymeric) radicals.[1-5] Direct detection of propagating radicals in radical polymerization by ESR has been difficult, mainly due to both the labile nature and the low concentration of the propagating radicals. About 20 years ago, Kamachi[3] observed the ESR spectra of propagating radicals of methacrylates under conditions similar to those in conventional polymerization by means of a specially designed flat cell and cavity. Starting in the early 1990s, relatively well-resolved ESR spectra of propagating radicals have been observed using a commercially available cavity and normal sample cell, due to both improvement in ESR spectrometers and careful optimization of sample preparation.[5,6] Well-resolved ESR spectra in the radical polymerizations of styrene and its derivatives, diene compounds, methacrylates, and vinyl esters, were obtained in benzene or toluene solutions under usual polymerization conditions.[6-13] Selected ESR spectra of these propagating radicals observed during polymerization reactions are shown in Fig. 1. Simulated spectra of the corresponding propagating radicals are also shown. Based on these spectra, a kinetic analysis, especially for estimating propagating rate constants (k_p) of the monomers, was performed. The ESR spectroscopy was also used to quantify radical concentration in the polymerization. However, the direct detection method did not reveal some aspects that are significant in radical polymerization chemistry; for example, the lifetime of the propagating chain is not known. The chain length of the observed propagating radicals was considered to be long enough for estimation of k_p. However, there has been no clear experimental information on the actual chain length of the propagating radicals observed by ESR. In order to detect well-resolved spectra, the initiator concentration in some ESR experiments was higher than in usual radical polymerizations. Under such conditions, the chain lengths of the propagating radicals could be shorter than under usual conditions. The chain length of the resulting polymers in the ESR samples can be estimated by size exclusion chromatography (SEC) experiments after ESR experiments. However, the resulting molecular weights, even if high, do not guarantee ESR detection of radicals with high molecular weights, because detection of shorter

INTRODUCTION

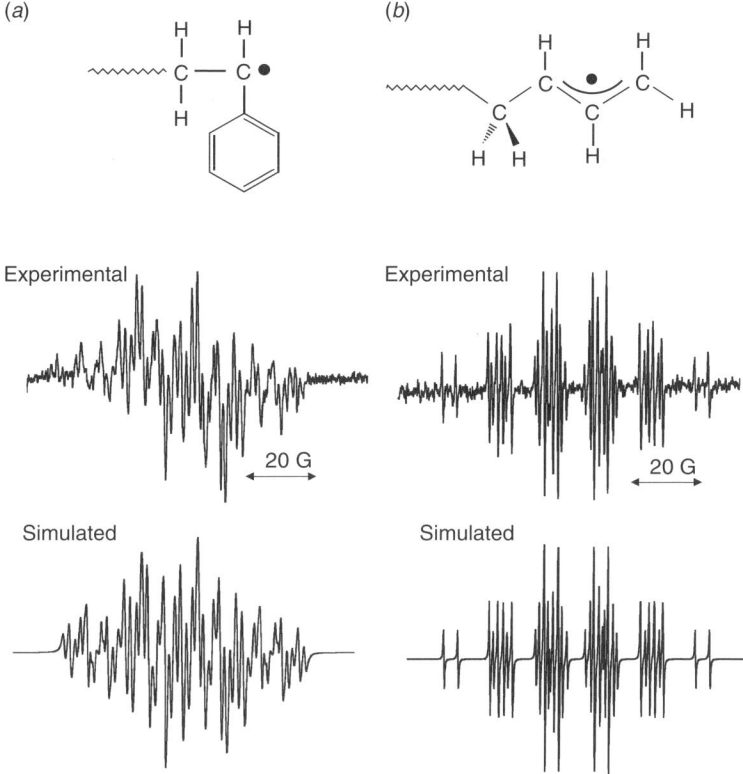

Fig. 1. Experimental and simulated ESR spectra of propagating radicals obtained during conventional radical polymerizations. (*a*) Styrene at 0°C. (*b*) Butadiene at −10°C.

chain radicals is easier than that of longer chain radicals. Therefore, the effect of the chain length of the propagating radicals should be examined carefully. Moreover, it is extremely difficult to investigate copolymerization by the direct detection method. During copolymerization, various kinds of propagating chain ends are formed, due to so-called penultimate unit effects. These problems have not yet been resolved. The development of controlled radical polymerization techniques, especially atom-transfer radical polymerization (ATRP), enables us to solve some of these problems.

Atom-transfer radical polymerization is one of the most widely applied techniques in the field of controlled/living radical polymerization, and allows the preparation of a wide range of polymeric materials with controlled molecular weights and well-defined architectures.[14–20] A general mechanism for ATRP is shown in Fig. 2. The radicals, or the active species, are generated through a reversible redox process catalyzed by a transition metal complex ($Mt^{z}-Y/L_nY$, where L is the ligand, n is the number of ligands, and Y may be another ligand or the counterion), which undergoes a one-electron oxidation with concomitant abstraction of a (pseudo)halogen atom, X, from a dormant species, R–X. This process occurs with a rate constant of activation, k_{act}, and of deactivation, k_{deact}. Polymer chains grow by addition of intermediate radicals to monomers in a manner similar to a conventional radical polymerization, with the rate constant of propagation k_p. The polymers obtained have narrow polydispersity and predetermined molecular weight, and can be precursors of model radicals with well-defined structures. The polymers formed in ATRP contain terminal carbon–halogen bonds (Fig. 2). Giese et al.[21] reported that these bonds can be homolytically cleaved by reaction with organotin compounds. Accordingly, various radicals that model the end groups in ATRP can be formed from the corresponding precursors prepared by atom-transfer radical addition (ATRA) and ATRP, and the generated radicals can be studied by ESR spectroscopy. The generation of model radicals from corresponding radical precursors by a reaction with organotin compounds for polymeric (meth)acrylates with various chain lengths is shown in Fig. 3. Variation of the chain length and composition of

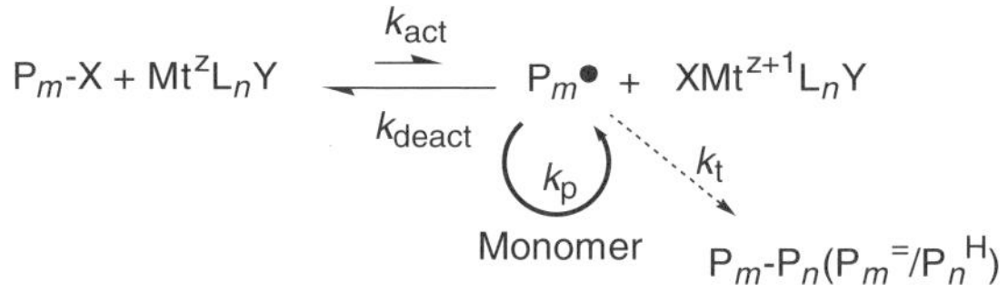

Fig. 2. Schematic diagram of atom-transfer radical polymerization. The resulting polymer (P_m–X) usually has a terminal carbon–halogen bond.

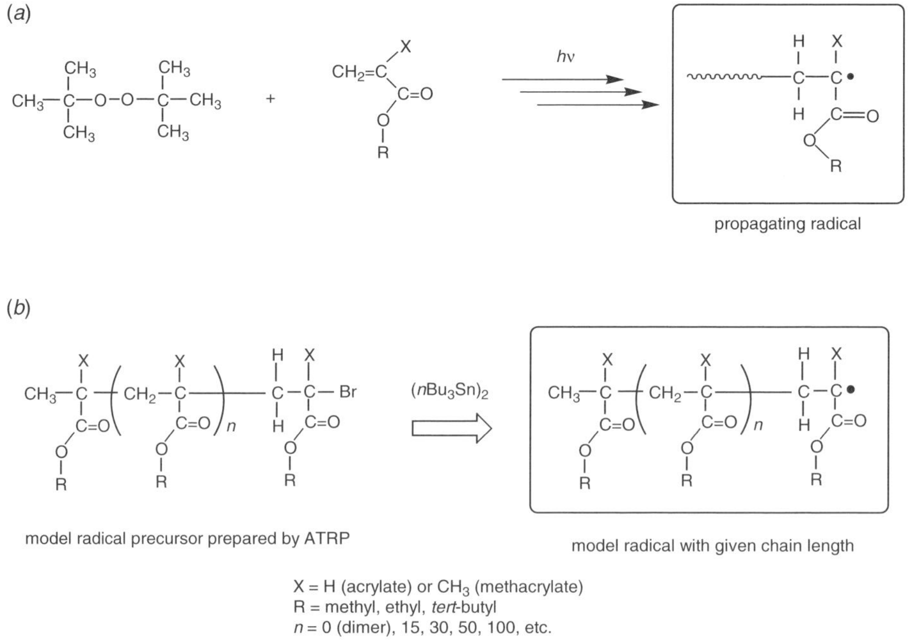

Fig. 3. Radicals observed by ESR. (*a*) Formation of propagating radicals during a standard free radical polymerization. (*b*) Generation of model radicals from radical precursors prepared by ATRP.

polymeric radical precursors elucidates the effect of chain length and penultimate units on the ESR spectra of the formed radicals, which is the basic topic of this chapter.

Investigations of radical reactions are difficult without model reactions. Poly(meth)acrylates with various chain lengths can be prepared by ATRP, as shown in Fig. 3b. Propagating radicals with various chain lengths can be generated from these model precursors. The ESR spectra at various temperatures of these model radicals are expected to show chain-length dependence, as detected for propagating radicals of methacrylates observed *in situ*.[22] This method can also be applied to investigate chain-transfer reactions for propagating acrylate radicals.

The ATRP is based on a sequence of ATRA reactions and ATRA can be used to prepare various dimeric species that model the expected chain ends in a copolymerization reaction. An ESR study of radicals generated from these mixed dimers can be used to clarify the penultimate unit effects relevant to copolymerization. The ESR spectra of monomeric radicals of (meth)acrylates were previously investigated by Fischer et al.,[23–25] Gilbert et al.,[26,27] and Matsumoto et al.[28,29] However, no ESR study has so far been conducted for dimeric radicals of (meth)acrylates, mainly due to the difficulty of obtaining clear ESR spectra.

In this chapter, three examples of the application of ESR to conventional radical polymerizations based on controlled/living radical polymerizations will be demonstrated. The first example is estimation of the effect of chain length on propagating radicals. The second example is the detection of chain-transfer reactions on the propagating radicals in polymerization of *tert*-butyl acrylate (*t*BA). The third example is investigation of penultimate unit effects using ESR analysis of dimeric model radicals of (meth)acrylates prepared by ATRA.

As shown in Fig. 3, when conducting ESR spectroscopy on radical polymerizations, direct detection in a conventional polymerization and use of model radicals based on controlled radical polymerization technique complement each other. Comparison of ESR spectra obtained by direct detection in a radical polymerization processes with those of model radicals generated from radical precursors prepared by ATRP leads to an understanding of each step in radical polymerizations, and to the reaction mechanism. Electron spin resonance spectroscopy has provided unambiguous proof of the intermediate reactions and products for several reactions involved in radical polymerization, aided by controlled radical polymerization techniques. This spectroscopy was also used to examine and quantify some intermediates in ATRP, and to provide a deeper understanding of the ATRP process, including identification of the structure and concentration of the paramagnetic species involved.[12, 13, 30–39] Although interesting and important for investigation of ATRP, the structural features of the paramagnetic transition metal compounds will not be discussed in this chapter.

2. CHAIN LENGTH DEPENDENCE

2.1. ESR Spectra of Propagating Radicals Formed in the Polymerization of *tert*-Butyl Methacrylate

Propagating radicals are formed (Fig. 3a, $X=CH_3$) when a mixture of monomer and radical initiator is heated or photoirradiated in an ESR sample cell. Well-resolved

ESR spectra of propagating radicals of *tert*-butyl methacrylate (*t*BMA) have been detected in such polymerization systems at various temperatures, as shown in Fig. 4. The 16-line spectrum is different compared to that previously reported for methacrylates (13-line or 9-line spectra),[3] and was assigned to propagating radicals. Since couplings due to the methyl protons of the *tert*-butyl group in the ester side chain did not appear, splittings were assigned to the three α-methyl protons and the two β-methylene protons. The temperature dependence of the spectra was interpreted by a hindered rotation model of two stable conformations.[3] The intensity of the inner eight lines increased with increasing temperature, indicating that there are two exchangeable conformations whose existence has been deduced from the ESR spectra of methacrylates.[3] At 150°C, the intensity of the inner eight lines increased and the ESR spectrum can be interpreted as a single conformation, indicating that the energy difference between the two conformers is small. The observed ESR spectrum of propagating radicals of *t*BMA at 150°C is shown in Fig. 5*a* along with the simulated spectrum [hyperfine splitting constants of 14.0 G for one methylene proton (1:1 doublet), 11.6 G for the other proton (1:1 doublet), and 21.7 G for three equivalent methyl protons (1:3:3:1 quartet)], as shown in Fig. 5*b*. A characteristic point of this result is estimation of different hyperfine splitting constants for the two methylene protons. This means that the rate of rotation of the end radical is not fast enough to make the methylene protons equivalent on the time scale of the ESR measurement. Thus, it leads to a 16-line spectrum (2 × 2 × 4). If the two β-methylene protons were equivalent, the total number of splitting lines would be 12 [4(CH_3−) × 3 (−CH_2−)]. The expected spectra for these two cases are shown in Fig. 6. This suggests the presence of a propagating radical with a long chain that hindered the rotation of the terminal bond to generate the 16-line spectrum, and also another oligomeric radical that may show a 12-line spectrum.

In order to clarify the chain length dependence, model radical precursors were prepared by the ATRP, a technique that can provide polymers with controlled molecular weights and low polydispersity, and with preserved terminal carbon–halogen bonds.[14–20] When the carbon–halogen bonds are cleaved homolytically by reaction with organotin compounds, model radicals of propagating chains with given chain length could be generated (Fig. 3*b*, X = CH_3).[21]

2.2. ESR Spectra of Model Radicals with Controlled Chain Length

First, a mixture of oligomers containing 2–7 monomer units (P_n = 2–7) was prepared by ATRP and model radicals with short-chain lengths were generated from the mixture without any further separation. Well-resolved ESR spectra of the model radicals were observed at various temperatures. The 12-line ESR spectrum observed at 150°C is shown in Fig. 7*a*; the two β-methylene protons are almost equivalent in small radicals at such high temperature. This finding indicates that rotation of the radical chain end is too fast to detect differences in methylene protons on the time scale of ESR spectroscopy. In order to estimate the critical chain length that would show the 16-line spectrum, model radical precursors with

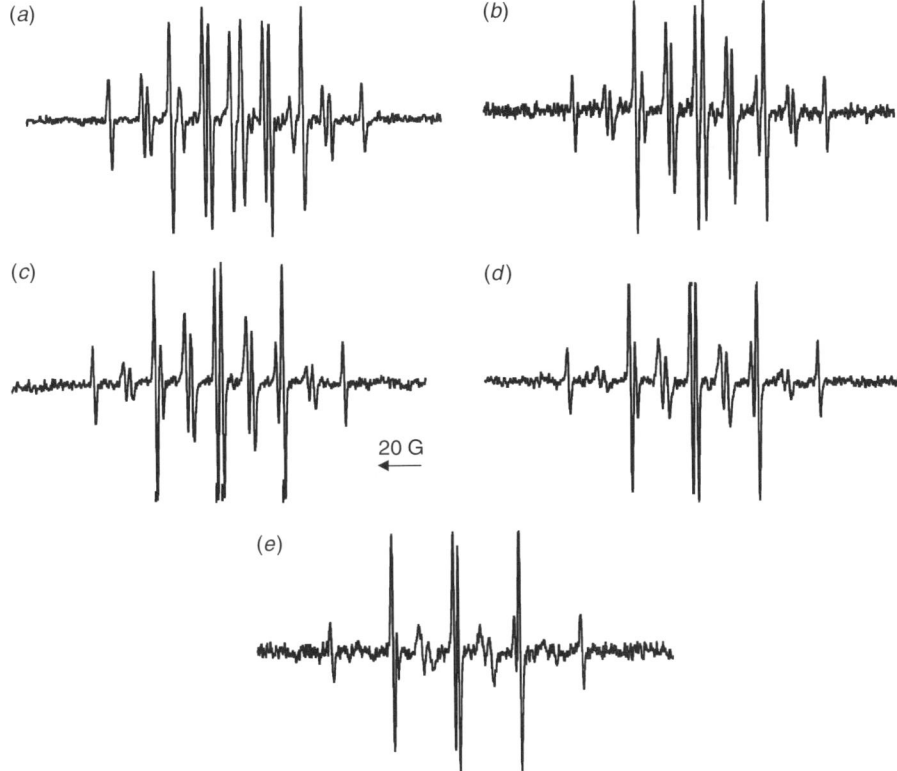

Fig. 4. The ESR spectra of propagating radicals of *t*BMA observed in polymerization systems initiated by *t*BPO under irradiation. Propagating radicals were observed *in situ*. (*a*) 150°C (in mesitylene). (*b*) 120°C (in mesitylene). (*c*) 90°C (in toluene). (*d*) 60°C (in toluene). (*e*) 30°C (in toluene). (From Ref. 22, with permission.)

degrees of polymerization (P_n) of 30, 50, and 100 were prepared by ATRP and polymers with calculated molecular weights and low polydispersities were obtained. The ESR spectra of radicals generated from these precursors were observed at various temperatures, as shown in Fig. 8. Although the lifetime of the model radicals were very short at 150°C, clear and well-resolved spectra were observed. These spectra showed similar temperature dependence to that shown in Fig. 4. In the case of $P_n = 100$, a 16-line spectrum was clearly observed at temperatures < 120°C.

Comparison of the ESR spectra of oligomeric radicals (Fig. 7a), of model radicals with $P_n = 100$ (Fig. 8), and of radicals in a polymerization system (Fig. 5a) at 150°C is shown in Fig. 9. From the comparison of the separation of the inner lines, the P_n of the propagating radical is considered to be >100. The hyperfine couplings measured from these spectra, plotted against chain lengths in Fig. 10, seemed to show a nearly linear correlation between hyperfine coupling constants

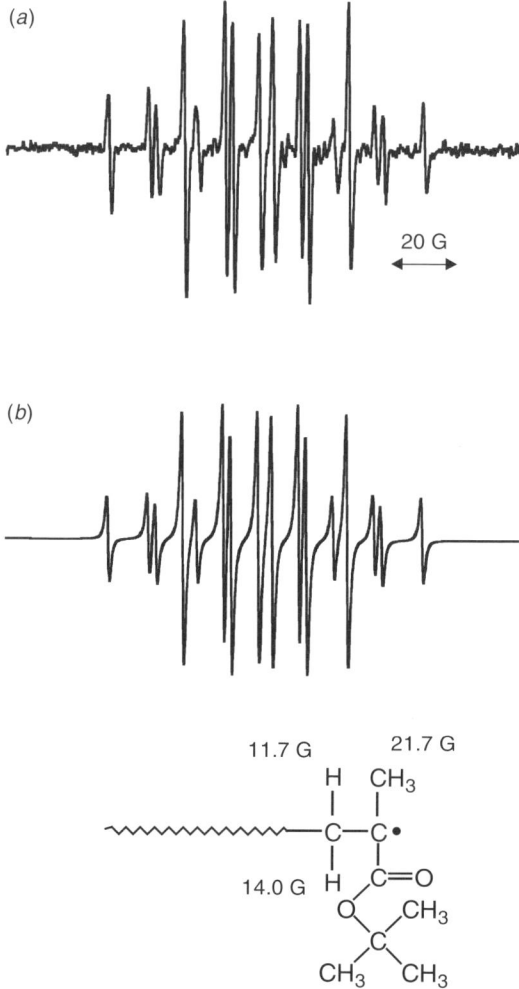

Fig. 5. Experimental (*a*) and simulated (*b*) ESR spectra of the propagating radical in polymerization of *t*BMA at 150°C. (From Ref. 22, with permission.)

and chain lengths in the range up to $P_n = 200$. The molecular weight (M_n) of the isolated polymers from the polymerization system (Fig. 9*c*) was determined to be 30,000 ($P_n = 210$) by SEC. Interpretation of the ESR spectra suggests that they correspond to "long" propagating radicals, in agreement with SEC. Before these experimental results, ESR spectra and overall SEC results did not correlate. However, more experimental results are needed for a comprehensive correlation of kinetic data with ESR spectra.

It can be concluded that the 16-line spectrum in ESR measurements is due to "polymeric" radicals with > 100 monomer units. Similar differences in ESR spectra

CHAIN-TRANSFER REACTIONS

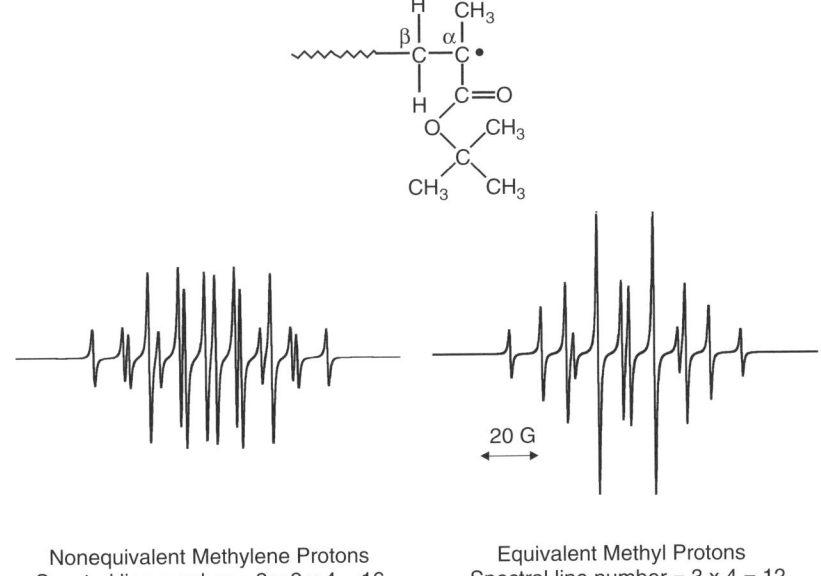

Fig. 6. Simulated spectra for nonequivalent β-methylene protons (16-line spectrum) and equivalent β-methylene protons (12-line spectrum) for propagating radicals of tBMA.

were observed for poly(methyl methacrylate) (polyMMA) radicals generated *in situ* and H-MA-MMA dimeric model radical (Fig. 23), which will be discussed in the section on the penultimate unit effect. Although they are more complicated, with a small quartet due to ester methyl protons, the ESR spectrum of polyMMA recorded during polymerization showed a basic 16-line spectrum and the dimeric model radical with terminal MMA radical showed a 12-line spectrum. The ESR spectroscopy has provided structural information of the propagating radicals at their chain ends. Direct information on the chain length of the radicals from ESR measurements are reported here for the first time.

Further information on the dynamic behavior of the propagating radicals was obtained by simulating the temperature dependence of the ESR spectra.[40] The average exchange time between the two conformers was calculated from the simulation of the spectra in Fig. 4. The activation energy for rotation of the terminal C_α–C_β bond was estimated to be 21.2 kJ mol^{-1}.[41]

3. CHAIN-TRANSFER REACTIONS

3.1. ESR Spectra Observed in Polymerization of *tert*-Butyl Acrylate

Interpretations of ESR spectra detected in acrylate radical polymerizations have been very difficult.[23,26,42–47] The ESR spectra observed for acrylates and

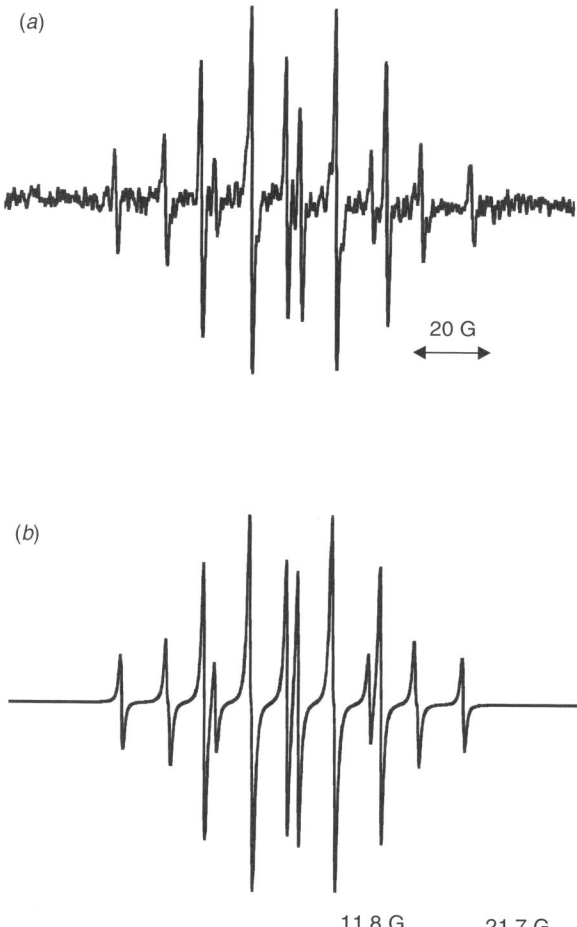

Fig. 7. Experimental (*a*) and simulated (*b*) ESR spectra of oligomeric model radical of *t*BMA (P_n = 2–7) at 150°C. (From Ref. 22, with permission.)

methacrylates are different, even under almost identical conditions, and it is difficult to interpret the spectra in terms of propagating radicals. The spectra measured during solution polymerization of *t*BA are shown in Fig. 11: A six-line spectrum or doublet of triplets with narrow line width (Fig. 11*a*) was observed at −30°C

CHAIN-TRANSFER REACTIONS   111

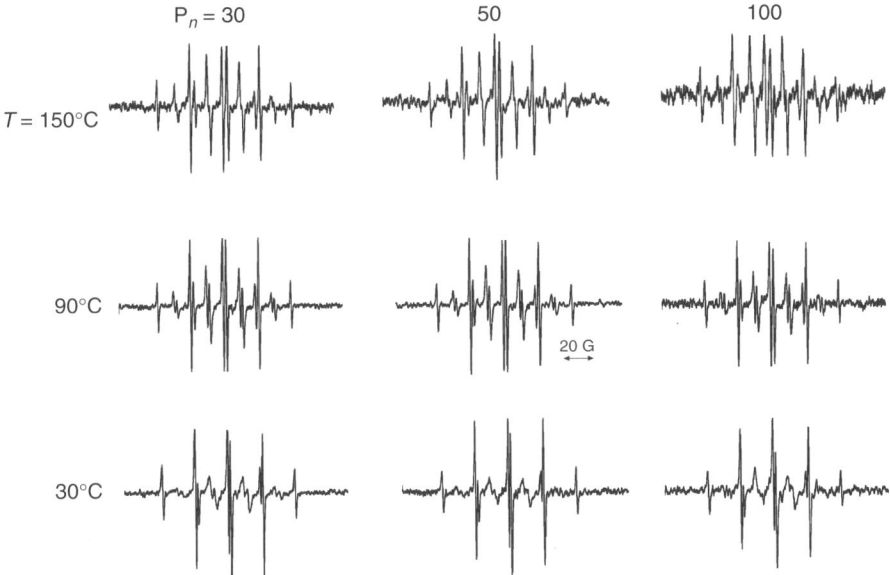

Fig. 8. The ESR spectra of model radicals of poly(*t*BMA) for the indicated chain length and temperature.

and assigned to be a propagating radical with two β-methylene protons (1:2:1 triplet) and one α-proton (1:1 doublet), as shown in Fig. 11*b*. A totally different, seven-line, spectrum with broader line width was observed at 60°C (Fig. 11*d*). Traces of the six-line spectrum can be seen in the spectrum at 60°C. At −10°C, overlapped spectra of the first and latter spectra were observed (Fig. 11*c*), suggesting that the spectrum observed at −30°C converted to the spectrum observed at 60°C. Two possible explanations for this result have been considered: A chain-length dependence of the spectra, and chemical reaction. These possibilities were examined by analysis of ESR spectra of model radicals with various chain lengths, as described in Section 3.2.[14–20]

3.2. ESR Study of Model *tert*-Butyl Acrylate Radicals with Controlled Chain Lengths

Precursors of $P_n\bullet$ with $n = 15$, 50, and 100 were prepared by ATRP. The model radicals were generated by a reaction with organotin compounds and observed by ESR.[21] The ESR spectra for $P_n = 15$, 50, and 100 are shown in Fig. 12. These radicals showed almost the same six-line (doublet of triplets) ESR spectrum of propagating radicals at low temperature (−30°C), as in Fig. 11*a*. This result indicates that there is no chain-length dependence in the ESR spectra of the propagating acrylate radicals. However, the six-line spectrum changed after raising the temperature to 60°C, and the change is in agreement with the ESR spectra observed directly during

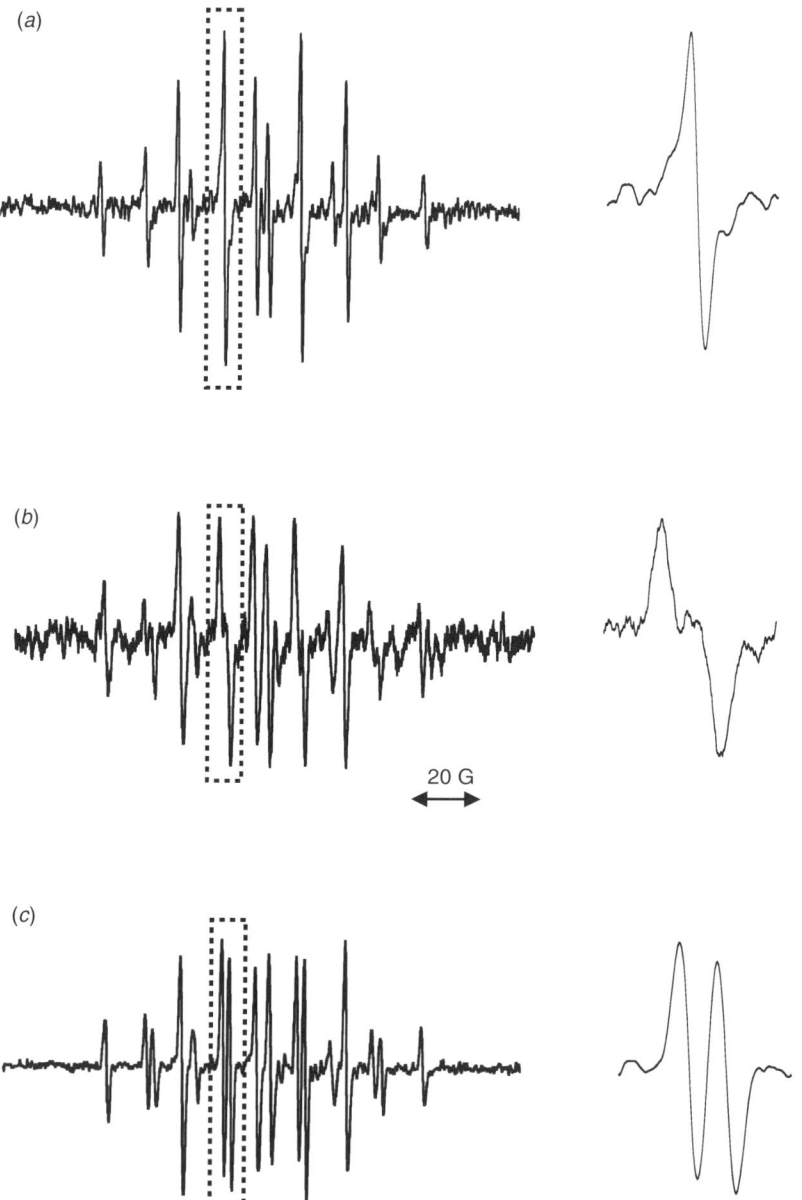

Fig. 9. Comparison of ESR spectra of radicals with various chain length at 150°C. (*a*) oligomeric model radical. (*b*) Model radical with $P_n = 100$. (*c*) Propagating radical. Characteristic lines (in dashed squares) were enlarged on the right-hand side. (From Ref. 22, with permission.)

CHAIN-TRANSFER REACTIONS

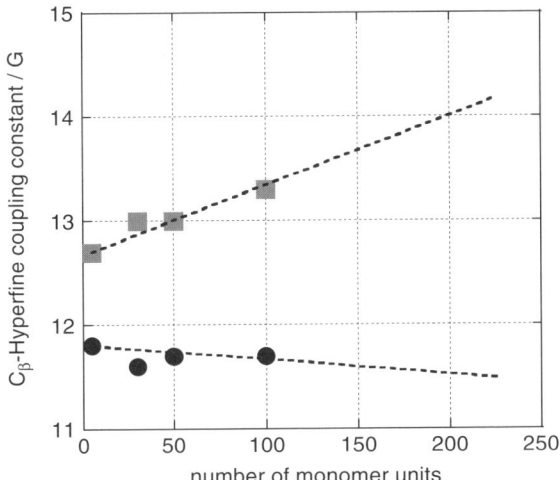

Fig. 10. The correlation between chain length and hyperfine coupling constants of C_β protons in oligomeric and polymeric tBMA radicals.

polymerization. A typical example of two kinds of ESR spectra for $P_n = 100$ is shown in Fig. 13. The model systems do not contain any monomer and the spectroscopic change occurred without propagation. This suggests that a chemical reaction is responsible for the observed changes.

In radical polymerization of acrylates, a 1,5-hydrogen shift to form a mid-chain radical has previously been considered.[43–47] Yamada and co-workers[45,46] observed ESR spectra similar to those in Figs. 11d and 13b, and reported that the spectra were due to a mixture of propagating and mid-chain radicals. However, there has been no clear experimental evidence or explanation for the formation of the mid-chain radicals. Here, a clear interpretation of the ESR spectra of propagating (Fig. 14a) and mid-chain radicals (Fig. 14b) was demonstrated with the aid of ATRP. The proposed mechanism for the 1,5-hydrogen shift is shown in Fig. 15. Formation of a six-membered ring enables the radical to migrate to the monomer located unit two units before the terminal unit. Since this shift needs at least three monomer units, it suggests that mid-chain radicals cannot be formed in a dimeric radical. A dimeric radical precursor was isolated from a mixture of oligomeric radical precursors, and the ESR spectrum of the dimeric radical was measured. The ESR spectra of the dimeric model radical showed no change even at relatively high temperature (40°C). From the same mixture of oligomers, a mixture of oligomers with tetramer to hexamer (4–6 monomer units) was also separated and radicals were generated from the oligomers. The ESR spectra of these radicals showed overlapped spectra of two kinds of spectra from the beginning of the reaction, as shown in Fig. 16. The absence of chain-length dependence and no changes with temperature for the dimer strongly suggested the formation of mid-chain radicals in radical polymerizations of acrylates via a 1,5-hydrogen shift reaction.

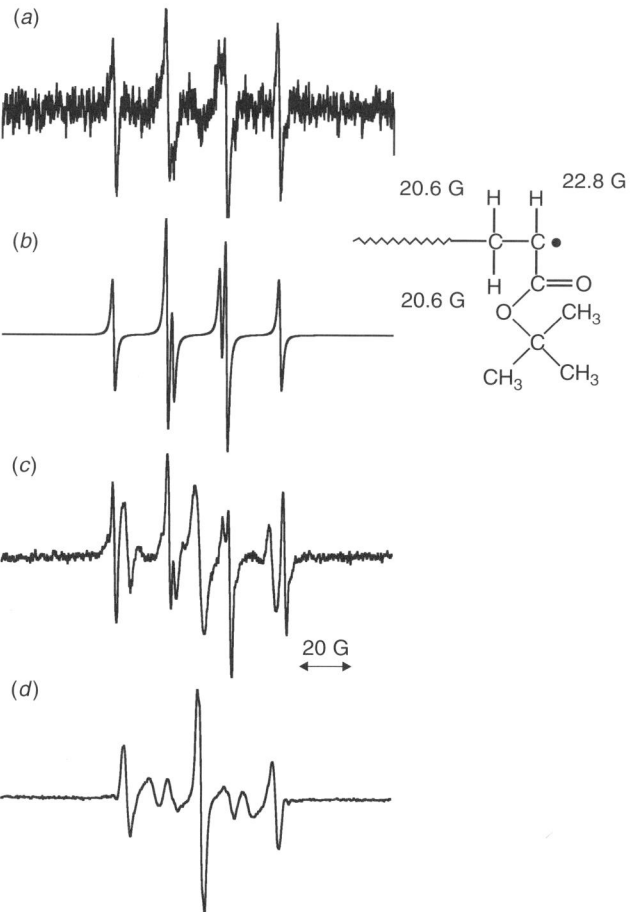

Fig. 11. The ESR spectra observed in radical polymerization of *t*BA initiated with *t*BPO under irradiation: At −30°C (*a*) and corresponding simulated spectrum (*b*). At −10°C (*c*). At 60°C (*d*) in toluene. (From Ref. 13, with permission.)

One additional piece of evidence supporting the formation of a mid-chain radical was obtained from an examination of hydrogen abstraction from polyacrylates. This mid-chain radical can be formed by hydrogen abstraction from polyacrylates by oxygen centered radicals. Poly*t*BA and *tert*-butyl peroxide (*t*BPO) were dissolved in benzene, and the mixture exhibited the ESR spectra shown in Fig. 17*a* under irradiation. The spectrum was similar to both the spectra observed in the polymerization system (Fig. 11*d*) and that reported by Westmoreland et al.[47] Furthermore, it was reasonably simulated by considering two sets of methylene protons with restricted rotation at both sides of the mid-chain radical, as shown in Fig. 17*b*. Consequently, the radical observed at high temperatures (Fig. 11*b*) is due to the formation of mid-chain radicals.

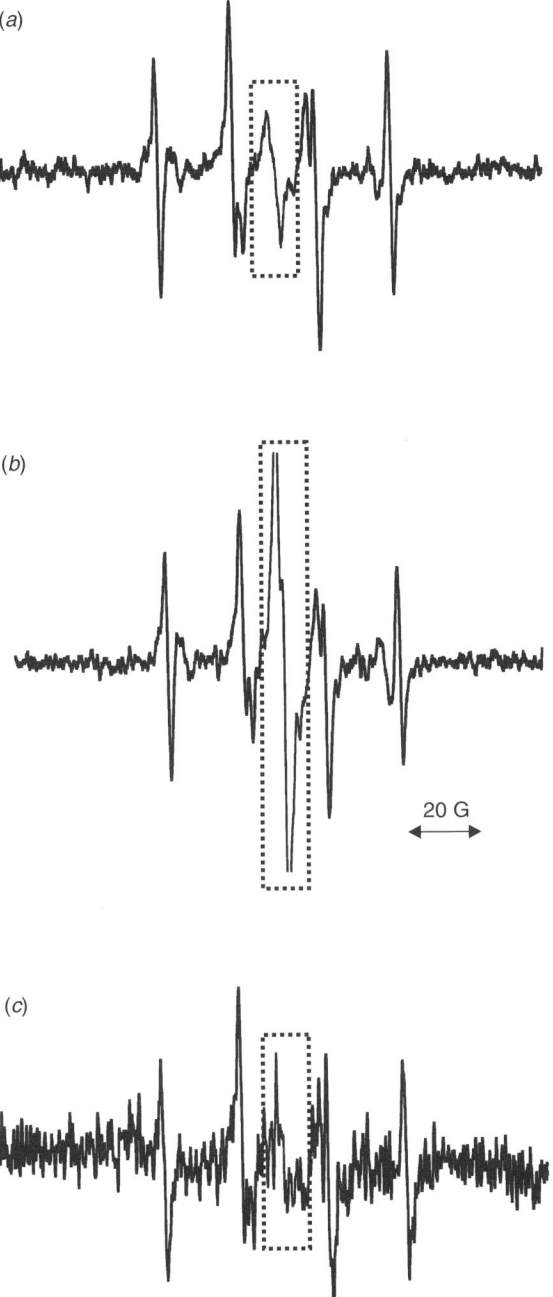

Fig. 12. The ESR spectra of model radicals of poly*t*BA with P_n = 15 (*a*), 50 (*b*), and 100 (*c*) observed at −30°C in toluene. No chain length dependent spectroscopic change was observed. (Center lines indicated in dashed squares are due to radicals of tin compounds.)

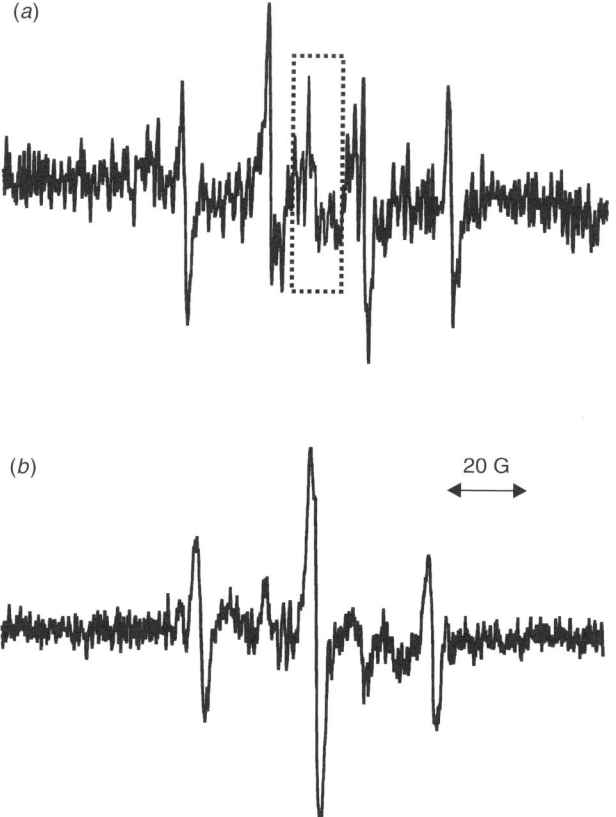

Fig. 13. The ESR spectra of model radicals of poly*t*BA with $P_n = 100$ at $-30°C$ (*a*), and at $60°C$ (*b*). Heating caused irreversible spectral changes. The chain length of the radical was fixed in this case. (Center lines indicated in dashed squares are due to radicals of tin compounds.)

It is known that polyacrylates prepared by conventional radical polymerization contain many branches.[48,49] The ESR study has now provided direct evidence for the origin of the branching. Note also that an estimate of k_p for acrylates is difficult by ESR, because it would provide the sum of the concentration of the active growing terminal radicals and the less active mid-chain radicals.

3.3. Polymeric Radicals and Chain-Length Dependent Changes in ESR Spectra

The ESR spectra of propagating radicals of *t*BA did not show any chain length dependence, due to much faster dynamics compared to methacrylates. However, a different type of chain-length dependence was observed for monomeric, dimeric, and polymeric radicals of *t*BA. The ESR spectra of H-*t*BA•, H-*t*BA-*t*BA•, and poly*t*BA• ($DP_n = 15$), where the radicals were generated from the corresponding alkyl bromides, are shown in Fig. 18 together with their simulations. In the monomeric and dimeric radicals, the hyperfine splittings of the α-proton are smaller than those of the

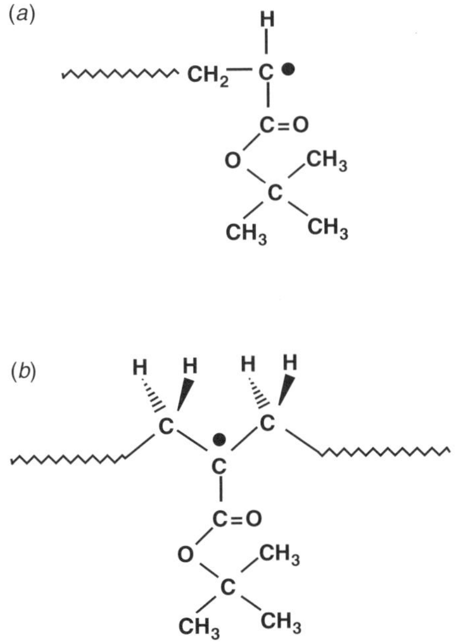

Fig. 14. Structure of propagating radicals for terminal tBA (*a*) and mid-chain radical (*b*).

β-protons. However, in the case of polymeric radicals with $DP_n = 15$, the hyperfine splitting for the α-proton is larger than that for the β-protons, as shown in Fig. 18c and Table 1. The ESR spectra of tBA polymeric radicals with $DP_n = 50$, $DP_n = 100$, and propagating radicals generated *in situ*, were very similar to that for $DP_n = 15$.[13] This result can indicate either a pen-penultimate unit effect, or another chain-length effect. The origin of the chain-length dependent variation of the ESR spectra is not yet explained. Fischer et al.[25] reported ESR spectra of polymeric radicals of acrylic acid by a flow method and detected two kinds of radical species with different splittings, termed polymeric radicals A and B, respectively. The coupling constants for α- and β-protons were reported to be 20.67 and 22.06 G for A and 22.62 and 21.34 mT for B. In the case of radical A, the coupling constant for the α-proton was smaller than that for the β-protons. In the case of radicals B, the value of the hyperfine coupling for the α-proton was larger than that for the β-methylene protons. A similar dependence was found for tBA derived radicals in this work. These results could indicate that radical A was an oligomeric radical (longer than dimeric), and that radical B had a chain length longer than radical A.

As mentioned above, chain length dependent ESR spectra were observed for methacrylate radicals. Apparently, the coupling constant of two β-methylene protons changed from equivalent to nonequivalent as the chain length increased. However, coupling constants of protons in the methyl group at the α-position do not show any chain length dependent change. The exchange of the magnitude of hyperfine coupling constants values between α- and β-protons in acrylates is therefore solely characteristic of acrylates.

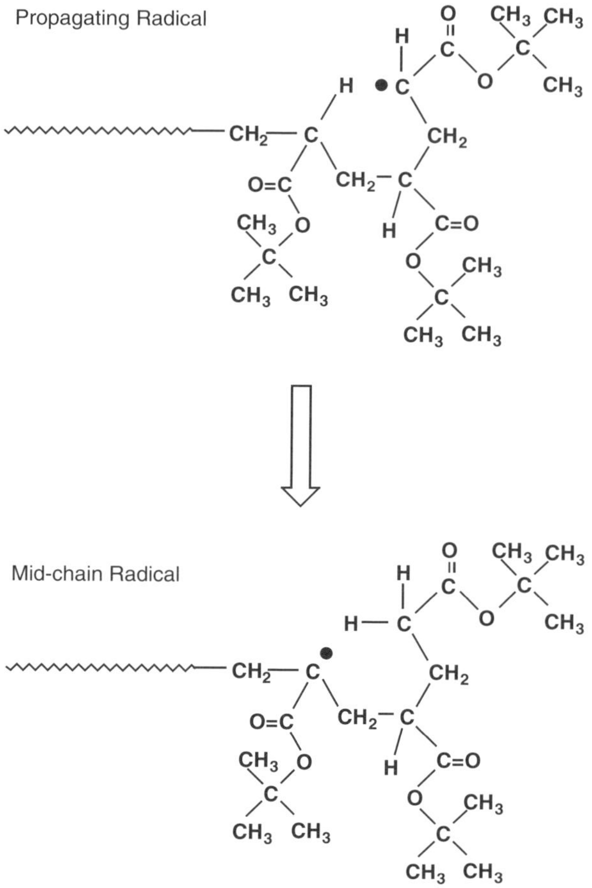

Fig. 15. Schematic diagram of potential 1,5-hydrogen shift for propagating *t*BA radical.

TABLE 1. Hyperfine Splitting Constants of *t*BA Radicals with Various Chain Length[a]

Radicals	Source	Hyperfine Splitting Constants, G			
		α	β	γ	Reference
H-*t*BA•	*t*BBrP	20.90	25.50		50
H-*t*BA-*t*BA•	*t*BA-*t*BA – Br	20.90	22.70	0.78	50
H-(*t*BA)$_n$•	(*t*BA)$_n$ – Br	22.8	20.6		13
poly A (poly(acrylic acid))	acrylic acid (*in situ*)	20.67	22.06		25
poly B (poly(acrylic acid))	acrylic acid (*in situ*)	22.62	21.34		25

[a] From Ref. 50, with permission.

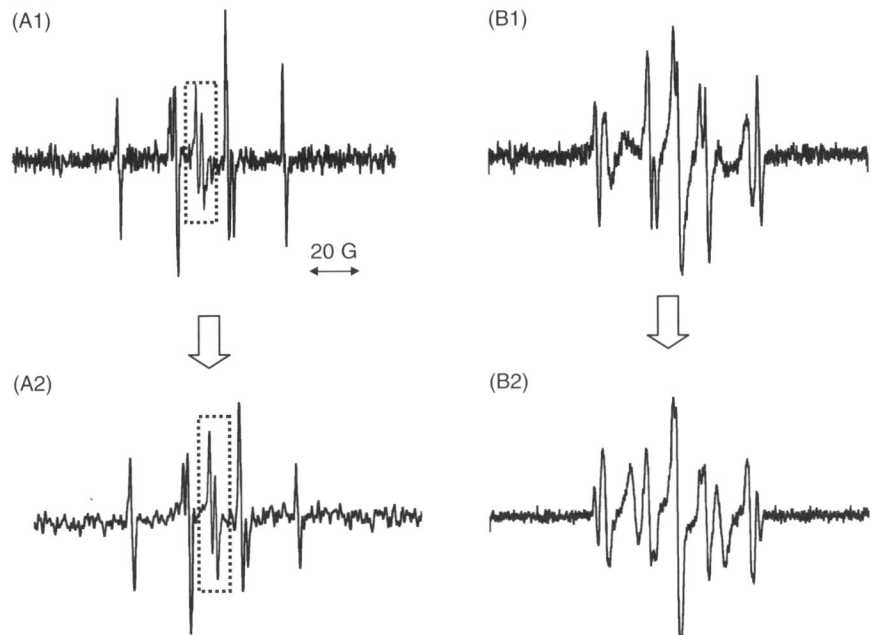

Fig. 16. Time-dependent ESR spectra of dimeric radicals of H–EA-tBA at 40°C [2 min (A1) and 6 min (A2) after irradiation], and of oligomeric tBA species (P_n = 3–7) at 40°C [2 min (B1) and 6 min (B2) after irradiation]. (Center lines indicated in dashed squares are due to radicals of tin compounds.)

In the case of tBA, experimental evidence for the formation of a significant amount of mid-chain radicals from propagating radicals via 1,5-hydrogen shift was shown. Progress in controlled radical polymerization enables us to investigate conventional radical polymerization systems in detail.

4. PENULTIMATE UNIT EFFECT

It has been known since ≈ 1980 that the terminal model for free-radical copolymerization sometimes fails, due to the penultimate unit effect. Direct detection of the penultimate unit effect by ESR has been unsuccessfully attempted many times. In this section, direct detection of the penultimate unit effect using dimeric model radicals generated from dimeric model radical precursors prepared by ATRA is discussed (Fig. 19).[50] The structures of the dimeric model radicals studied are summarized in Fig. 20. For a detailed discussion of the penultimate unit effect, dimeric, monomeric, and polymeric model radicals were examined. The radicals were generated by three methods: homolytic cleavage of carbon–bromine bonds of alkyl bromides with hexabutyldistannane, photodecomposition of an azo-initiator, and radical polymerization performed directly in a sample cell in a cavity.

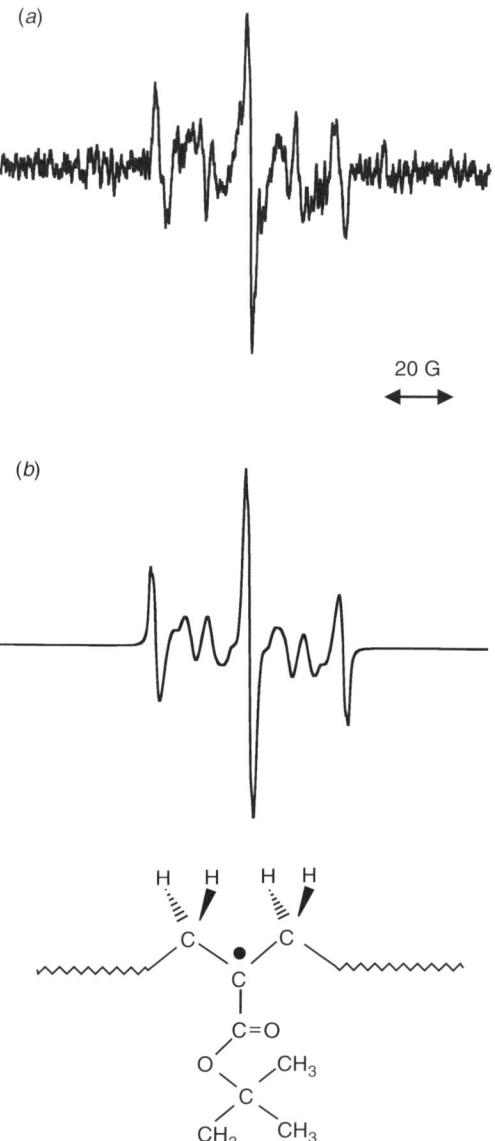

Fig. 17. Experimental (*a*) and simulated (*b*) ESR spectra of mid-chain radical generated by hydrogen abstraction from poly*t*BA by peroxide radicals (*b*). The spectrum was reasonably simulated based on the structure shown in the figure.

4.1. Monomeric Model Radicals

In order to understand the penultimate unit effect in dimeric model radicals, ESR spectra of monomeric model radicals were studied for comparison. The ESR spectra

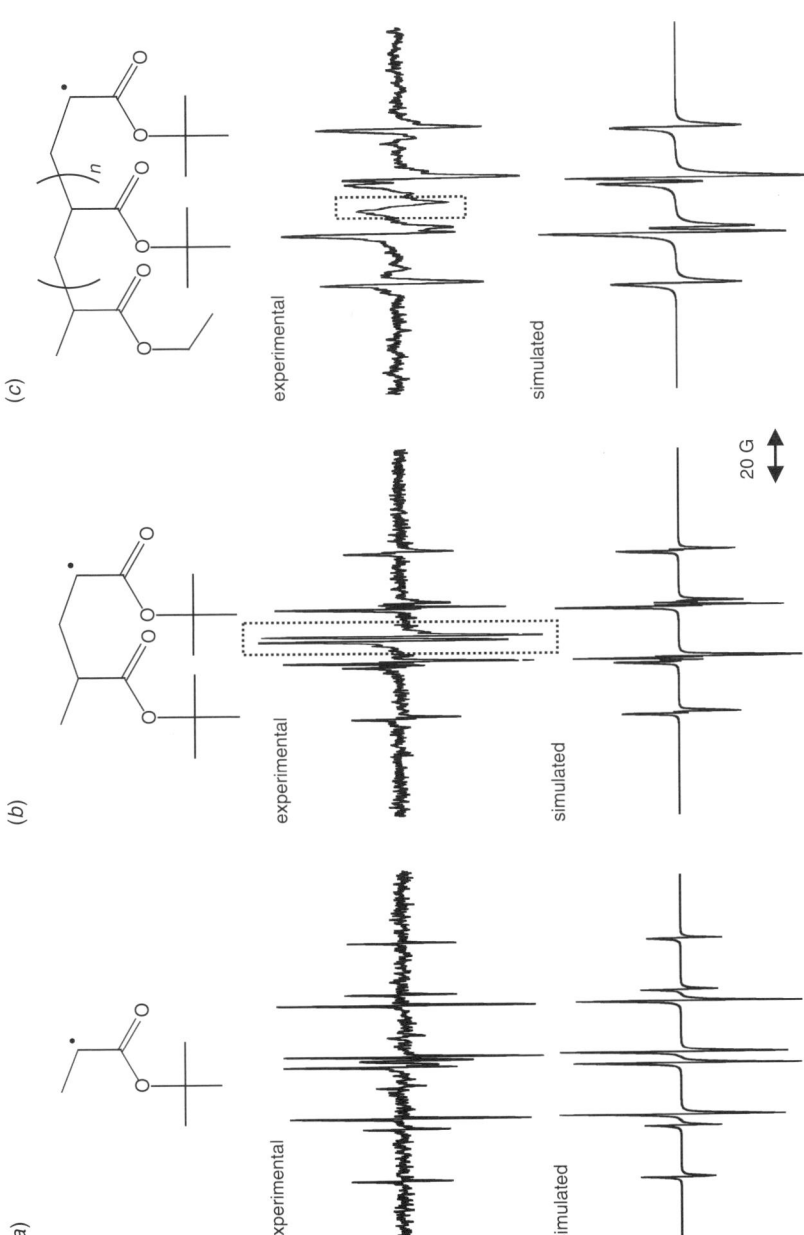

Fig. 18. Experimental and simulated ESR spectra of *t*BA radicals with various chain lengths observed at 30°C: (*a*) H-*t*BA• (monomeric), (*b*) H-*t*BA-*t*BA• (dimeric), and (*c*) H-(*t*BA)$_n$-*t*BA• [polymeric model radicals, $n \sim 15$ (DP$_n$ = 15)]. (Center lines indicated in dashed squares are due to radicals of tin compounds). (From Ref. 50, with permission.)

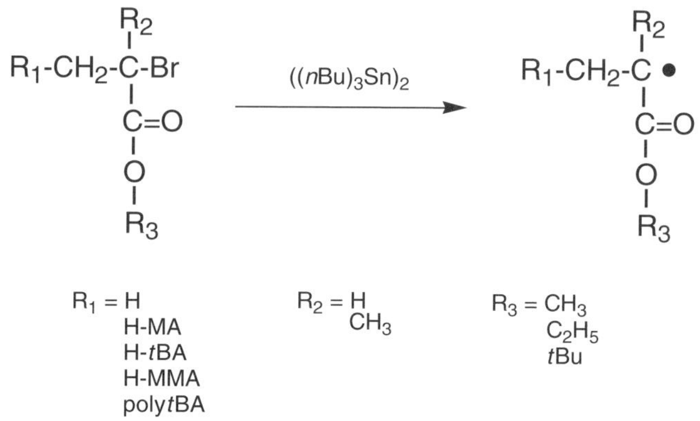

$$R_1\text{-}CH_2\text{-}\underset{\underset{\underset{R_3}{O}}{\underset{O}{C=O}}}{\overset{R_2}{C}}\text{-}Br \xrightarrow{((n\text{Bu})_3\text{Sn})_2} R_1\text{-}CH_2\text{-}\underset{\underset{\underset{R_3}{O}}{\underset{O}{C=O}}}{\overset{R_2}{C}}\bullet$$

$R_1 = H$
H-MA
H-*t*BA
H-MMA
poly*t*BA

$R_2 = H$
CH_3

$R_3 = CH_3$
C_2H_5
*t*Bu

Fig. 19. Generation of monomeric and dimeric model radicals for investigation of the penultimate unit effect.

of radicals generated from monomeric alkyl bromides were studied first. Reaction of methyl 2-bromopropionate (MBrP), ethyl 2-bromopropionate (EBrP), and *tert*-butyl 2-bromopropionate (*t*BBrP) with organotin compounds provided radicals that model monomeric methyl acrylate (H–MA•), ethyl acrylate (H–EA•), and *tert*-butyl acrylate (H-*t*BA•), respectively. As shown in Fig. 21, well-resolved spectra were observed and simulated. The hyperfine couplings for these and other model radicals are summarized in Table 2. Two sets of quartets were observed in each spectrum, indicating three equivalent protons and one more proton with a different coupling constant. For H–MA• and H–EA•, each spectroscopic line has an additional splitting in the form of a fine quartet (1.45 G) and triplet (1.38 G), respectively, due to the methyl and methylene protons in the ester side groups, indicating spin density delocalization through the ester bonds. In the case of H-*t*BA•, the splitting due to methyl groups in the *tert*-butyl group is not resolved. These findings indicate the formation of radicals with the expected structures, as shown in Fig. 22.

The hydrogenated monomeric model radical of poly(ethyl methacrylate) (H–EMA•) was formed in the reaction of ethyl 2-bromoisobutyrate (EB*ri*B) with organotin compounds. A well-resolved ESR spectrum was observed, where interpretation confirms the formation of a radical with the structure shown in Fig. 22. The hydrogenated monomeric methacrylate type radical (H–MMA•) was generated through photodecomposition of MAIB under irradiation. The ESR spectra of monomeric methacrylate radicals led to the hyperfine couplings summarized in Table 2.

The ESR spectra of H–EA•, H–EMA•, and H–MMA• were reported previously and the values of hyperfine coupling constants in that study are comparable to the results obtained in the present study.[28, 51]

4.2. Dimeric Model Radicals

The dimeric model compounds studied can be classified into two groups. The terminal radical can have either an acrylate or a methacrylate structure. Model dimeric radicals with an acrylate terminal group were generated in the reaction of the corresponding alkyl bromides (H-tBA-tBA–Br, H–MA-tBA–Br, H–EMA-tBA–Br, and H–MA–MA–Br) with an organotin compound under irradiation. The resulting radicals had the structure of hydrogenated dimeric radicals, that is H-tBA-tBA•, H–MA-tBA•, H–EMA-tBA•, and H–MA–MA•, respectively. The ESR spectra of these radicals led to the hyperfine couplings included in Table 2. A doublet of triplets was observed in each spectrum and can be assigned to the α-proton and two equivalent β-methylene protons, respectively.

The model dimeric H–MA–MMA• radicals were generated by reacting the corresponding alkyl bromide (H-MA-MMA-Br) with an organotin compound under irradiation. The ESR spectra were obtained at 30, 60, and 90°C. The spectrum of

TABLE 2. Hyperfine Splitting Constants of Radicals at 30°C[a]

Radicals	Source	Hyperfine Splitting Constant, G				
		α	β	γ	ester	Line width[a,b]
H-MA-MMA•[c]	MA-MMA-Br		22.60 (CH$_3$)	0.80	1.28	0.55
			14.80 (CH$_2$)			
polyMMA•[c]	MMA (*in situ*)		22.2 (CH$_3$)		1.1	0.50
			14.4 (β-CH)			
			9.7 (β-CH)			
oligo$t$BMA•[d]	oligo$t$BMA-Br		21.7 (CH$_3$)		1.0	
			12.7 (β-CH)			
			11.8 (β-CH)			
poly$t$BMA•[d]	tBMA (*in situ*)		21.7 (CH$_3$)		1.2	
			14.0 (β-CH)			
			11.6 (β-CH)			
H-tBA-tBA•	tBA-tBA-Br	20.90	22.70	0.78		0.47
H-MA-tBA•	MA-tBA-Br	20.70	22.70	0.78		0.40
H-EMA-tBA•	EMA-tBA-Br	20.90	22.10			0.47
H-MA-MA•	MA-MA-Br	20.7	22.9	0.8	1.0	0.39
poly$t$BA•[e]	poly$t$BA-Br	22.8	20.6			1.5

[a] From Ref. 50, with permission.
[b] Line width at half-height.
[c] Measured at 90°C.
[d] Measured at 150°C.
[e] Measured at −30°C.

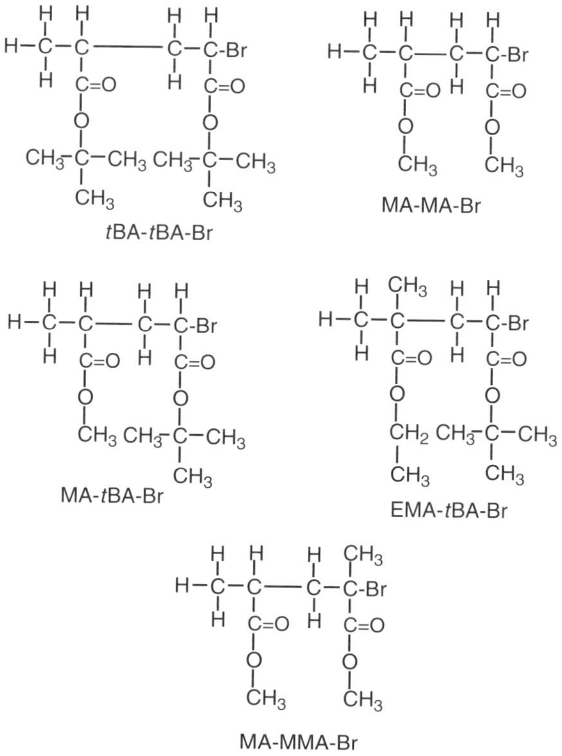

Fig. 20. Model radical precursors of dimeric (meth)acrylates. (From Ref. 50, with permission.)

H-MA-MMA• at 90°C, along with its simulation, is shown in Fig. 23. A 12-line spectrum with a very small quartet in each spectroscopic line was detected. The 12-line spectrum is caused by a quartet of triplets (4 × 3) from three equivalent methyl protons and two methylene protons. An additional very small splitting of a doublet (0.80 G) due to the γ-proton in the MA moiety can also be observed. The signal intensity of each spectroscopic line displays a temperature dependent change due to hindered rotation around the $C_\alpha-C_\beta$ bond.[3,12,22] At 90°C, the bond rotates freely and the simulated spectrum is shown in Fig. 23. The ESR spectra of such model dimeric radicals, with well-defined structures, have not been observed before. Analyses of the values obtained from the hyperfine coupling constants clearly show the dimeric radicals have the structures indicated in Fig. 22.

4.3. Dimeric Radicals and the Penultimate Unit Effect

Penultimate unit effects were observed in the ESR spectra of model dimeric radicals. The ESR spectra of dimeric radicals with acrylate terminal units will be discussed first. The effect of a methacrylate penultimate unit (due to the presence of a methyl group at the γ-position) is seen by comparing the spectrum of H–EMA-tBA• with

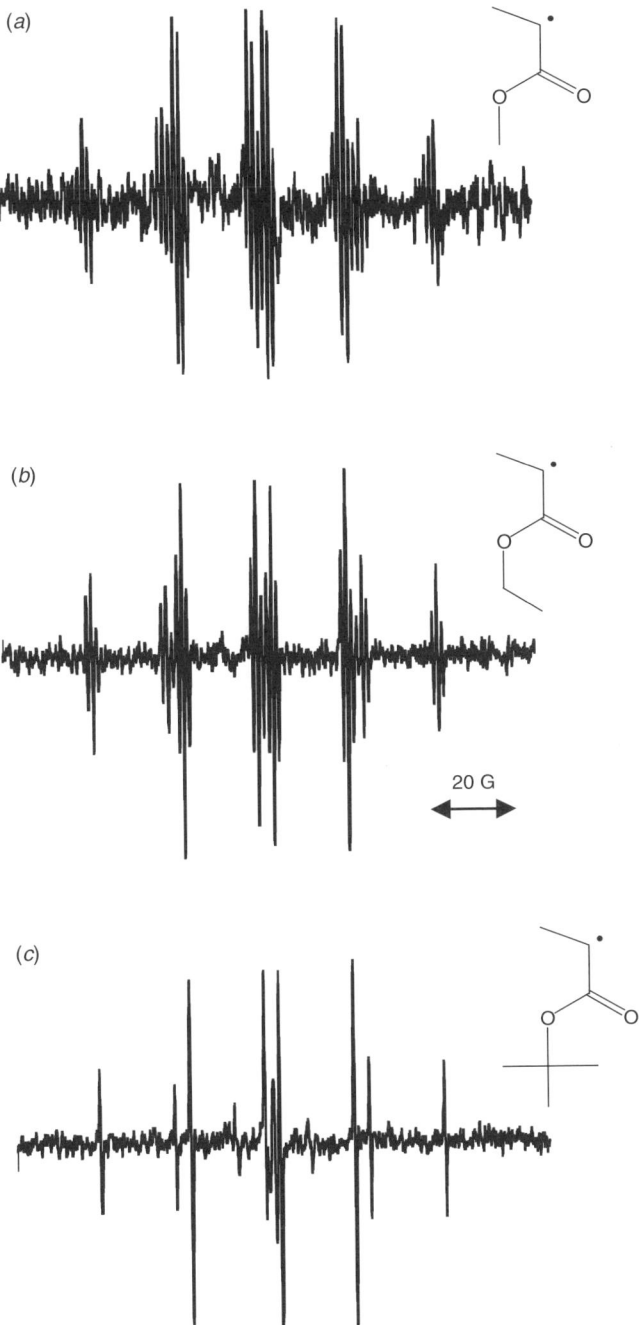

Fig. 21. The ESR spectra of radicals generated from corresponding alkyl bromides observed at 30°C: (*a*) H-MA•, (*b*) H-EA•, and (*c*) H-*t*BA•. (From Ref. 50, with permission.)

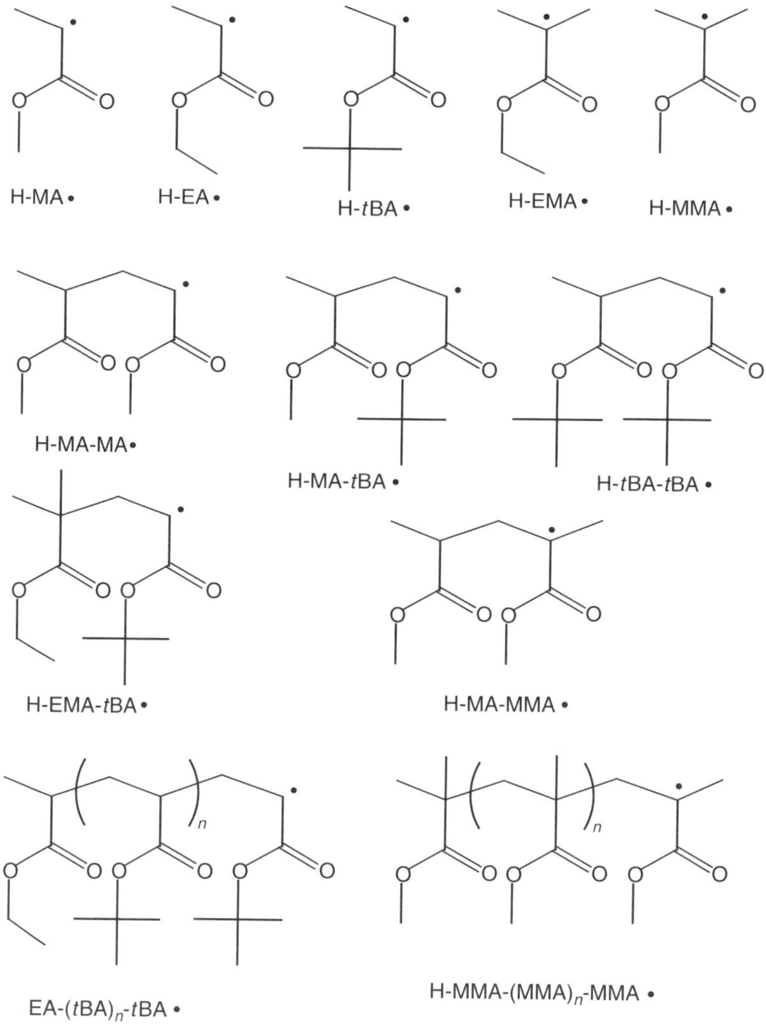

Fig. 22. Generated monomeric, dimeric, and polymeric model radicals of (meth)acrylates. (From Ref. 50, with permission.)

that of H–MA-tBA• or H-tBA-tBA•. The presence of an EMA unit shows only small variations in the coupling of the α-proton. Obviously, the spin density at the α-carbon atom does not change appreciably by β-substitution. On the other hand, the coupling of the β-methylene protons varies considerably with the presence of an EMA penultimate unit, as shown in Table 2: The value for H-EMA-tBA• (22.10 G) is smaller than that of H-MA-tBA• (22.70 G) or for H-tBA-tBA• (22.70 G). The ESR spectra of monomeric acrylic acid radicals with various substituents, that is, HO–, NH_2–, CH_3–, HO–CH_2–, were discussed previously by Fischer et al.[23] and the reported values of the hyperfine coupling constants of the β-protons varied

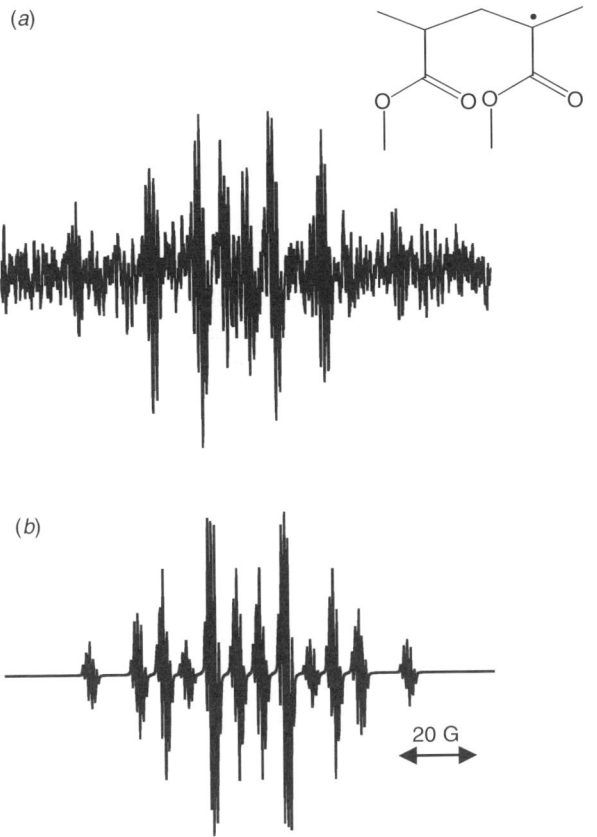

Fig. 23. Experimental and simulated ESR spectra of H–MA–MMA• observed at 90°C. (From Ref. 50, with permission.)

considerably with the substituents. Judging from the reported results, the electronic effect of a penultimate MMA unit is larger than those of substituents. These findings show the electron-withdrawing effect of the presence of the MMA unit. The steric effect due to the γ-methyl group may appear in the broader line width than that seen in the case of H-MA-tBA•. In comparison with the ESR spectrum of H-MA-tBA•, the steric effect of H-tBA-tBA• appeared solely in the line width of the spectrum, although the electronic effect of replacement of a methyl group by a *tert*-butyl group is very small. Sterically, the small methyl ester group of MA caused less hindrance to the rotation of C_β-C_α bond. As a result, a clear separation of spectroscopic lines, due to narrower line width, was observed. This difference in ESR spectra between H–MMA-tBA• and H-tBA-tBA• suggests that the reactivity of the terminal tBA radical may depend on the penultimate unit in copolymerizations of MMA and tBA.

Radical copolymerization of MA (M_1) and MMA (M_2) was investigated: The copolymer reactivity ratios r_1 and r_2 were determined to be 0.4 and 2.15, respectively.[52]

However, these results were obtained without considering the penultimate unit effect. The present ESR results for H-MA-MMA•, H-MA-MA•, and H-MMA-*t*BA• suggest the presence of a penultimate unit effect on the acrylate terminal unit.

Thus, electronic and steric penultimate effects influence not only radical reactivity,[53–62] but also their ESR spectra.

4.4. Activation Rate Constants for Monomeric and Dimeric Alkyl Bromides and ESR Spectra

The penultimate unit effect may play a very important role in ATRP. The rate constants of activation of monomeric and dimeric alkyl bromides with a CuBr–bpy (bpy=2,2'-bipyridine) complex as activator were determined.[63–65] The ATRP relies on the reversible activation of a dormant alkyl halide through halogen abstraction by a transition metal complex to form a radical that participates in the classical free-radical polymerization figure (Fig. 2) prior to deactivation. In this equilibrium, the alkyl radical (P_m•) is formed in an activated process, with a rate constant k_{act}, by the homolytic cleavage of an alkyl halogen bond (P_m-Z) catalyzed by a transition metal complex in its lower oxidation state (Cu^I). The relative values of k_{act} of the alkyl bromides were determined for CuBr/bpy catalyst systems in acetonitrile at 35°C. These systems followed the order EBriB (30)>>MBrP (3)>*t*BBrP (1) for monomeric initiators[29, 64] and MMA-MMA-Br (100)>>MA-MMA-Br (20) > MMA-MA-Br (5) > MA-MA-Br (1) for dimeric initiators.[65]

The relative values of k_{act} for alkyl bromides follow the order EBriB (30)>>MBrP (3)>*t*BBrP (1). This order indicates not only the higher thermodynamic stability of the tertiary radicals (H–EMA•), but also the lower reactivity of the secondary radical with a bulkier *tert*-butyl group rather than the smaller methyl group. This was explained by suggesting steric hindrance in the atom-transfer process, where the catalyst complex must approach the halogen atom. However, judging from the values of hyperfine coupling constants for H-*t*BA•, H-MA•, and H-EA•, the electronic effect from the ester side groups seems negligibly small. Replacement of a methyl group (MBrP) by a *tert*-butyl group (*t*BBrP) has no significant effect on the values of the hyperfine coupling constants determined in the ESR spectra, but the steric effect in the transition state for atom transfer leads to three fold difference in k_{act}.

The values of k_{act} for the dimeric bromides follow the order: MMA–MMA–Br (100), MA–MMA–Br (20), MMA–MA–Br (5), and MA–MA–Br (1). Differences between MMA–MMA–Br and MA–MMA–Br, and between MMA–MA–Br and MA–MA–Br, may be explained in the same manner as the MA-*t*BA and MMA-*t*BA systems. When a MMA unit is located in the penultimate position, the values of hyperfine splitting constants of β-protons of *t*BA were smaller than with MA as the penultimate unit. These electronic effects and steric effects manifest themselves in different values of k_{act}.

It is anticipated that a combination of ESR analysis and kinetic studies lead to a better understanding of both the structure and the reactivity of the radicals. Such information is very important for basic research, not only for ATRP, but also for conventional radical polymerization processes.

5. CONCLUSIONS

Development of controlled radical polymerization techniques has stimulated basic research on radical chemistry in conventional radical polymerizations. Information on the effect of chain lengths on propagating radicals, chain-transfer reactions to polymers, and penultimate unit effects has been obtained from ESR observation of model radicals generated from radical precursors prepared by ATRP. Previously, it has been extremely difficult, even impossible, to obtain such information from ESR spectra during conventional radical polymerizations. The ESR study of radical polymerizations has made remarkable progress as a result of the combination of study of radicals formed as a result of various kinds of controlled radical polymerization techniques.

ACKNOWLEDGMENTS

Financial support from the Japan Society for the Promotion of Science (JSPS) for Japan-U.S. Cooperative Science Program and National Science Foundation (CHE 04-05627) are gratefully acknowledged.

REFERENCES

1. Fischer, H. *Adv. Polym. Sci.*, **1968**, *5*, 463.
2. Fischer, H. In *Polymer Spectroscopy*; Hummel, D. O., Ed., Springer Verlag:Berlin, 1974; pp. 289–354.
3. Kamachi, M. *Adv. Polym. Sci.* **1987**, *82*, 207.
4. Yamada, B.; Westmoreland, D.G.; Kobatake, S.; Konosu, O. *Prog. Polym. Sci.* **1999**, *24*, 565.
5. Kamachi, M. *J. Polym. Sci., Part A:. Polym. Chem.* **2002**, *40*, 269.
6. Kamachi, M.; Kajiwara, A. *Macromol. Symp.* **2002**, *179*, 53.
7. Kamachi, M.; Kajiwara, A. *Macromolecules* **1996**, *29*, 2378.
8. Kamachi, M.; Kajiwara, A. *Macromol. Chem. Phys.* **1997**, *198*, 787.
9. Kamachi, M.; Kajiwara, A. *Macromol. Chem. Phys.* **2000**, *201*, 2165.
10. Kajiwara, A.; Kamachi, M. *Macromol. Chem. Phys.* **2000**, *201*, 2160.
11. Kamachi, M. In *Controlled Radical Polymerization*, ACS Symposium series 685; Matyjaszewski, K., Ed.; American Chemical Society: Washington, DC, 1998; Chapt. 9, pp. 145–168.
12. Kajiwara, A.; Matyjaszewski, K.; Kamachi, M. In *Controlled/Living Radical Polymerization*, ACS Symposium series 768; Matyjaszewski, K., Ed; American Chemical Society: Washington, DC., 2000; Chapt. 5, pp. 68–81.
13. Kajiwara, A.; Kamachi, M. In *Advances in Controlled/Living Radical Polymerization*, ACS Symposium series 854; Matyjaszewski, K., Ed.; American Chemical Society: Washington, DC., 2003; Chapt. 7, pp. 86–100.
14. Wang, J.-S.; Matyjaszewski, K. *J. Am. Chem. Soc.* **1995**, *117*, 5614.

15. Wang, J.-S.; Matyjaszewski, K. *Macromolecules* **1995**, *28*, 7901
16. Patten, T.E.; Matyjaszewski, K. *Adv. Mater.* **1998**, *10*, 901.
17. Coessens, V.; Pintauer, T.; Matyjaszewski, K. *Prog. Polym. Sci.* **2001**, *26*, 337.
18. Matyjaszewski, K.; Xia, J. *Chem. Rev.* **2001**, *101*, 2921.
19. Kamigaito, M.; Ando, T.; Sawamoto, M. *Chem. Rev.* **2001**, *101*, 3689.
20. Davis, K.A.; Matyjaszewski, K. *Adv. Polym. Sci.* **2002**, *159*, 2.
21. Giese, B.; Damm, W.; Wetterich, F.; Zeitz, H.-G. *Tetrahedron Lett.* **1992**, *33*, 1863.
22. Kajiwara, A.; Maeda, K.; Kubo, N.; Kamachi, M. *Macromolecules* **2003**, *36*, 526.
23. Fischer, H.; Giacometti, G. *J. Polym. Sci. C* **1967**, *16*, 2763.
24. Fischer, H.Z. *Naturforsch.* **1964**, *19a*, 267.
25. Fischer, H.Z. *Naturforsch.* **1964**, *19a*, 866.
26. Gilbert, B.C.; Lindzay Smith, J.R.; Milne, E.C.; Whitwood, A.C.; Taylor, P. *J. Chem. Soc. Perkin Trans. 2* **1993**, 2025.
27. Gilbert, B.C.; Lindzay Smith, J. R.; Milne, E. C.; Whitwood, A. C.; Taylor, P. *J. Chem. Soc. Perkin Trans. 2* **1994**, 1759.
28. Matsumoto, A.; Giese, B. *Macromolecules* **1996**, *29*, 3758.
29. Spichty, M.; Giese, B.; Matsumoto, A.; Fischer, H.; Gescheidt, G. *Macromolecules* **2001**, *34*, 723.
30. Knuehl, B.; Pintauer, T.; Kajiwara, A.; Fischer, H.; Matyjaszewski, K. *Macromolecules* **2003**, *36*, 8291.
31. Zhou, W.; Zhu, S. *Ind. Eng. Chem. Res.* **1997**, *36(4)*, 1130.
32. Matyjaszewski, K.; Kajiwara, A. *Macromolecules* **1998**, *31*, 548.
33. Kajiwara, A.; Matyjaszewski, K. *Macromol. Rapid Commun.* **1998**, *19*, 319.
34. Kajiwara, A.; Matyjaszewski, K.; Kamachi, M. *Macromolecules* **1998**, *31*, 5695.
35. Kajiwara, A.; Matyjaszewski, K. *Polym. J.* **1999**, *31*, 70.
36. Yu, Q.; Zeng, F.; Zhu, S. *Macromolecules* **2001**, *34*, 1612.
37. Wang, A. R.; Zhu, S. *Polym. Prepr. (Am. Chem. Soc., Div. Polym. Chem.)* **2002**, *43(2)*, 11.
38. Wang, A.R.; Zhu, S. *Macromolecules* **2002**, *35*, 9926.
39. Wang, A.R.; Zhu, S.; Matyjaszewski, K. In *Advances in Controlled/Living Radical Polymerization*, ACS Symposium series 854; Matyjaszewski, K., Ed.; American Chemical Society: Washington, DC., 2003; Chapt. 12, pp. 161–179.
40. Rockenbauer, A. *Mol. Phys. Rep.* **1999**, *26*, 117.
41. Kajiwara, A.; Kamachi, M.; Rockenbauer, A., to be published.
42. Best, M.E.; Kasai, P. H. *Macromolecules* **1989**, 22, 2622.
43. Doetschman, D.C.; Mehlenbacher, R. C.; Cywar, D. *Macromolecules* **1996**, 29, 1807.
44. Sugiyama, Y. *Bull. Chem. Soc. Jpn.* **1997**, 70, 1827.
45. Azukizawa, M.; Yamada, B.; Hill, D. J. T.; Pomery, P. J. *Macromol. Chem. Phys.* **2000**, *201*, 774.
46. Yamada, B.; Azukizawa, M.; Yamazoe, H.; Hill, D. J. T.; Pomery, P. J. *Polymer* **2000**, *41*, 5611.
47. Chang, H. R.; Lau, W.; Parker, H. -Y.; Westmoreland, D. G. *Macromol. Symp.* **1996**, 111, 253.
48. Heatley,F; Lovell, P. A.; Yamashita, T. *Macromolecules* **2001**, 34, 7636.

49. Ahmad, N. M.; Heatley, F.; Lovell, P. A. *Macromolecules* **1998**, 31, 2822.
50. Kajiwara, A.; Nanda, A. K.; Matyjaszewski, K. *Macromolecules* **2004**, *37*, 1378.
51. Tanaka, H.; Kagawa, T.: Sato, T. *Macromolecules* **1986**, *19*, 934.
52. Zubov, V. P.; Valuev, L. I.; Kavanov, V. A.; Kargin, V. A. *J. Polym. Sci. A-1* **1971**, *9*, 833.
53. Giese, B.; Engelbrecht, R *Polym. Bull.* **1984**, *12*, 55.
54. Fukuda, T.; Ma, Y.-D.; Inagaki, H. *Makromol. Chem., Suppl.* **1985**, *12*, 125.
55. Fukuda, T.; Ma, Y.-D.; Inagaki, H. *Macromoleculs* **1985**, *18*, 17.
56. Ma, Y.-D.; Fukuda, T.; Inagaki, H. *Macromoleculs* **1985**, *18*, 26.
57. Cywar, D.A.; Tirrell, D.A. *Macromolecules* **1986**, *19*, 2908.
58. Prementine, G.S.; Tirrell, D.A. *Macromolecules* **1987**, *20*, 3034.
59. O'Driscoll, K.F.; Davis, T.P. *J. Polym. Sci., C* **1989**, *27*, 417.
60. Fukuda, T.; Kubo, K.; Ma, Y.-D. *Prog. Polym. Sci.* **1992**, *17*, 875.
61. Davis, T.P.; *J. Polym. Sci., A, Polym. Chem.* **2001**, *39*, 597.
62. Heuts, J.P.A.; Coote, M.L.; Davis, T.P.; Johnston, L.P.M. In *Controlled Radical Polymerization*, ACS Symposium series 685; Matyjaszewski, K., Ed.; American Chemical Society: Washington, DC., 1998; p. 120.
63. Nanda, A.K.; Matyjaszewski, K. *Macromolecules* **2003**, *36*, 599.
64. Nanda, A.K.; Matyjaszewski, K. *Macromolecules* **2003**, *36*, 1487.
65. Nanda, A.K.; Matyjaszewski, K. *Macromolecules* **2003**, *36,* 8222.

6

LOCAL DYNAMICS OF POLYMERS IN SOLUTION BY SPIN-LABEL ESR

JAN PILAŘ

Academy of Sciences of the Czech Republic, Prague, Czech Republic

Contents

1. Introduction	134
2. Local Dynamics in Polymers by NMR and Fluorescence Techniques	134
2.1. Segmental Mobility of Polymers	134
2.2. NMR and Fluorescence Studies	135
3. Effect of Nitroxide Rotational Dynamics on ESR Line Shapes	136
3.1. Basic Principles	136
3.2. Fast-Motional Spectra	138
3.3. Slow-Motional Spectra and Ordering	139
4. ESR Studies of Segmental Dynamics of Polymers in Solution	140
4.1. Dynamics of Nitroxide Spin Labels Attached to Polymer Chain in Dilute Solution	140
4.2. Segmental Mobility of Poly(methyl methacrylate)	142
4.3. Segmental Mobility of Poly(methacrylic acid) and Poly(acrylic acid)	142
4.4. Segmental Mobility of Polystyrene	147
4.5. Segmental Mobility of Poly(2-hydroxyethyl methacrylate)	155
4.6. Comparison with Other Techniques	161
5. Conclusions	162
Acknowledgments	162
References	162

Advanced ESR Methods in Polymer Research, edited by Shulamith Schlick.
Copyright © 2006 John Wiley & Sons, Inc.

1. INTRODUCTION

The study of local segmental dynamics in polymers resulting from thermal motion plays a significant role for an understanding of structure–property relationships in polymers. These relationships represent an important subject in polymer research. It is useful to study local chain motions of an isolated chain in dilute solutions before chain–chain interactions are taken into account. The conformation of a polymer chain in dilute solution, resulting, among others, from polymer–solvent interactions, depends strongly on the dynamics on the length scale of a few monomer units. This chapter demonstrates that ESR is a important tool that is capable to provide information on the local polymer dynamics, complementary to the information obtained by other advanced methods.

2. LOCAL DYNAMICS IN POLYMERS BY NMR AND FLUORESCENCE TECHNIQUES

Orientational relaxation of a polymer chain segment is the most elementary process for conformational change among a number of relaxation modes of the polymer backbone. Corresponding energy barriers and polymer–solvent interactions primarily affect conformational transitions. In recent years, great efforts have been devoted to understanding the local dynamics of polymers by measuring the correlation time, τ_c, for segmental rotational diffusion in dilute solution. Measurement of the fluorescence depolarization of the anthracene chromophore bonded to the main chain of the polymer, and measurement of ^{13}C NMR (nuclear magnetic resonance) relaxation times and nuclear Overhauser effect enhancements (NOE) have been the main methods used for the purpose.

2.1. Segmental Mobility of Polymers

It was shown that Kramers' theory,[1] in the high-friction limit resulting in Eq. 1a with $\alpha=1$, cannot accurately describe the segmental dynamics of a synthetic polymer in dilute solutions.

$$\tau_c = K[\eta(T)]^\alpha \exp(E_a/RT) = K' \exp(E_{exp}/RT) \tag{1a}$$

$$E_{exp} = E_a + \alpha E_\eta \tag{1b}$$

$$\eta(T) = \eta_0 \exp(E_\eta/RT) \tag{1c}$$

For some polymers in solvents with high viscosity and large activation energy of viscosity, the theory yields nonphysical negative values for the height of the potential barrier for segmental rotation.[2] Hence, the prediction that the experimental correlation times should scale linearly with viscosity at constant temperature (Kramers' behavior) is not generally fulfilled. Provided that the temperature dependence of viscosity exhibits Arrhenius behavior in the temperature range investigated, a power law relationship (Eq. 1a) between τ_c and the solvent viscosity η with exponent $0 \leq \alpha \leq 1$ was suggested[3] on a theoretical basis.[4] The prefactor K in this equation is independent of temperature and viscosity, E_{exp} is the activation energy determined from the Arrhenius plot of a parameter the characterizing local segmental mobility (e.g., τ_c), and E_η is the activation energy of viscosity for the solvent. The parameter K' is independent of temperature, E_a is the height of the potential barrier for local segmental motions, T is the absolute temperature, and R is the gas constant. The exponent α is assumed to depend on the moment of inertia and the size of the unit involved in the conformational transition and on the curvature at the top of the potential barrier for the transition. A larger size, a larger moment of inertia, and a lower barrier prolong the time required to get across the barrier, and thereby lead to a larger value of α. When this time period is sufficiently long, the high-frequency behavior of friction is less significant, and a behavior close to Kramers' high-friction limit should be observed. The value of the exponent α is expected to vary from polymer to polymer. The height of the intramolecular potential barrier, E_a, should not depend on the solvent viscosity, but it may vary with the polymer conformation in solution; it follows that it should be solvent independent for all good nonpolar solvents of a particular polymer, but may change in poor solvents and in theta solutions.[3]

2.2. NMR and Fluorescence Studies

In the frame of the model resulting in Eq. 1a, the values $E_a = 13 \pm 2$ kJ mol^{-1} and $\alpha = 0.41 \pm 0.02$ were found for polyisoprene by ^{13}C NMR,[3] and $E_a = 10 \pm 1$ kJ mol^{-1} and $\alpha = 0.75 \pm 0.06$ by fluorescence depolarization.[2] The difference in E_a is small and can be related to a modification of the potential barrier by the presence of the anthracene chromophore, because the fluorescence measurement senses the chain motion in the vicinity of anthracene. A larger size of the unit subjected to conformational transition and a larger moment of inertia due to the presence of the anthracene chromophore in the labeled chain would explain the higher α value determined by fluorescence depolarization. The ^{13}C NMR study of segmental mobility of poly (1-naphthylmethyl acrylate) (PNMA) in chlorinated solvents (1,1,2,2-tetrachloroethane-d_2, pentachloroethane and CDCl$_3$) has shown[5] that, in the range of viscosities studied, the polymer exhibits nearly Kramers' behavior with $\alpha=1.07$; the value $E_a = 11.3$ kJ mol^{-1} for the segmental PNMA dynamics was deduced. Kramers' theory has also succeeded in explaining results of ^{13}C NMR study of poly(N-vinylcarbazole) (PNVC) in five solvents covering viscosities differing by a factor of 5, leading to $E_a = 9.1 \pm 0.3$ kJ mol^{-1} for PNVC chain local dynamics.[6] A similar study of poly(vinyl chloride) (PVC) in three solvents (chloroform, dioxane, and dimethyl sulfoxide) has revealed[7] deviation from Kramers' theory, with $\alpha = 0.6$ and $E_a = 19.5 \pm 2.5$ kJ mol^{-1}. A fluorescence depolarization study of local chain dynamics of poly(ethylene oxide) (PEO) in dimethylformamide (DMF) and cyclohexane has shown non-Kramers' behavior with $\alpha = 0.69$ and $E_a \div 5$ kJ mol^{-1}.[8] The temperature dependence of the correlation times for rotational diffusion of the anthracene chromophore bonded in the polystyrene main chain in various solvents, measured by fluorescence depolarization, resulted in $\alpha = 0.90 \pm 0.05$, which was considered a reasonable agreement with Kramers' theory.[9] The larger α observed for labeled polystyrene in comparison with labeled polyisoprene was rationalized by the presence of bulky benzene rings in polystyrene. The height of the potential barrier, $E_a = 11 \pm 3$ kJ mol^{-1}, has been found for polystyrene in good solvents and higher values in theta solutions ($E_a = 21 \pm 2$ kJ mol^{-1} in cyclohexane).

In all earlier papers dealing with polystyrene local dynamics, data were analyzed in the frame of Kramers' theory. Heatley and Wood[10] measured ^{1}H NMR spin lattice relaxation times for polystyrene in four solvents at 5% concentration. They determined the activation energy for the correlation time characterizing conformational jumps of polystyrene chains and deduced $E_a = 11-16$ kJ mol^{-1} for polystyrene in chlorinated solvents (CDCl$_3$, CCl$_4$, and hexachlorobuta-1,3-diene) and $E_a = 19 \pm 5$ kJ mol^{-1} for polystyrene in 90/10 (v/v) solution of cyclohexane-d_{12} and toluene-d_8, which is a theta solvent for polystyrene at 15°C. Gronski et al.[11] performed ^{13}C spin lattice relaxation and NOE measurements of 1% solutions of ^{13}C-enriched polystyrene and determined heights of potential barriers for conformational transitions of polystyrene: $E_a = 15$ and 10.5 kJ mol^{-1}, in benzene and toluene solutions, respectively. Yokotsuka et al.[12] measured the mean relaxation time related to local conformational transitions of the anthracene-labeled polystyrene by the fluorescence depolarization and deduced $E_a = 6.3, 6.7, 2.9, 3.3$ and -8.0 kJ mol^{-1} in butyl acetate, cyclohexane, dioxane, cyclohexanone, and more viscous glycerol tripropionate, respectively, using the original Kramers' equation. When applying the generalized equation combining both high- and low-friction limits suggested by Helfand,[13] they deduced $E_a = 6.7, 7.1, 5.0, 5.0$, and 8.0 kJ mol^{-1}, respectively. Ono et al.[14,15] used the same technique and determined the heights of potential barriers for local conformational transitions in polystyrene chain in various solvents including theta solvents: $E_a = 5.9, 7.5$, and 10.9 kJ mol^{-1} in toluene, butyl acetate, and in theta solvent cyclohexane, respectively, using the classic Kramers' approach. Horinaka et al.[16] studied the influence of a fluorescent probe on local relaxation times and their dependence on molecular weight and tacticity of the polystyrene used and deduced $E_a = 7.1$ kJ mol^{-1} for $M_w > 10^4$ by the same approach.

3. EFFECT OF NITROXIDE ROTATIONAL DYNAMICS ON ESR LINE SHAPES

3.1. Basic Principles

In nitroxide free radicals characterized by a hyperfine tensor, A, and g-tensor, g, the interaction between the magnetic moment of an unpaired electron and the magnetic moment of nitrogen nucleus is highly anisotropic. The anisotropy determines the line shape of electron spin resonance (ESR) spectrum of completely immobilized nitroxides (Fig. 1), and can be completely averaged out for fast thermal Brownian rotational diffusion in low-viscous media. For the ^{14}N nucleus with $I = 1$ and no other interacting nuclei in the nitroxide, three equidistant lines of equal intensities and widths should be observed in ESR spectra in such media. In fact, even in media of

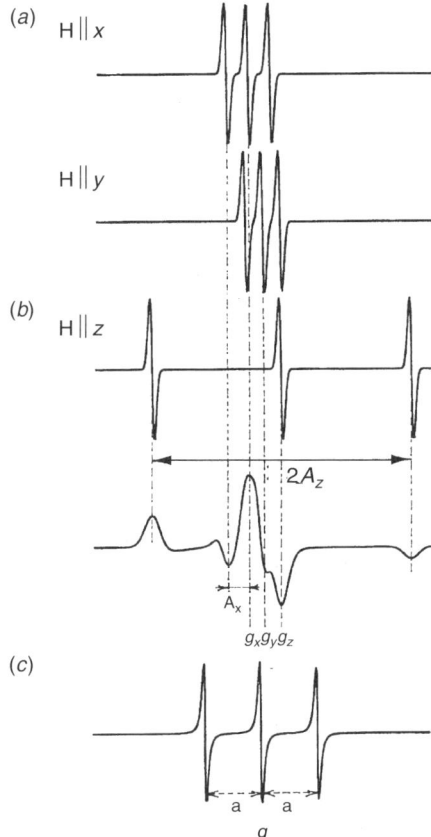

Fig. 1. Effect of A- and g-tensor anisotropy on the line shape on the ESR spectrum of a typical nitroxide: (*a*) Immobilized and perfectly oriented nitroxides taken at the indicated orientations of the external magnetic field with respect to the nitroxide axis system. (*b*) Immobilized nitroxides in a polycrystalline sample (rigid limit spectrum). (*c*) Fast-motional spectrum in a solvent of very low viscosity (*a* and *g* are isotropic hyperfine splitting and isotropic g-factor, respectively).

very low viscosity, the averaging is not complete and the anisotropy affects the ESR line shape.

In real media, nitroxides change their orientations with respect to the external magnetic field due to the Brownian thermal rotational mobility usually characterized by the rotational correlation time, τ_R. The anisotropic hyperfine interaction between the unpaired electron and nitrogen nucleus is modified by these changes with a frequency dependent on τ_R. In this way, frequency-dependent perturbations are generated, which modify the energy levels and transition probabilities in the system. As a result, the line shape of ESR spectra of nitroxides (and of other free radicals in which anisotropic magnetic interactions occur) depends on the correlation time τ_R.

The line shape of the ESR spectrum can be calculated using the spin Hamiltonian $\hat{H}_S(t)$ for the nitroxide radical. The time dependence of the spin Hamiltonian describes the orientation-dependent part $\hat{H}_1(t)$, which contains terms characterizing the anisotropy of the A- and g-tensors, and terms characterizing the rotational reorientation of the free radical as a classic stochastic process using the Wigner rotation matrix elements. A comprehensive theory of ESR spectra of nitroxides has been described step-by-step in contributions to two monographs[17,18] and papers cited therein. Computer programs[19,20] suitable for calculation of theoretical line shapes of ESR spectra measured in continuous wave (CW) and pulse experiments have become indispensable tools for practical applications. The ESR spectra of nitroxides are also described in chapters 1 and 3 in this volume.

The "ESR working window" in which the ESR line shapes are sensitive to rotational reorientation depends on the anisotropy of the A- and g-tensors of the radical; for nitroxides, the range for τ_R is $10^{-12}-10^{-5}$ s, which means that the technique is applicable to polymer solutions, polymer gels, and solid polymers at temperatures close to or above T_g. At the X-band ESR frequency (≈ 9 GHz), the rotational dynamics in most liquids is usually sufficiently fast, that is, $\tau_R \Delta\omega \ll 1$ [where $\Delta\omega$ is a measure of the magnitude of the orientation-dependent part of the spin Hamiltonian $\hat{H}_1(t)$]; such motions ($\tau_R < 10^{-8}$ s) fall within the motional narrowing (fast-motional) regime. For slower dynamics ($\tau_R \Delta\omega \geq 1$), more complicated slow-motional spectra are observed.[17]

3.2. Fast-Motional Spectra

The fast-motional nitroxide ESR spectrum is a superposition of three Lorentzian lines of different widths and, consequently, with different amplitudes when recorded as first derivative of the absorption. Intrinsic first-derivative peak-to-peak widths of lines, ΔH_{pp}, of the three components of the nitrogen hyperfine splitting in the fast-motional nitroxide spectra can be expressed as

$$\Delta H_{pp}(M) = (2/3^{1/2})(A + BM + CM^2) \tag{2}$$

where $M = -1, 0, +1$ is the spectral index number. The experimental values of parameters A, B, and C can be determined by using three experimental widths of lines $\Delta H_{pp}(M)$ and Eq. 2. The calculation of the A, B, and C values from the A- and g-tensors and parameters for axially symmetric nitroxide rotational reorientation was described by Goldman et al.,[19] and for a more general orientation of the symmetry axis of nitroxide rotational reorientation in one of our papers.[21] In principle, parameters for nitroxide rotational reorientation could be found by fitting experimental A, B, and C parameters with calculated theoretical values. Unfortunately, it was found that three line widths cannot be fully and unambiguously related to the nitroxide rotational parameters without additional assumptions about the motion.

Deuterated spin labels significantly simplify analysis of both fast- and slow-motional spectra. The contribution of unresolved hyperfine interaction due to deuterium nuclei is much smaller than that of protons (Fig. 2) due to a smaller nuclear magnetic moment.[22]

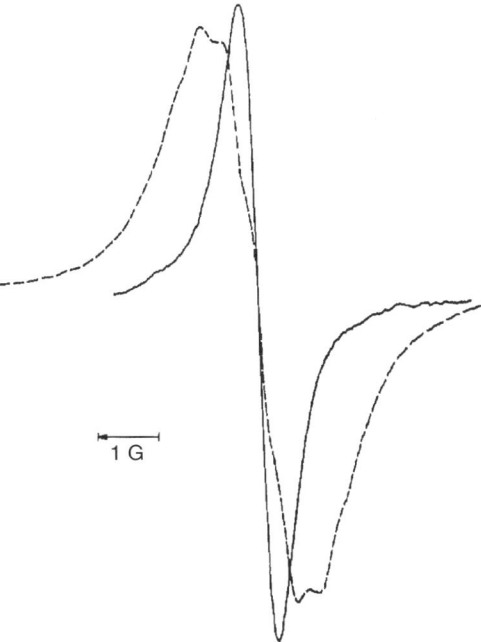

Fig. 2. Narrowing of the central component of the typical three line fast-motional spectrum of a nitroxide attached to a polymer chain, after replacing all protons with deuterium; central line in the room temperature ESR spectrum of the protonated nitroxide exhibiting partially resolved proton hyperfine structure (dashed line) and of the deuterated nitroxide exhibiting no visible hyperfine structure (solid line). (From Ref. 22, with permission.)

Pioneering applications of ESR to the study of dynamics of nitroxides attached to polymers was described by Cameron and Bullock,[23] who analyzed fast-motional solution ESR spectra of simple nitroxides attached to polystyrene. The ESR spectroscopy of nitroxides in the fast-motional regime based on even simplified versions of Eq. 2 has been widely used for solving various problems mainly in the fields of biochemistry and medicine. Unfortunately, critical assessment of the model, and the accuracy and validity of results is frequently missing.

3.3. Slow-Motional Spectra and Ordering

For slower motions in more viscous media where $\tau_R \Delta\omega \geq 1$, the ESR spectrum depends dramatically on the combined influence of molecular motion and magnetic interactions. The slow-motional ESR line shapes, in principle, provide a more detailed picture of rotational dynamics compared with fast-motional line shapes, and can be fully analyzed using a theoretical approach based on numerical solution of the stochastic Liouville equation.[17,18] The theoretical approach resulting in the Schneider–Freed set of programs[18] enables the calculation of theoretical line shapes of ESR spectra of nitroxides subjected to anisotropic rotational reorientation using various models (Brownian rotation, free rotation, jump diffusion). In addition, the set allows for constraints in rotational reorientation. The tendency of the nitroxide to order is modeled by a restoring potential that is defined relative to a director axis. The MOMD (Microscopic Order with Macroscopic Disorder) model describes the case, when there is local ordering described by the potential, but its directors are isotropically distributed so that there is no macroscopic ordering. Such a case is extremely important for nitroxides attached to polymers (spin-labeled polymers). The model requires integration of spectral line shapes calculated in the presence of a particular orienting potential over the director orientations. Later on, a nonlinear least-squares (NLSL) fitting program[20] (including its PC version) based on the modified Levenberg–Marquart minimization algorithm was developed, which is able to deduce parameters for nitroxide rotational reorientation by iterating the simulation until a minimum least-squares fit to the experiment is reached.

Other theoretical approaches to the theory of slow-motional ESR spectra of nitroxides were also developed,[24–27] aimed at solving specific problems. However, none of these affords a comprehensive treatment of the subject comparable with the approach described above.

4. ESR STUDIES OF SEGMENTAL DYNAMICS OF POLYMERS IN SOLUTION

4.1. Dynamics of Nitroxide Spin Labels Attached to Polymer Chain in Dilute Solution

Information on rotational diffusion of nitroxides was used to deduce and characterize the rotational mobility of polymer chain segments. A model for rotational diffusion of a nitroxide spin label randomly distributed along the polymer chain was proposed, based on theoretical considerations of Campbell et al.[28] The rotational diffusion of nitroxides was approximated as superposition of the isotropic rotational diffusion of the polymer chain segment with rotational parameter, R_S, and the internal rotation of the spin label about the tether through which it is attached to the polymer chain segment, with the rotational parameter, R_I.[21] The contribution of rotational diffusion of the polymer coil as a whole was neglected for high-molecular-weight polymers. When the spin label is attached to the polymer chain with a tether containing only one single bond through which conformational transitions of the tether can occur, the axis of internal rotation of the spin label is identical to this bond axis. This approximate model gives rise to axially symmetric rotational diffusion with two components of the rotational tensor, $R_{prp} = R_S$ and $R_{pll} = R_S + R_I$. In addition, the axis of internal rotation can be tilted by angle β relative to the z axis in the xz plane of the nitroxide axis system (in regard with the symmetry plane of the piperidine ring). This is shown below for the 2,2,6,6-tetramethylpiperidin-1-yloxyl-type label. In this case the tether involves three single bonds (C–CO, CO–NH, and NH–SL, where SL stands for 2,2,6,6-tetramethylpiperidin-1-yloxyl). Given that the amide bond can be considered to be fixed during the time window of the relevant motions, and that there are steric constraints around the first bond, the only bond through which conformational changes of the tether are expected to occur is the NH–SL bond. The applicability and adequacy of the model was tested in practice by critical assessment of consistency of the rotational parameters determined.

Local segmental dynamics in spin-labeled poly(methyl methacrylate) (PMMA), poly(methacrylic acid) (PMA), poly(acrylic acid) (PAA), polystyrene (PS), and poly(2-hydroxyethyl methacrylate) (PHEMA) was studied using the model described above. These polymers were spin labeled by incorporating a spin-labeled chain unit in the polymer chain at randomly distributed sites using various synthetic techniques (polymer reaction or copolymerization with a suitable monomer carrying an amine precursor of the spin label and subsequent oxidation). Free unreacted (unattached) low-molecular-weight nitroxides were removed from solutions of spin-labeled polymers by long-term dialysis or repeated precipitation until the ESR signal from free nitroxides (fast-motional three-line ESR spectrum) could not be observed. The weight-average molecular weights (M_w) of the spin-labeled polymers exceeded 20,000 in all cases, which is considered as a rough limit above which the contribution of the rotational dynamics of polymer coil as a whole in solution to the rotational dynamics of the spin label can be neglected. The molar concentration of the spin-labeled chain units ranged between 1 and 3 mol%.

The ESR studies of local segmental dynamics of polymers benefited from the progress achieved in theoretical description of the effects of dynamics on line shape of nitroxide ESR spectra and in resulting software. The Schneider–Freed set of programs[18] made possible the analysis of slow-motional spectra of nitroxides subjected to anisotropic rotational diffusion with rotation symmetry axis oriented quite arbitrarily with respect to the nitroxide axis system. The simulation of one such ESR spectrum with computers available in the 1980s required ~ 20 min of computer time and the best fits to experimental spectra were found by visual comparison simulated spectra with experimental ones.

4.2. Segmental Mobility of Poly(methyl methacrylate)

Randomly spin-labeled PMMA carrying the spin label units shown above and end spin-labeled PMMA were prepared. The ESR spectra of dilute solutions (1%) of both polymers were measured in ethyl acetate and dibutyl phthalate as solvents[21,31] in a suitable temperature range, considering that full determination of rotational parameters can only be performed by analysis of slow-motional ESR spectra. The components of both A- and g-tensors for the spin label were determined by analysis of spectra of the immobilized nitroxides with random orientations in the system under study (rigid-limit ESR spectra). Spectra in most solvents taken at temperatures close to 120 K usually fulfill these requirements. The distribution of the unpaired electron in nitroxides can be affected by interaction with the solvent and by the structure of the chemical bond through which it is attached to the polymer. Therefore the A- and g-tensors must be determined for every nitroxide – polymer – solvent combination.[29]

Consistent data for segmental dynamics of the randomly labeled PMMA were obtained by simulating the experimental spectra. Best fits were obtained for $\beta \approx 51°$, a value close to that determined for the orientation of the NH–SL bond for the chair form of piperidine nitroxides determined by diffraction methods on nitroxide single crystals.[30] The Arrhenius plot of the values for the segmental rotational diffusion (R_S) in ethyl acetate was divided into two parts by a break at room temperature. The low-temperature part of the plot had a higher activation energy (Fig. 3). The sudden decrease in the activation energy at ≈ 300 K was explained by the conformational transition of PMMA chain. This transition of PMMA (and some other polymethacrylates) above room temperature is characterized by an increase in flexibility and a decrease of 8–25% in unperturbed dimensions of the polymer coil K_Θ, as observed by Katime et al.[32] in studies of molecular weight and temperature dependences of the intrinsic viscosity of polymethacrylates in nonpolar organic solvents. For PMMA in ethyl acetate, the transition temperature range was 318–328 K. The presence of both conformations of PMMA coils was found at temperatures close to this range (Fig. 3). No such break was observed when plotting data for PMMA in the more viscous dibutyl phthalate (Fig. 4), probably because the transition temperature range in this solvent, if it exists, is shifted outside the temperature range studied. Analysis of the end-labeled PMMA spectra in both solvents revealed much faster rotational diffusion of the nitroxide compared with the randomly labeled chain, where most spin labels are expected to be attached to the inner-chain segments. The higher value, $\beta \approx 65°$, observed for end-labeled PMMA in ethyl acetate solvent indicates participation of more bonds in the internal rotation of the spin label attached to the end of the chain, which probably shifts β toward the average value for all-trans chains ($\approx 90°$).

4.3. Segmental Mobility of Poly(methacrylic acid) and Poly(acrylic acid)

Polyelectrolytes offer an excellent opportunity to study the effect of chain conformation on local segmental motions. A polyelectrolyte chain in solution can expand due to the electrostatic repulsion between polar groups. This effect leads to a high intrinsic

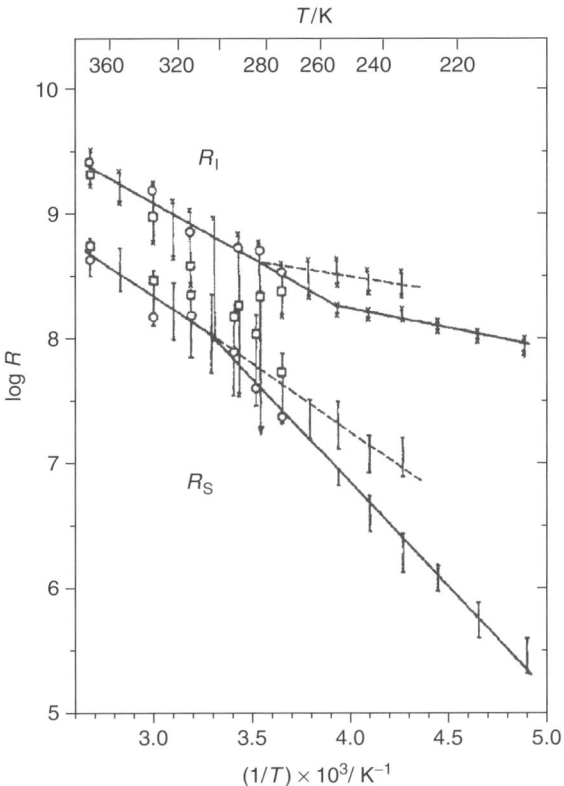

Fig. 3. Arrhenius plots of the rotational parameters R_S (+) and R_I (×) in s^{-1} determined in randomly spin-labeled PMMA in ethyl acetate. Dashed lines show plots of the parameters for the more mobile spin label attached to the PMMA coils in more flexible conformation present in solution, simultaneously with the coils in less flexible conformation at temperatures below the break temperature. The symbols ○ and □ are used for corresponding pairs of R_S and R_I values representing acceptable accuracy limits of spectral analysis. (From Ref. 31, with permission.)

viscosity. The size of the polyelectrolyte random coil is, among other things, a function of the degree of neutralization, α. The reduced viscosity of a typical polyelectrolyte in aqueous solution, which exhibits the so-called "normal polyectrolyte behavior" (e.g., PAA), increases with α increasing up to 0.4 and than slowly decreases due to the saturation effect.[33] The PMA deviates from the normal polyelectrolyte behavior: it is known to undergo a conformational transition in aqueous solution upon ionization. It exists in a very compact conformation stabilized by short-range interactions at low α values, whereas at $\alpha > 0.3$, long-range electrostatic interactions become predominant. Its compact structure is converted to an extended one, which then shows a normal polyelectrolyte behavior upon further increase in α. This conformational transition (at 293 K 50% of the very compact conformation disappears at $\alpha \approx 0.17$[34]) is responsible for the steep increase in reduced viscosity at $\alpha = 0.15$.[33]

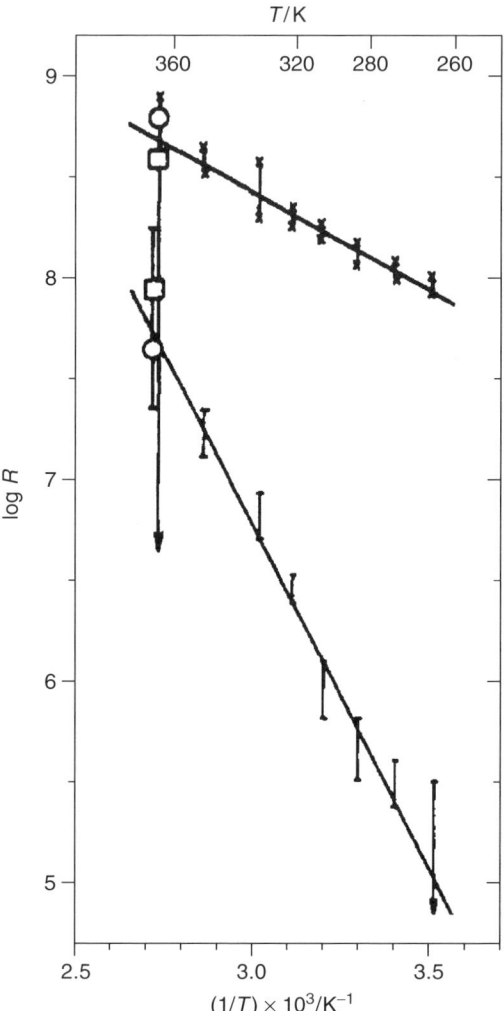

Fig. 4. Arrhenius plots of the rotational parameters R_S (+) and R_I (×) in s^{-1} determined in randomly spin-labeled PMMA in dibutyl phthalate. For the symbols, see Fig. 3. (From Ref. 31, with permission.)

Randomly spin-labeled PMA[35] and PAA[36] carrying the spin label units shown above were prepared. Dilute solutions of both polyacids neutralized to the required degrees were prepared by mixing a polyacid stock solution ($\alpha = 0.0$) in quartz-bidistilled water with an appropriate amount of a polysalt stock solution ($\alpha = 1.0$) or of 0.4 M NaOH. Extreme care was taken to prevent accidental neutralization of the polyacids during preparation. The degree of neutralization α achieved was checked by titration. Components of A- and g-tensors for the spin label attached to each of the acids were determined by analysis of the rigid-limit ESR spectra measured at 113 K. In both cases, the line shapes of the rigid-limit spectra were practically independent of the degree of neutralization α. The ESR spectra of the solutions were measured in the temperature range 273–333 K and their line shapes were analyzed using the model described above.

For PMA, data suggest a slower segmental rotational mobility with lower activation energy for the very compact conformation that prevails at low values of α, compared to that for the extended conformation, which prevails at higher α.[35] Both conformations were found to exist simultaneously in the α range 0.0–0.2 at higher temperatures (293–333 K) in agreement with spectrophotometric data.[34] Corresponding spectra were analyzed in terms of a two-site model, as a superposition of the slow-motional spectrum of the spin labels bound to the PMA coils in the very compact conformation and of the fast-motional spectrum of the spin labels bound to the extended PMA coils. Plots of R_S determined by the simulation procedure are presented in Figs. 5 and 6. It follows that an adequate representation of the α dependence of PMA segmental mobility at each

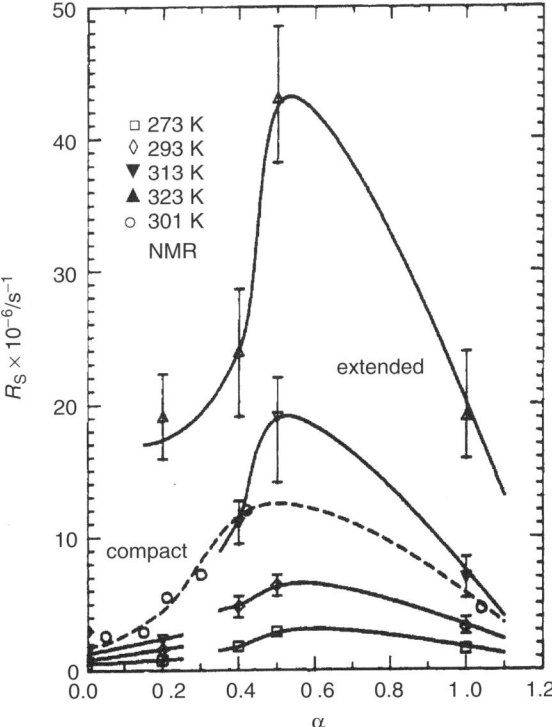

Fig. 5. Dependences of the rotational parameter R_S determined in randomly spin-labeled PMA in dilute aqueous solutions at indicated temperatures on the degree of PMA neutralization α. Data determined by the NMR study[38] are given for comparison. (From Ref. 35, with permission.)

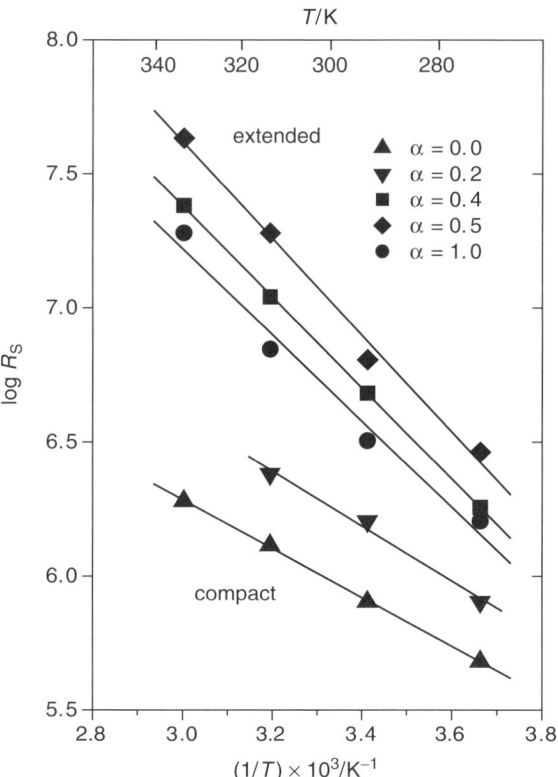

Fig. 6. Arrhenius plots of the rotational parameter R_S in s^{-1} determined in analysis of ESR spectra of randomly spin-labeled PMA in dilute aqueous solutions measured at various degrees of neutralization. Data characterizing both the very compact and extended conformations are given. (From Ref. 35, with permission.)

temperature from the range studied (Fig. 5) should consist of two curves, for the very compact and for the extended conformations. In Fig. 5, the data given for $\alpha = 0.0$ at the four temperatures and the data given for $\alpha = 0.2$ at 273, 293, and 313 K represent the very compact conformation, the rest of the data represent extended PMA conformation. A low proportion of extended PMA coils for $\alpha = 0.0$ and for $\alpha = 0.2$ at 273, 293, and 313 K prevented reliable determination of R_S for the extended PMA conformation. Linear Arrhenius plots of parameters R_S are presented in Fig. 6. The slopes yielded experimental activation energies of the segmental rotational mobility, E_{exp}, of ~ 20 and 30 kJ mol^{-1} for the very compact and extended PMA conformations, respectively.

In the range of α 0.2–0.5, the segmental rotational mobility increases with coil expansion, a result that may be explained by a generally accepted increase in the persistence length in polyelectrolytes. This finding is closely related to the number of segments in the moving sequence of the polymer chain.[37] The value $\beta \approx 40°$ determined when fitting all experimental spectra is somewhat lower than $\beta \approx 51°$ determined for

PMMA,[31] indicating that the conformation of the chain unit carrying the spin label depends on the host polymer and solvent. Our results are in good agreement with NMR data,[38] which are presented in Fig. 5 by dashed line. To our knowledge, such quantitative agreement between the data of two different magnetic resonance methods for polymer dynamics has not been reported yet.

In contrast, quite different results were obtained for PAA,[36] which is expected to exhibit normal polyelectrolyte behavior. The very slow rotational mobilities for PMA in the very compact conformation were not observed for PAA. In the entire α and temperature (293–353 K) ranges studied, the segmental rotational mobility of PAA at corresponding α and temperature was found to be higher than segmental rotational mobilities of PMA in the extended conformation and nearly independent of α. On the other hand, experimental activation energies of segmental rotational mobility, E_{exp}, close to that for PMA in the extended conformation, were found for PAA. The absence of methyl groups is probably responsible for a higher segmental rotational mobility at all degrees of neutralization and in the temperature range studied, compared with PMA under the same conditions.

4.4. Segmental Mobility of Polystyrene

As mentioned in Section 3.1, in more viscous media, where the motion of the nitroxide label is slower and $\tau_R \Delta\omega \geq 1$, the ESR spectrum depends more dramatically on the combined influences of molecular motion and magnetic interactions. Thus, the slow-motional ESR line shapes provide, in principle, a more detailed picture of rotational dynamics compared to the fast-motional line shapes. This idea has motivated the extension of molecular dynamics studies by ESR to higher magnetic fields requiring resonant radiation at wavelengths of the order of $\lambda \cong 1.2$ mm (250 GHz), corresponding to far-infrared (FIR) ESR.[39,40] Because of the greatly increased $\Delta\omega$ at such frequencies, which require magnetic fields close to 90,000 G compared with 3300 G at X band, rotational motions encountered in liquid phases are more likely to fall into the slow-motional regime, thereby permitting a more detailed analysis. Another important feature of high-frequency studies is the enhanced g-factor resolution: regions of the rigid-limit spectrum corresponding to the magnetic x, y, and z components are clearly discerned and the components of the g-tensor may be determined from the rigid-limit spectra with higher accuracy. This enhanced orientational resolution also has the potential of providing better resolution of microscopic details of the motions from the slow-motional ESR studies.

A successful MOMD model for analyzing ESR line shapes for spin-labeled polymers was introduced by Meirovitch et al.[18,41] This model, mentioned in Section 3.3, allows for constraints in the rotational diffusion. More precisely, the polymer chain segmental motion sensed by the tethered nitroxide, that is, the wobbling motion of the effective axis of internal rotation, was considered as constrained by an orienting potential, typically given by Eq. 3,

$$-U(\theta,\varphi)/kT = \sum_{L=2,4} \{c_0^L D_{00}^L(\theta,\varphi) + c_2^L [D_{02}^L(\theta,\varphi) + D_{0-2}^L(\theta,\varphi)]\} \quad (3)$$

where the c_0^L terms refer to the strength and shape of the restricting potential for the wobbling motion and the c_2^L to its asymmetry, and $D_{0K}^L(\theta,\varphi)$ are Wigner rotation matrix elements. The angles θ and φ are polar and azimuthal angles, respectively, which describe the orientation of the effective axis of internal rotation relative to the director expressed in the principal axes of nitroxide rotational diffusion tensor. Calculation of simulated line shapes in the frame of the MOMD model requires more computer time and fitting of additional parameters c_K^L. The revolution in the power and availability of computer hardware together with improvements in computational methodology has made it possible to quickly perform spectral calculations on small laboratory computers (PC) using the NLSL program[20] (Section 3.3) and made it feasible to incorporate sophisticated models, such as MOMD. The input parameters for the NLSL calculation are the components of A- and g-tensors for the spin label attached to the polymer chain and determined in the solvent used. The parameters of the fit for the MOMD model and axially symmetric Brownian rotational diffusion of the spin label include rotational parameters $R_{prp} = R_S$, $R_{pll} = R_S + R_I$, angle β, parameters for line widths, and potential parameters c_K^L. A PC computer with 2-GHz processor is able to find the best-fit parameters for slow-motional spectrum in the frame of this model overnight.

Styrene copolymers with spin-labeled acrylic or methacrylic acid units distributed randomly along the main chain were also synthesized. Local segmental dynamics of the styrene copolymer with spin-labeled acrylic acid units was studied by comparing data obtained by analysis of spectra of dilute toluene solutions of the copolymer measured using X band (9 GHz) and FIR (250 GHz) ESR spectrometers. The high-frequency spectra were measured on the FIR ESR spectrometer in Professor Jack Freed's laboratory at Cornell University. All the experimental spectra were analyzed in the frame of the MOMD model using NLSL program (Fig. 7). The MOMD model, when applied to the system of nitroxide spin labels attached via short side chains (tethers) to the polystyrene main chain in solution, implies a preferred orientation for the axis of internal rotation of each attached nitroxide label with respect to the polymer main chain, but there is an isotropic distribution of these preferred orientations in the macroscopic sample due to the random distribution of orientations of polymer coils. The nature of the preferred orientational distribution within each polymer coil is determined by the shape of the orienting potential. The best-fit shape can be optimized during the fitting process by varying the potential parameters c_0^2, c_2^2, c_0^4, and so on. Although a reasonable agreement was found[42] between the results obtained by NLSL analysis at 9 and 250 GHz, there were systematic discrepancies such that the orienting potentials obtained from the 250-GHz spectra were almost twice (or more) as high as those from the 9-GHz spectra, and the rotational diffusion tensor components from the 250-GHz spectra were at least twice as high as those from the 9-GHz spectra (Fig. 8). This implies a greater sensitivity of the 9-GHz spectra to slower tumbling motions. For the faster "time scale" of the 250-GHz spectra, such motions are likely "frozen out", consistently with the MOMD model. The effective motion at 9 GHz is a combination of both faster internal and slower segmental motions. Since the slower segmental motions are unconstrained, the effect of the faster and constrained internal rotation prevails in 250-GHz spectra. Nevertheless, the results at both frequencies

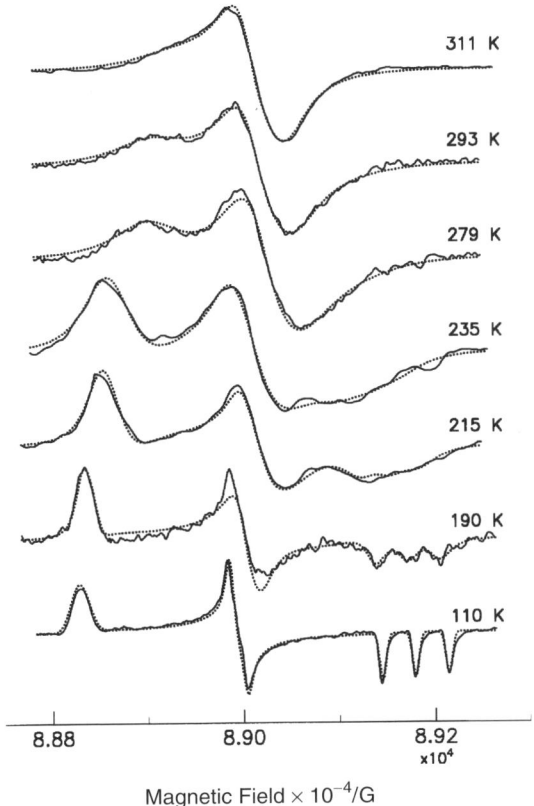

Fig. 7. The FIR (250-GHz) ESR spectra of the random copolymer of styrene with spin-labeled acrylic acid units measured in toluene at the given temperatures; experimental and best-fitting simulated spectra are given with full lines and dotted lines, respectively. (From Ref. 42, with permission.)

yielded a common activation energy, $E_{exp} = 20.7 \pm 1.5$ kJ mol^{-1} (Fig. 8), which, when corrected for the contribution of activation energy of toluene viscosity according to classic Kramers' theory, yielded $E_a = 11.9 \pm 1.5$ kJ mol^{-1}.

The ESR spectra of styrene copolymer with spin-labeled methacrylic acid units were measured at X band (9 GHz) in dilute solutions (~ 1 wt%) over a broad temperature range in four solvents differing in thermodynamic quality, viscosity, and activation energy of viscosity.[43] Theta solvent dioctyl phthalate (DOP), marginal solvent dibutyl phthalate (DBP), and good solvents toluene (TOL) and DMF were used. Components of A- and g-tensors of the nitroxide spin label attached to polystyrene chains in the solvents used have been determined in a usual way from rigid-limit ESR spectra of the spin-labeled copolymer measured in frozen solutions at 120 K. The rigid-limit spectra were successfully fitted using the NLSL program and assuming a very slow isotropic Brownian rotational diffusion of the spin label ($R_{prp} = R_{pll} = 1 \times 10^3 \text{s}^{-1}$). Best fits required axially symmetric Lorentzian inhomogeneous broadening and a Gaussian inhomogeneous

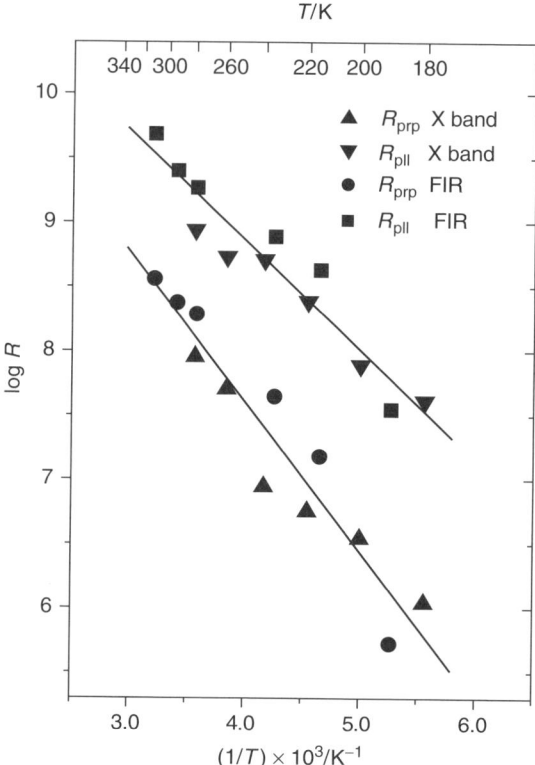

Fig. 8. Arrhenius plots of the rotational parameters R_{prp} and R_{pll} in s^{-1} determined in analysis of the X band (9-GHz) and FIR (250-GHz) ESR spectra taken in dilute toluene solution of the random copolymer of styrene with spin-labeled acrylic acid units. (From Ref. 42, with permission.)

broadening. This corresponds to a combined Lorentzian–Gaussian line shape frequently observed in rigid limit spectra. Note that the line shapes of the well-resolved spectra are very sensitive to the A_{xx} and A_{zz} values, but less sensitive to the A_{yy} values. The uncertainty in A_{yy} was resolved by adopting the values calculated from the isotropic nitrogen hyperfine splitting determined from the fast-motional three-line spectra.

The parameters R_S and the other parameters characterizing the local dynamics in polystyrene chains were determined by fitting experimental ESR spectra to the MOMD model using the NLSL program. The first three potential parameters c_0^2, c_2^2, c_0^4 were found to be sufficient for potential characterization in this study. The best-fit values of the fitting parameters confirmed the consistency of the model. The angle β was found to be solvent dependent; it increased with increasing solvent viscosity and slightly decreased or remained practically constant with increasing temperature in each of the solvents. The parameter c_0^2 characterizes the component of the orienting potential symmetric in the plane perpendicular to the direction of the preferred rotational tensor symmetry axis orientation specified by angle β. This parameter increased with increasing temperature in all solvents up to a maximum ≈ 1.5 reached at intermediate temperatures, at which a fast-motional character of the spectra begins to prevail, and it sharply decreased toward zero at higher temperatures. The parameter c_0^4, which affects the shape of orienting potential significantly, behaves in a similar way and is negative at low temperatures. The parameter c_2^2 characterizing the rhombic distortion of the orienting potential keeps a relatively high value (> 1.0) at temperatures at which the slow-motional character of the spectra prevails; at intermediate and higher temperatures, it sharply decreases toward zero. Fits of similar quality have been obtained in all four solvents. Some of the experimental spectra (Fig. 9) clearly indicate the presence of a small amount (< 5 % of total nitroxide concentration in the sample) of a much more mobile spin label in the sample. Such a superposition in the ESR spectra of spin-labeled polymers in solution, which is probably due to a slow detachment of some of the side chains through which the spin label is bonded to the main chain, was frequently observed. This results in the appearance of the free nitroxide spectrum. All such spectra have been analyzed using a two-site model.

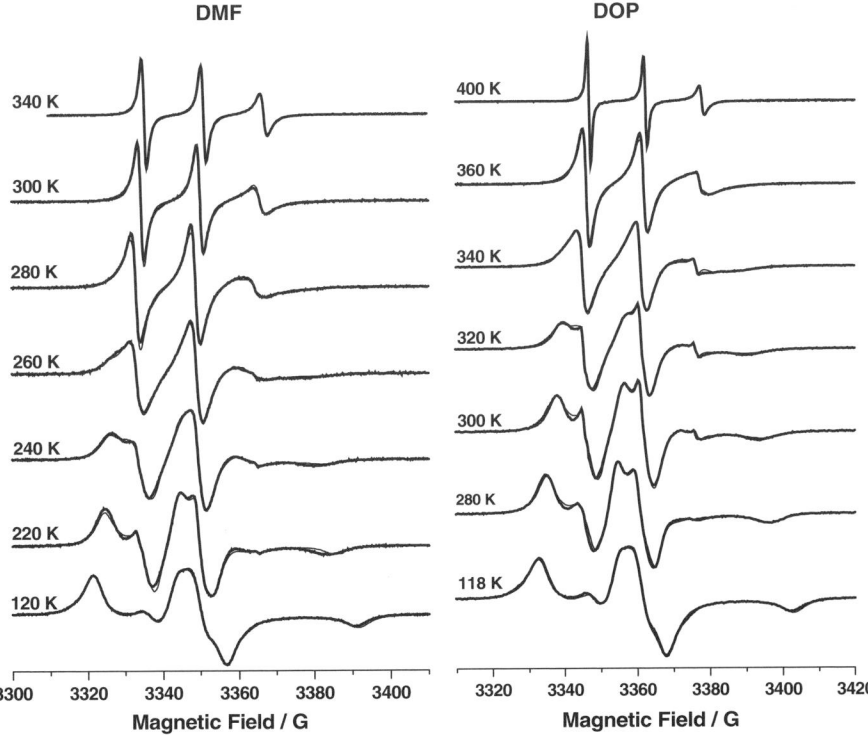

Fig. 9. The ESR spectra of the random copolymer of styrene with spin-labeled methacrylic acid units measured in DMF and in DOP at the indicated temperatures; experimental, and best-fitting simulated spectra are given with full lines and dotted lines, respectively. (From Ref. 43, with permission.)

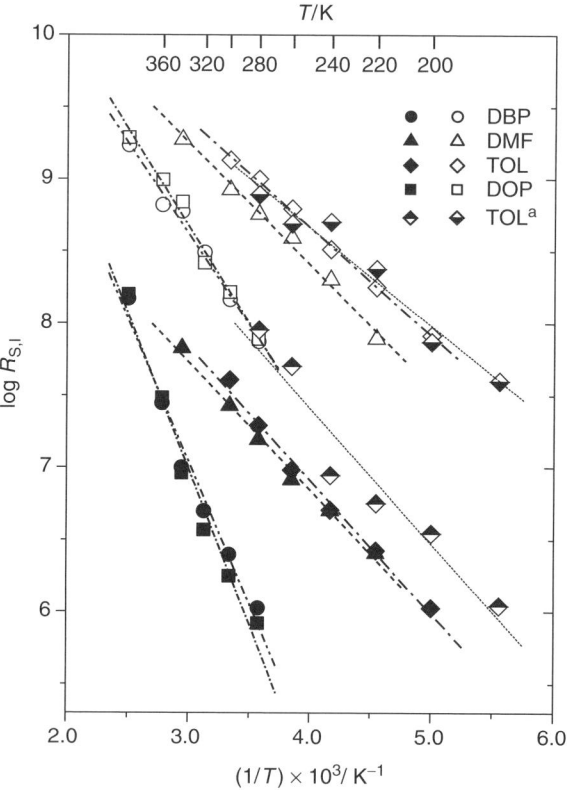

Fig. 10. Arrhenius plots of rotational parameters R_S (full symbols) and R_I (empty symbols) in s^{-1} determined in the random copolymer of styrene with spin-labeled methacrylic acid units in four solvents and their best fits to Eq. 4. The X-band data for the styrene copolymer with spin-labeled acrylic acid units in toluene[42] (TOL[a]) are given for comparison. (From Ref. 43, with permission.)

Arrhenius plots of the best-fit values determined for the rotational parameters R_S and R_I are given in Fig. 10. The data for both parameters in all four solvents have been fitted by Eq. 4,

$$\log R_{S,I} = K'' - (E_{exp}^{S,I}/RT) \log e \qquad (4)$$

which represents a linear dependence in $\log R$ versus $1/T$ coordinates and is analogous to Eq. 1a (K'' is independent of temperature). Higher mobilities and lower activation energies $E_{exp}^{S,I}$ for both less viscous solvents (TOL and DMF) compared with both more viscous ones (DBP and DOP) follow from Fig. 10.

The temperature dependence of viscosity of the solvents used does not exhibit Arrhenius behavior even in the temperature ranges studied. This follows from Fig. 11, which clearly shows nonlinearity of the plots in $\log \eta$ versus $1/T$ coordinates. Consequently, the activation energy of the viscous flow of the solvent, E_η, cannot be a relevant parameter for further discussion. For this reason, the approach introduced by Glowinkowski et al.[3] was used. Eq. 4 can be rewritten as shown in Eq. 5,

$$\log R_S = K''' - \alpha \log[\eta(T)] - (E_a/RT)\log e \qquad (5)$$

where K''' is independent of temperature and viscosity. In the inset of Fig. 12, $\log R_S$ versus $\log \eta$ is plotted at three temperatures at which measurements were performed at least in three of the four solvents. At a constant temperature, the slopes of these plots determine the values of the coefficient α, as follows from Eq. 5. Linear fits to the presented data led to the value $\alpha = 0.73 \pm 0.02$ at all three temperatures. Equation 1 can be rewritten as shown in Eq. 6,

$$\log (R_S [\eta(T)]^\alpha) = A''' - (E_a/RT)\log e \qquad (6)$$

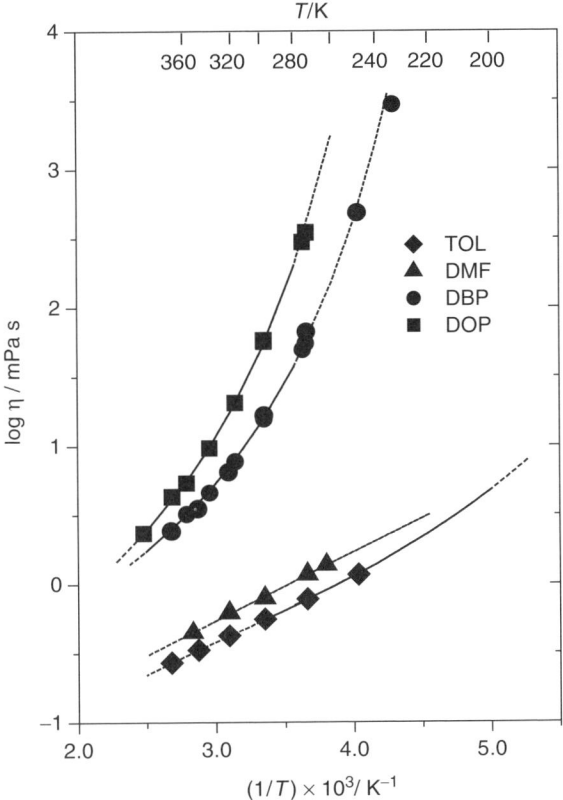

Fig. 11. Temperature dependence of viscosity of the solvents used. Literature data and calculated fits[43] are presented; solid parts of the curves indicate the temperature ranges studied. (From Ref. 43, with permission.)

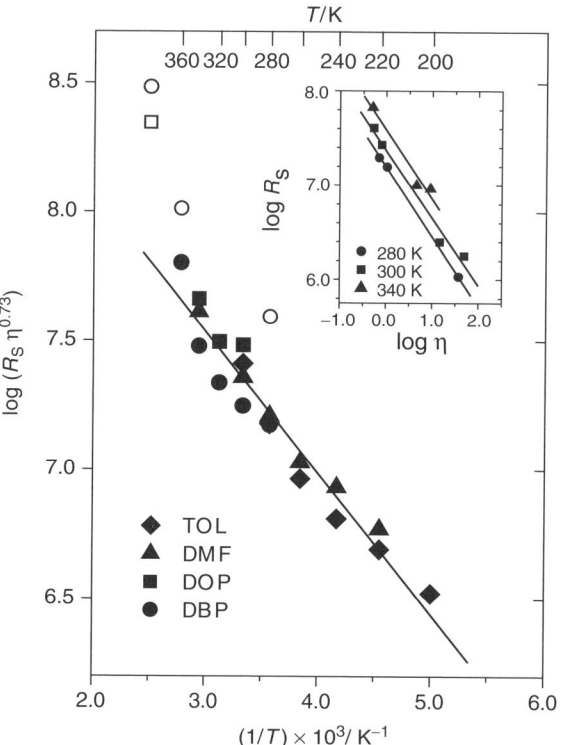

Fig. 12. Arrhenius plot of reduced rotational parameter $R_S \times \eta^{0.73}$. The line is the best fit by Eq. 6 and the slope gives $E_a = 10.5 \pm 0.6$ kJ mol^{-1}. The points excluded from the fit are given by hollow symbols. Log R_S versus log η dependences at the indicated temperatures are given in the inset; the lines are best fits to Eq. 5, and their slopes at all temperatures give $\alpha = 0.73 \pm 0.02$. (From Ref. 43, with permission.)

Finally, $\log(R_S \times \eta^{0.73})$ versus $1/T$ was plotted in Fig. 12. The linear fit of Eq. 6 to the data characterized by a satisfactory correlation coefficient of 0.9712 has yielded the value $E_a = 10.5 \pm 0.6$ kJ mol^{-1} for the solvent-independent height of the potential barrier for the local dynamics of polystyrene chains. The four points indicated in Fig. 12 have been excluded from the data set before performing the fit. In the case of DOP and DBP high-temperature data, the reason is that corresponding ESR spectra could not be analyzed using a two-site model because their line shapes were practically insensitive to the second site parameters. The one-site model fit cannot give parameters consistent with the two-site one due to disregarding the second-site contribution. The value determined in DOP at 280 K was excluded because of increased error in its determination due to a low sensitivity of the line shape of the corresponding ESR spectrum to the R_{prp} parameter. The error in determination of R_S parameters by fitting all other ESR spectra has been estimated at 10%.

For the nitroxide spin label chemically attached to polymer chains in solution, the conclusions presented above generally confirm the importance of microscopic ordering for the dynamics resulting in slow-motional ESR spectra and its decreasing effect on the dynamics in the motional narrowing region. It was concluded that polystyrene in the solvents used (theta solvent DOP, marginal solvent DBP and good solvents TOL and DMF) regardless of their thermodynamic quality, exhibits non-Kramers' behavior characterized by parameter $\alpha = 0.73 \pm 0.02$ and by the height of the potential barrier for local conformational transitions $E_a = 10.5 \pm 0.6$ kJ mol^{-1}. The value agrees with the value published by Waldow et al.[9] for good solvents ($E_a = 11 \pm 3$ kJ mol^{-1}) within the given error limits. The difference between the value of parameter α presented in this chapter and the value determined by Waldow et al.[9] using the fluorescence depolarization technique and anthracene chromophore attached to the polystyrene chain ($\alpha = 0.9$) is minor. Nevertheless, a higher α value has been found by fluorescence depolarization in the presence of the bulky chromophore label than by ESR in the presence of the nitroxide spin label affecting the chain geometry to a small extent. A similar case was observed with polyisoprene when comparing the values determined by fluorescence depolarization[2] and NMR[3] (0.75 and 0.41, respectively).

4.5. Segmental Mobility of Poly(2-hydroxyethyl methacrylate)

Hydrogels based on the 2-hydroxyethyl methacrylate (HEMA) monomer exhibit excellent biocompatibility and physical properties similar to those of living tissue. For these reasons, they have found large-scale applications in medical practice; in particular, materials based on such hydrogels have been used for manufacturing soft contact and intraocular lenses. Recently, investigation of the local segmental dynamics of spin-labeled PHEMA in methanolic solution (linear PHEMA is insoluble in water) was performed over a wide concentration range.[44] A copolymer of HEMA with spin-labeled methacrylic acid units distributed randomly along the main chain (**I**) at concentration ≈ 3 mol% was synthesized. The ESR spectra of methanolic solutions of the copolymer at concentrations between 2 and 50 wt% were measured at X-band (9 GHz) over a broad temperature range (Fig. 13). The temperature dependence of the R_S parameter characterizing the local segmental dynamics in PHEMA and of the other parameters for local dynamics and

I

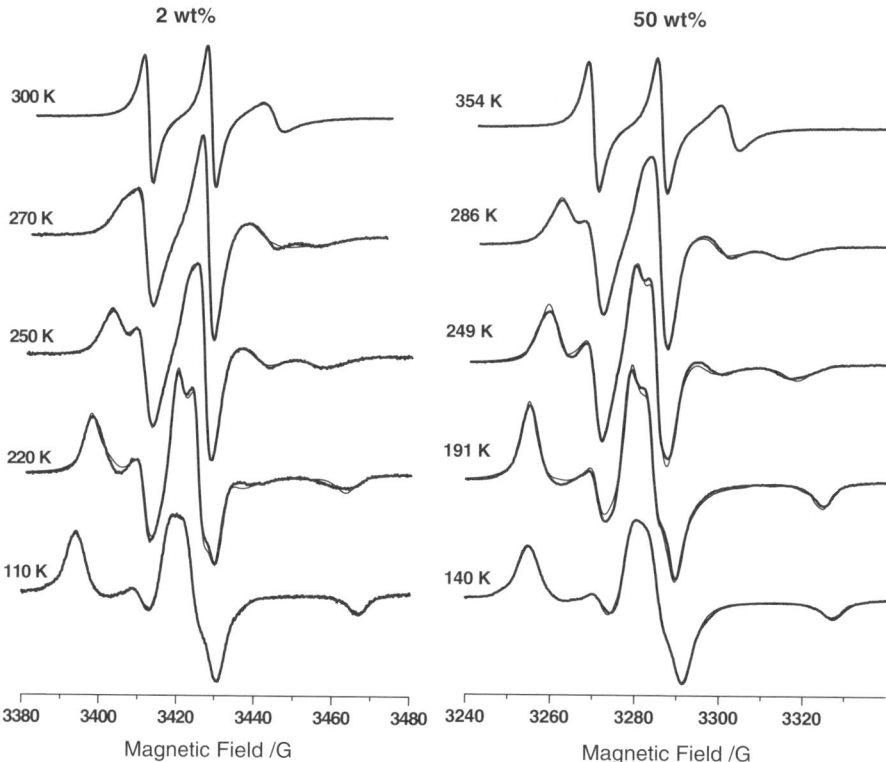

Fig. 13. The ESR spectra of spin-labeled PHEMA in methanolic solution at given temperatures and concentrations; experimental and simulated spectra are given with thick and thin lines, respectively. (From Ref. 44, with permission.)

conformation of the copolymer were determined by fitting experimental ESR spectra to the theoretical spectra calculated using the MOMD model and NLSL program. The best-fit values of the fitting parameters (Fig. 14) again demonstrate the general consistency of the model. At all the concentrations investigated, the angle β first increases with temperature increasing, reaches maximum at ~ 250 K, and then decreases. The isotropic inhomogeneous Lorentzian broadening w^L reaches minimum at temperatures close to 230 K, and then increases with increasing temperature. The best-fit values of the three potential parameters employed determine the shape of the ordering potential at a given temperature and concentration. The parameters c_0^2 and c_0^4 characterize the component of the ordering potential independent of the θ coordinate, and the parameter c_2^2 characterizes its rhombic distortion (See also Chapter 3). Best-fit values of parameters c_0^2 and c_2^2 are positive and decrease with increasing temperature at all concentrations studied. Best-fit values of parameter c_0^4 reach negative minimum values at temperatures close to 250 K, and then increase toward zero with increasing

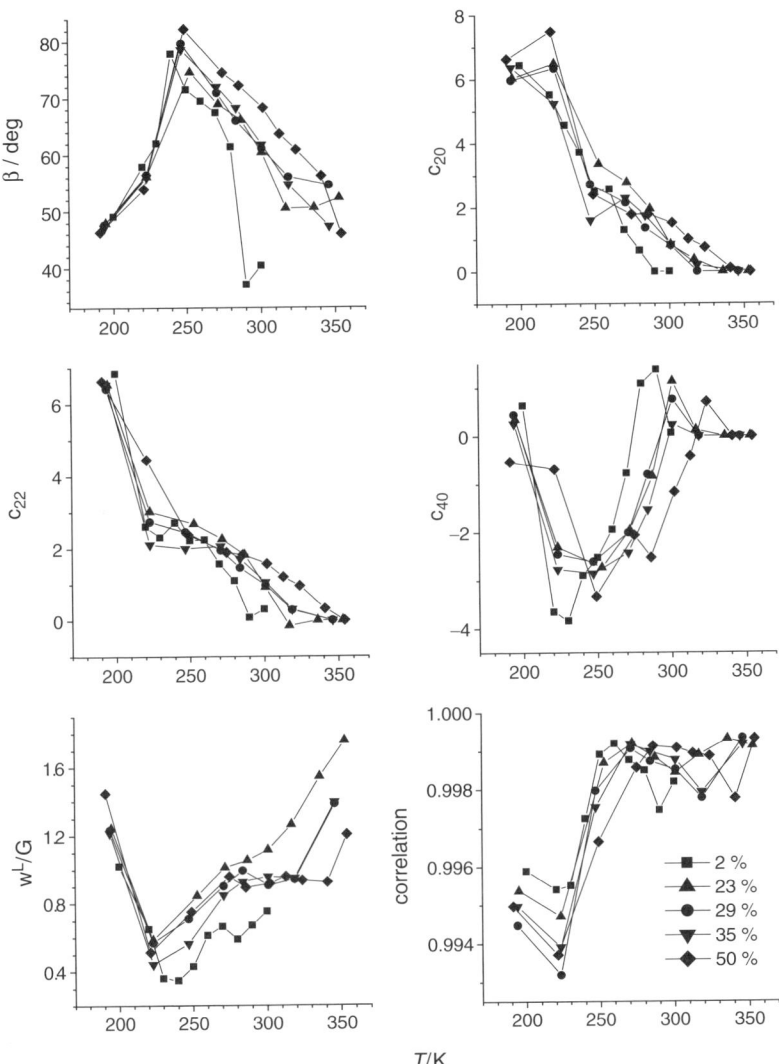

Fig. 14. Temperature dependence of the fitting parameters determined in spin-labeled PHEMA in methanol for the concentrations studied. (From Ref. 44, with permission.)

temperature. Typical shapes of the orienting potential $U(\theta,\varphi)$ in the coordinates θ and φ ($\theta = 0°$ for the directions in the symmetry plane of the piperidine moiety — in the xz plane of the nitroxide axis system defined in Chart 1 — and $\varphi = 0°$ for the direction along the director) show the probability distribution of effective orientation of the nitroxide rotational tensor symmetry axis with respect to the director (Fig. 15). The appearance of two clearly visible maxima of the distribution separated by an angle of $2\varphi_{max}$ increasing from 60 to 100° ($\varphi_{max} \approx 30-50°$) with

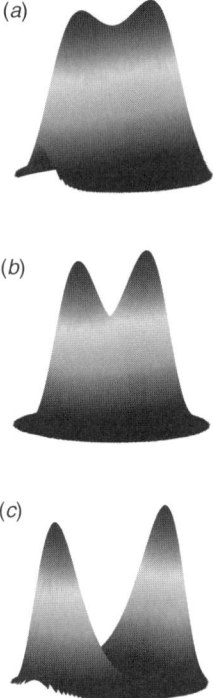

Fig. 15. Population distribution of the instant orientation of the effective nitroxide rotational tensor symmetry axis with respect to the director as characterized by the ordering potential parameters c_0^2, c_2^2, and c_0^4 determined from the ESR spectra of spin-labeled PHEMA in methanol at 50 wt% concentration measured at (a) 191 K, (b) 221 K, and (c) 249 K. The magnitude of the ordering potential decreases with increasing temperature as seen in Fig. 14. (From Ref. 44, with permission.)

temperature in this range is typical of negative values of parameter c_0^4 and positive values of c_2^2. The ordering potential keeps the plane of symmetry of the piperidine moiety, with the maxima appearing at $\theta = 0$ and at $180°$, which follows from the positive values of the parameter c_2^2.

Arrhenius plots of rotational parameters R_S and R_I for the segmental rotational dynamics of PHEMA, as determined by the fitting process at all the concentrations studied (Fig. 16), differ at first sight by their nonlinearity for both rotational parameters from the published plots for segmental dynamics of a number of polymers in dilute solution. In particular, they differ from the above mentioned plots characterizing the local dynamics of polystyrene in dilute solution (Fig. 10), which are linear for both rotational parameters regardless of the thermodynamic quality of the solvent. The plots for PHEMA exhibit an atypical nonlinear behavior characterized by two breaks at specific temperatures in the R_S plot. The difference between the lower break temperature (≈ 250 K for all the concentrations studied) and the upper break temperature was found to increase with increasing concentration of PHEMA in methanol. Local dynamics of the polymer was found to be practically independent of temperature in the temperature

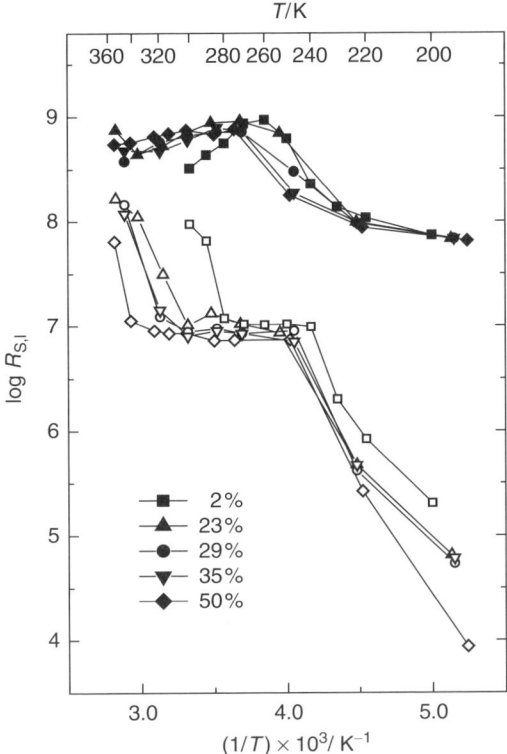

Fig. 16. Arrhenius plots of rotational parameters R_S (empty symbols) and R_I (full symbols) in s^{-1} determined in spin-labeled PHEMA in methanol for all the concentrations studied. (From Ref. 44, with permission.)

range between both breaks. It was concluded that, similarly to the behavior of PMMA and some other polymethacrylates in a number of nonpolar solvents,[32] PHEMA undergoes a conformational transition from a less compact to more compact conformation of the polymer chain in methanol above the lower break temperature. The PHEMA, a polymethacrylate, differs from the polymers studied by Katime et al.[32] in its higher hydrophobicity (and consequently insolubility in non-polar solvents) due to the presence of the 2-hydroxyethyl groups. The decrease in the polymer globule radius observed in 2% methanolic solution of PHEMA with temperature increasing in the range in which the saddle part for this PHEMA concentration exists, as determined by dynamic light scattering (DLS) and small-angle X-ray scattering (SAXS) experiments (Fig. 17), supports this conclusion. Intramolecular interactions in a dilute solution (2 wt%) in the range between both break temperatures, when PHEMA exists in a compact conformation, hinder local polymer dynamics. The contribution of the intermolecular interactions in entangled undiluted solutions (at polymer concentrations higher than ~ 20 wt%) is probably responsible for the same effect in more concentrated solutions and for the observed increase in the upper break temperature with increasing PHEMA concentration.

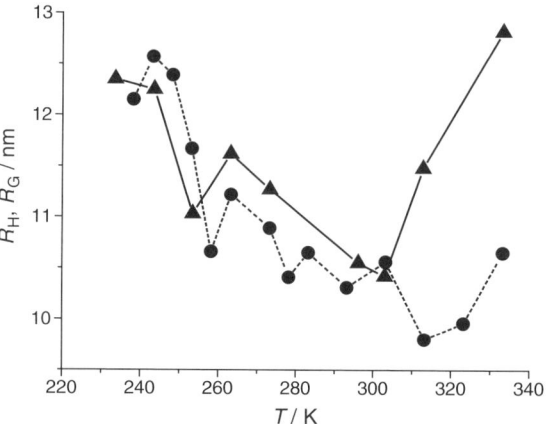

Fig. 17. Hydrodynamic (R_H, ▲) and gyration (R_G, ●) radius of PHEMA coils in 2% methanolic solution determined by DLS and SAXS, respectively. (From Ref. 44, with permission.)

Analysis of the shape of the orienting potential, which constrains the preferred orientation of the effective axis of internal rotation of the tethered nitroxide, revealed the presence of two different conformations of the spin-label moiety. In order to explain these results, the quantum chemical procedure was utilized. The spin-label moiety in the polymer system was reduced to 4-acetamido-2,2,6,6-tetramethylpiperidine-1-yloxyl (**II**) (spin label plus tether). The molecule was subjected to a search for stable conformers, which resulted in the two minima corresponding to the structures **IIa** and **IIb**.[44]

IIa   **IIb**

The minima can be roughly described as structures containing different chair-like conformations of the piperidine ring. For the sake of simplicity, the x axis of the nitroxide axis system was considered to be oriented in the direction of the >N–O• bond. This is justified by the natural bond orbital analysis of molecule **II** performed, which revealed that the p_z orbital of the nitrogen atom is oriented nearly perpendicular to the >N–O• bond. Considering that the axis of nitroxide internal rotation is oriented in the direction of the NH–SL bond of the tether, it is easy to calculate the angle between the x axis of the nitroxide axis system and the symmetry axis of the rotational tensor of the nitroxide using coordinates of the atoms involved. The angles

$\beta_1{}^Q = 11°$ and $\beta_2{}^Q = 77°$ for each of the conformers were found by subtracting this angle from 90°, which are in reasonable agreement with the values $\beta_1{}^S = 15°$ and $\beta_2{}^S = 85°$ obtained by MOMD simulations at temperatures close to 190 K, with the inclusion of the orienting potential. In particular, the value of the angle separating the two most probable orientations determined by MOMD simulation of the spectra measured for the lowest temperatures (close to 200 K), $2\varphi_{max} = 60°$, agrees very well with the value of this angle $\beta_2{}^Q - \beta_1{}^Q = 66°$ determined by quantum chemical calculations performed for 0 K.[44]

We also tried to verify the relative probabilities of occurrence of each conformation. The MOMD simulation suggests that each conformation occurs with similar probability. This is due to the symmetry of the potential $U(\theta,\varphi)$ resulting in equal probability for each of the extremes. The absolute energy of the optimized conformers as provided by quantum chemical calculations was converted to the relative probability of occurrence of each of the conformers using the Boltzmann exponential law. This calculation suggested that conformer **IIa** is far more preferred than the more tightly packed conformer **IIb**. When putting relative energy for conformer **IIa** equal to zero, the relative energy of conformer **IIb** equals ≈ 17 kJ mol^{-1}. This prediction is strongly affected by the reduced size of the model system subjected to quantum chemical calculations: intra- and intermolecular interactions between polymer chain segments and the effects of interactions with solvent were not taken into account. These effects could increase the probability of occurrence of the more tightly packed conformer **IIb**. The appearance of two well-separated maxima of the orienting potential indicates that the conformational transitions between **IIa** and **IIb** in the temperature range 190–280 K are slow with respect to the ESR time scale.

At temperatures above the upper break temperature, the strength of steric interaction decreases and the probability of occurrence of the more tightly packed conformation **IIb** decreases. The orienting potential diminishes and the nitroxide rotation can be characterized by a simple Brownian axially symmetric rotational diffusion characterized by the rotational parameters given in Fig. 16 and by the angle β close to 40°. The presence of the conformations revealed by these calculations can be accepted as an explanation of the experimental data only, provided that the effects omitted in the calculations, in particular the neglect of solvent effect and of interactions mediated by polymer chain(s), would increase the probability of occurrence of the more tightly packed nitroxide conformation.

4.6. Comparison with Other Techniques

When comparing various mentioned techniques used for characterization of local segmental dynamics in polymers, only NMR techniques do not require the presence of any label that can affect the chain dynamics; on the other hand, its applicability requires appearance of suitable lines in the NMR spectrum of the polymer (^{13}C enrichment is frequently needed). The presence of a bulky chromophore label bonded in the main chain, which was required when applying the fluorescence technique, affects both main-chain conformation and dynamics in the vicinity of the "signal" group significantly. The ESR technique requires the presence of a spin label, which is usually attached to a short side chain; the volume and structure of the spin-label-bearing chain unit is usually very similar to the volume and structure of the main-chain units. Information on the local main-chain dynamics is extracted from the parameters characterizing the anisotropic rotational diffusion of the label. The ESR spin-label technique is applicable to polymer systems (copolymers, gels) regardless of their complex chemical structure, provided that a suitable method of spin-label attachment is found.

5. CONCLUSIONS

Investigation of local dynamics in some polymer systems using the ESR spin-label technique at various stages of development has been reviewed. The review shows that interpretation of experimental ESR data, based on the currently available theoretical models, can provide both data complementary to some other experimental techniques and data that clarify some newly observed effects.

ACKNOWLEDGMENTS

This research was supported by the Grant Agency of the Academy of Sciences of the Czech Republic (project A4050306). The author is grateful to Jiří Labský from the Institute for long-term cooperation and for performing all the chemical syntheses, including preparation of deuterated nitroxides and spin-labeled polymers, to Jack H. Freed, Cornell University, Ithaca, and his co-workers for long-term collaboration and for providing access to unique experimental equipment available in his laboratory, and to the most recent versions of the software needed, and to my Ph.D. student Antonín Marek for performing creative computer and experimental work.

REFERENCES

1. Kramers, H.A. *Physica* **1940**, *7*, 284.
2. Adolf, D.B.; Ediger, M.D.; Kitano, T.; Ito, K. *Macromolecules* **1992**, *25*, 867.
3. Glowinkowski, S.; Gisser, D.J.; Ediger, M.D. *Macromolecules* **1990**, *23*, 3520.
4. Courtney, S., H.; Fleming, G.R. *J. Chem. Phys.* **1985**, *83*, 215.
5. Spyros, A.; Dais, P.; Heatley, F. *Macromolecules* **1994**, *27*, 6207.
6. Karali, A.; Dais, P.; Heatley, F. *Macromolecules* **2000**, *33*, 5524.
7. Tylianakis, E.I.; Dais, P.; Heatley, F. *J. Polym. Sci., Part B: Polym. Phys.* **1997**, *35*, 317.
8. Horinaka, J.; Amano, S.; Funada, H.; Ito, S.; Yamamoto, M. *Macromolecules* **1998**, *31*, 1197.
9. Waldow, D.A.; Ediger, M.D.; Yamaguchi, Y.; Matsushita, Y.; Noda, I. *Macromolecules* **1991**, *24*, 3147.
10. Heatley, F.; Wood, B. *Polymer* **1978**, *19*, 1405.
11. Gronski, W.; Schäfer, T.; Peter, R. *Polym. Bull.* **1979**, *1*, 319.

REFERENCES

12. Yokotsuka, S.; Okada, Y.; Tojo, Y.; Sasaki, T.; Yamamoto, M. *Polym. J.* **1991**, *23*, 95.
13. Helfand, E. *J. Chem. Phys.* **1971**, *54*, 4651.
14. Ono, K.; Ueda, K.; Yamamoto, M. *Polym. J.* **1994**, *26*, 1345.
15. Ono, K.; Ueda, K.; Sasaki, T.; Murase, S.; Yamamoto, M. *Macromolecules* **1996**, *29*, 1584.
16. Horinaka, J.; Ito, S.; Yamamoto, M.; Tsujii, Y.; Matsuda, T. *Macromolecules* **1999**, *32*, 2274.
17. Freed, J.H. In *Spin Labeling*; Berliner, L.J. Ed., Academic Press: New York, 1976; pp. 53–132.
18. Schneider, D.J.; Freed, J.H. In *Biological Magnetic Resonance*; Berliner, L. J., Reuben, J. Eds., Plenum: New York, 1989; Vol. 8, pp. 1–76.
19. Goldman, S.A.; Bruno, G.V.; Polnaszek, C.F.; Freed, J.H. *J. Phys. Chem.* **1972**, *56*, 716.
20. Budil, D.E.; Lee, S.; Saxena, S.; Freed, J.H. *J. Magn. Reson. Ser. A* **1996**, *120*, 155.
21. Pilař, J.; Labský, J. *J. Phys. Chem.* **1984**, *88*, 3659.
22. Pilař, J.; Labský, J.; Freed, J.H. *J. Chem. Phys.* **1979**, *83*, 1907.
23. Bullock, A. T.; Cameron, G.G.; Smith, P.M. *J. Chem. Soc., Faraday Trans.* **1974**, *70*, 1202, and references cited therein.
24. Wassmer, K.-H.; Ohmes, E.; Portugall, M.; Ringsdorf, H.; Kothe, G. *J. Am. Chem. Soc.* **1985**, *107*, 1511.
25. Hustedt, E.J.; Cobb, C.E.; Beth, A.H.; Beechem, J.M. *Biophys. J.* **1993**, *64*, 614.
26. Usova, N.; Westlund, P.-O.; Fedchenia, I.I. *J. Phys. Chem.* **1995**, *103*, 96.
27. Usova, N.; Persson, L.; Westlund, P.-O. *Phys. Chem. Chem. Phys.* **2000**, *2*, 2785.
28. Campbell, R.F.; Meirovitch E.; Freed, J.H. *J. Phys. Chem.* **1979**, *83*, 525.
29. Labský, J.; Pilař, J.; Löwy, J. *J. Magn. Reson.* **1980**, *37*, 515.
30. Lajzerowicz-Bonneteau, J. In *Spin Labeling*; Berliner, L.J. Ed., Academic Press: New York, 1976; pp. 239–249.
31. Pilař, J.; Labský, J. *J. Phys. Chem.* **1986**, *90*, 6038.
32. Katime, I.A.; Garay, M.T., François, J. *J. Chem. Soc., Faraday Trans. 2* **1985**, *81*, 705 and references cited therein.
33. Braud, C.; Muller, G.; Fenyo, J.; Selegny, E. *J. Polym. Sci., Polym. Chem. Ed.* **1974**, *12*, 2767.
34. Mandel, M.; Leyte, J.C.; Stadhouder, M.G. *J. Phys. Chem.* **1967**, *71*, 603.
35. Pilař, J.; Labský, J. *Macromolecules* **1991**, *24*, 4188.
36. Pilař, J.; Labský, J. *Macromolecules* **1994**, *27*, 3977.
37. Bahar, I.; Erman, B.; Monnerie, L. *Macromolecules* **1990**, *23*, 1174.
38. Mulder, C.W.R.; Schriever, J.; Leyte, J.C. *J. Phys. Chem.* **1985**, *89*, 475.
39. Budil, D.E.; Earle, K.A.; Freed, J.H. *J. Phys. Chem.* **1993**, *97*, 1294.
40. Earle, K.A.; Budil, D.E.; Freed, J.H. *J. Phys. Chem.* **1993**, *97*, 13289.
41. Meirovitch, E.; Nayeem, A.; Freed, J.H. *J. Phys. Chem.* **1984**, *88*, 3454.
42. Pilař, J.; Labský, J.; Marek, A.; Budil, D.E.; Earle, K.A.; Freed, J.H. *Macromolecules* **2000**, *33*, 4438.
43. Pilař, J.; Labský, J. *Macromolecules* **2003**, *36*, 913.
44. Marek, A.; Czernek, J.; Steinhart, M.; Labsky, J.; Stepanek, P.; Pilař, J. *J. Phys. Chem. B* **2004**, *108*, 9482.

7

SITE-SPECIFIC INFORMATION ON MACROMOLECULAR MATERIALS BY COMBINING CW AND PULSED ESR ON SPIN PROBES

GUNNAR JESCHKE

Max Planck Institute for Polymer Research, Mainz, Germany

Contents

1. Introduction	166
2. Strategies for Spin Labeling and Probing	167
2.1. When to Label and When to Probe	167
2.2. Spin Labeling	168
2.3. Spin Probing	168
2.3.1. Hydrophobic Probes	169
2.3.2. Hydrogen-Bonding Probes	169
2.3.3. Ionic Probes	169
2.3.4. Amphiphilic Probes	170
3. Macromolecules in Solution	171
3.1. Choice of Solvent and Temperature	171
3.2. Frozen Solutions	172
3.3. Polyelectrolyte–Counterion Interactions	173
3.4. Coconformation and Motional Modes in [2]Catenanes	178
4. Materials with Mesoscopic Structure: Interfaces and Nanostructure	182
4.1. Polymer Latices	182
4.2. Ionomers	187
5. Conclusions	192
Acknowledgments	193
References	193

Advanced ESR Methods in Polymer Research, edited by Shulamith Schlick.
Copyright © 2006 John Wiley & Sons, Inc.

1. INTRODUCTION

Since the development of stable free nitroxide radicals by Hoffmann and Henderson[1] and the Rozantsev group[2] in the 1960s, spin labeling has found its main field of application in the biosciences.[3–5] However, the potential of spin labeling in materials science was already recognized early, as evidenced by book chapters on liquid crystals by Seelig[3] and on synthetic polymers by Miller.[4] Later work on synthetic polymers has been summarized in several reviews.[6–9] Nowadays polymer-related materials science often concentrates on systems that are not fully described by classical polymer physics. Such systems exhibit self-organization to mesoscopic structures with features on length scales between a few nanometers and a few micrometers. Concepts of colloid science are then of interest for their description. Many of these systems are hybrid materials made up of incompatible components that would normally segregate, but form dispersions when compatibilizers, such as surfactants, stabilize the interface. Self-organization and the structure at the interface usually depend on a balance between several supramolecular interactions whose energies are close to thermal energy. Examples are hydrogen bonds, electrostatic interactions, and π stacking of aromatic systems. If solvents are present or used during preparation of the material, their interaction with the macromolecular component may also play a role.

This wealth of relevant interactions leads to a complex behavior. To optimize material properties for a given application, we need to understand which interactions dominate, which structures form, and which dynamical processes take place under certain conditions; this goal is not easily achieved by techniques that obtain average information from the bulk of the material. Needed are techniques to probe specific sites that are known or suspected to govern dynamics and structure formation. This is a task for which spin labels and spin probes are well suited.

The complexity of the materials also requires that information other than rotational dynamics of the nitroxide radical be collected. In particular for spin probes, which are not covalently bound to the site of interest, one needs additional information to verify their location. In general, information on the spatial distribution of the probes or labels is of interest. Method development therefore aims to provide tools for obtaining as much information as possible about the vicinity of a nitroxide radical. In application work, the goal is to obtain exactly the information required for solving the given problem with minimum effort. Analyzing the line shape of continuous wave electron spin resonance (CW ESR) spectra as in classical spin-labeling methodology[3–5, 10] remains the starting point for all such work. Explicit consideration of exchange broadening[11] can yield additional information from such spectra. Where necessary, CW ESR at X-band frequencies of $\approx$ 9.6 GHz is complemented by high-field ESR[12] or by pulsed ESR[13] and double-resonance techniques (see also Chapter 2).

This chapter describes the basic concepts used when applying such a combination of ESR techniques to spin labels and probes in complex macromolecular materials. Illustrative examples are taken from our own work. The chapter is organized as follows: Section 2 discusses which systems should be studied with covalently bound labels and which systems with spin probes. Also considered are the choice of the tether that connects the spin label to the macromolecule and the optimum choice of spin probes. Section 3 is devoted to macromolecules in solution. Solution ESR sometimes cannot provide all the required information, hence also considered are how the distribution of conformations in frozen solution is related to the dynamic equilibrium between conformations in liquid solution. The concepts are illustrated by work on polyelectrolytes and on concatenated macrocycles that consist of both flexible and rigid parts. Section 4 is focused on characterizing the nanostructure and addressing interfaces in mesoscopically structured materials by surfactant spin probes or ionic spin probes. This is a versatile approach that was applied to polymer latices, ionic clusters in block copolymer-based ionomers, and nanoporous polymers.

2. STRATEGIES FOR SPIN LABELING AND PROBING

2.1. When to Label and When to Probe

Spin probes are free radicals or paramagnetic transition metal ions that are admixed to the system of interest, while spin labels are stable free radicals that are covalently bound to a macromolecule of interest. Spin labeling thus involves modification of the synthesis of the material and some modification of its structure. Suitable spin *probes* are often commercially available, so that spin probing typically requires less effort. If a question can be answered either by labeling or probing, probing is thus the technique of choice.

However, in many cases the problem dictates which of the two approaches is more appropriate. Generally, covalently bound labels are suited for characterizing structure and dynamics of individual macromolecules. Examples are dynamics of a polymer chain end or the distribution of end-to-end distances of a polymer chain. Spin probes are suited for characterizing the collective behavior of molecules or, to use a more fashionable term, supramolecular behavior. For example, in a pure polymer, spin probes reside in the voids that make up the free volume.[6] The characteristic temperature of nitroxide mobility, $T_{50\,G}$ (see Chapter 1) for spin probes thus depends on the size of the probe and the size distribution of free volume voids. As this size distribution in turn depends on the structure of the polymer, the relation of $T_{50\,G}$ to the glass transition temperature, T_g, of the polymer is not linear, even when comparing the same spin probe in different polymers. However, if the same probe is used, $T_{50\,G}$ does exhibit some correlation with T_g. Therefore, it is often (somewhat incorrectly) being referred to as an ESR T_g.

The main fields of application for spin probes are complex materials that consist of different components. Often, suitably selected spin probes can be used as tracers for one or even several of these components. Examples of such components are the counterions of polyelectrolytes[14–19] or ionomers[20–24] and surfactants.[25–30] This approach may also involve synthetic effort, as the spin probe should mimic the component of interest as closely as possible; otherwise it might distribute in the material in a different way or it might even perturb the structure. To avoid the latter complication, one should use the smallest fraction of probe molecules that provides a sufficiently large signal in the intended experiments. It is good practice to consider, estimate, and discuss possible perturbations introduced by probing or labeling.

2.2. Spin Labeling

A spin label is usually a nitroxide radical with one of the structures shown in Fig. 1 that is linked to the macromolecule by a tether. Nitroxides of TEMPO type (Fig. 1a) are most affordable and are commercially available with the broadest range of linker groups R. Their disadvantage is the conformational freedom of the saturated six-membered ring, which leads to a distribution of the position of the N–O bond from the point of attachment to the macromolecule. In addition, it introduces a motional process that is characteristic for the label itself rather than for the macromolecule. These disadvantages are negligible if disorder of the structure implies spin-to-spin distance distributions that are much broader than the one arising from label conformation and if dynamics is studied in a temperature range where large-angle excursions of the N–O bond occur on the time scale of CW ESR ($\approx$ 3.5 ns). If high precision of distance distributions or stronger coupling of nitroxide motion to motion of the macromolecule are required, nitroxides of the PROXYL type (Fig. 1b) and, in particular, of the dehydro-PROXYL type (Fig. 1c) are better suited. The most rigid attachment of a nitroxide to an alkyl chain is achieved in the DOXYL type (Fig. 1d) for which the z axis of the molecular frame of the nitroxide is fixed and almost parallel to an all-trans alkyl chain. Unfortunately, DOXYL labeling is synthetically much more involved than TEMPO or PROXYL labeling and is thus restricted to special cases.

For TEMPO and PROXYL labels, there is some choice of linker groups and tether lengths. A minimum tether length may be required in proteins to allow for packing of the nitroxide side group into the folded protein structure. In contrast, structures of synthetic materials are not usually so closely packed that they are easily perturbed by a side group with a size of $\approx 6 \times 6 \times 6$ Å^3. To provide best coupling to the macromolecule, the tether should thus be as short and rigid as possible. Ester and amide linkages combine this requirement with relatively simple and reliable synthetic procedures. The amide linkage allows for less conformational flexibility of the tether and thus leads to a simplification of the spectra.[31]

An important consideration in spin-labeling strategies is the radical functionality of the nitroxide, which complicates further synthesis steps and product characterization (by introducing considerable line broadening in NMR). Nitroxide radicals are stable under ambient conditions, but may cause side reactions in many standard approaches of organic or polymer synthesis. For example, it is well known that nitroxides interfere with radical polymerization by forming thermolabile coupling products with chain radicals.[32] Nitroxides are also sensitive to reduction; for example, they are reduced to ESR-silent hydroxylamines by ascorbic acid. In most cases, the hydroxylamines can be reoxidized by air or Cu^{2+} ions to nitroxides, however, DOXYL-type hydroxylamines may undergo irreversible hydrolysis to ketones.[33] These complications strongly suggest that spin labels be introduced as late as possible in the synthesis of a material.

2.3. Spin Probing

Many nanostructured polymer materials form by self-assembly based on noncovalent interactions. The very same interactions can be used for directing a spin probe to the site of interest. The coupling of the spin probe to its vicinity is then dominated by the same interactions that govern structure formation and structural dynamics, and a spin probe is particularly well suited to characterize these interactions. By using a set of different probes, it is furthermore possible to study different sites and thus different aspects of the same material.[26] For this purpose, probes will be classified according to their size and polarity.[9]

2.3.1. Hydrophobic Probes. The nitroxide functionality itself is hydrophilic, while the five- or six-membered hydrocarbon rings with their four methyl substituents are hydrophobic. In the absence of further substituents (e.g., TEMPO, structure in Fig. 1a with R = H), a nitroxide is thus slightly amphiphilic. TEMPO is therefore soluble in water at a concentration sufficient for recording a good CW ESR spectrum in 1 min. However, hydrophobicity dominates, as demonstrated by the strong preference of TEMPO for the hydrophobic polymer over the aqueous serum in polymer dispersions.[28] By enlarging the hydrophobic part, this tendency can be increased and water solubility may be lost. A possible application would be selective addressing of the hydrophobic block in an amphiphilic block copolymer.

2.3.2. Hydrogen-Bonding Probes. All nitroxides are hydrogen-bond *acceptors*. Hydrogen bonding to the N–O group causes changes in the energies of molecular orbitals, so that the g-tensor changes, with the g_x component being most affected. The polarity of the matrix also influences g_x, but in addition has an effect on the hyperfine coupling along the z axis of the molecular frame. The two contributions can be unraveled by high-field ESR.[34–36] Probes, such as TEMPOL (R = OH in Fig. 1a), can act as hydrogen-bond donors. Such probes may thus have a slight preference for attachment to hydrogen-bond acceptors,[26] and may therefore be coupled to the relaxation modes of a macromolecule.[36]

2.3.3. Ionic Probes. Ionic probes (Fig. 2) are generally better suited than hydrogen-bonding probes for selectively addressing a site in a complex material, because the energy gain in forming a single hydrogen bond is always smaller than the thermal energy, while the interaction of multivalent ions with a sufficiently strong electrostatic potential may exceed the thermal energy. Strong electrostatic potentials are found near charged surfaces and interfaces[20–26] as well as in the vicinity of polyelectrolyte chains.[14–19] Of the commercially available ionic probes, Fremy's salt dianion (Fig. 2a) is particularly well suited for studies in solution, as the lack of protons in the molecule

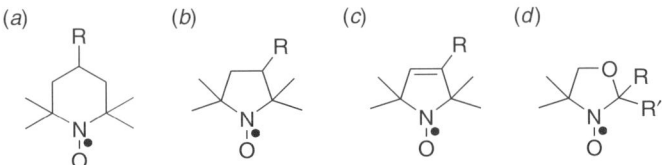

Fig. 1. Structures of the most common classes of nitroxide spin probes and labels. (*a*) TEMPO-type nitroxides. (*b*) PROXYL-type nitroxides. (*c*) Dehydro-PROXYL-type nitroxides. (*d*) DOXYL-type-nitroxides.

leads to narrow lines and thus to a higher sensitivity of the line shape to intermolecular interactions. Further advantages are the smaller size compared to other nitroxides, and the high charge density and low tendency for specific complexation of the charge-carrying sulfonate groups. Disadvantages of Fremy's salt are a low solubility in most solvents other than water, and a low stability at pH < 8 as well as at very high pH.

Other suitable anionic spin probes are TEMPO-4-carboxylate (Fig. 2b) and TEMPO-4-phosphonoxylate (Fig. 2c). Note that these probes exist in their charge-neutral acid form at too low pH. They are more easily incorporated into solid macromolecular materials than Fremy's salt. Negatively charged interfaces or polyelectrolyte chains can be addressed by the probe Cat-1 (Fig. 2d). In solid materials consisting of separated charged and unpolar domains, spin probes of either charge attach to the charged domains. In solution, it is necessary to use probes with a charge opposite to the one of the site of interest.

2.3.4. Amphiphilic Probes. Self-organization of surfactant molecules and compatibilization of polar and unpolar components by surfactants are widely used concepts in materials science. An understanding of the resulting materials critically depends on an understanding of the structure and dynamics of the surfactant layer or of surfactant aggregates. Spin-labeled surfactants, as shown in Fig. 3, are good tracers for unlabeled surfactants and can thus yield such information. The *n*-DOXYL-stearates (e.g., 5-DOXYL-stearate in Fig. 3a) provide the most informative CW ESR line shapes, as the label is rigidly attached to the alkyl chain and the *z* axis of the nitroxide molecular frame is within 5° deviation parallel to the axis of an all-trans alkyl chain. Ordering and anisotropic rotational diffusion of these surfactant probes is thus easily detectable.[10] Up to a point, *n*-DOXYL-stearates are universally applicable as tracers for both anionic and cationic surfactants and even for lipids, as they insert into chemically different surfactant or lipid layers and aggregates. However, there is a danger to overinterpret the precise information that can be obtained on dynamics and ordering of an *n*-DOXYL-stearate molecule by assuming that dynamics and ordering of the native surfactants are analogous.

This problem is particularly acute for cationic surfactants attached to negatively charged surfaces, as in polymer–clay nanocomposites.[30] In this situation, negatively charged stearates are poor tracers, as they are held in place by the neighboring cationic surfactants rather than by interaction with the surface charges. The

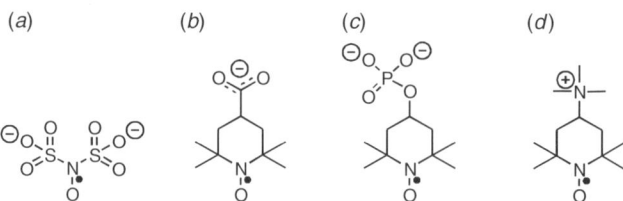

Fig. 2. Structures of ionic spin labels. (*a*) Fremy's salt dianion. (*b*) TEMPO-4-carboxylate. (*c*) TEMPO-4-phosphonooxylate. (*d*) Cat-1.

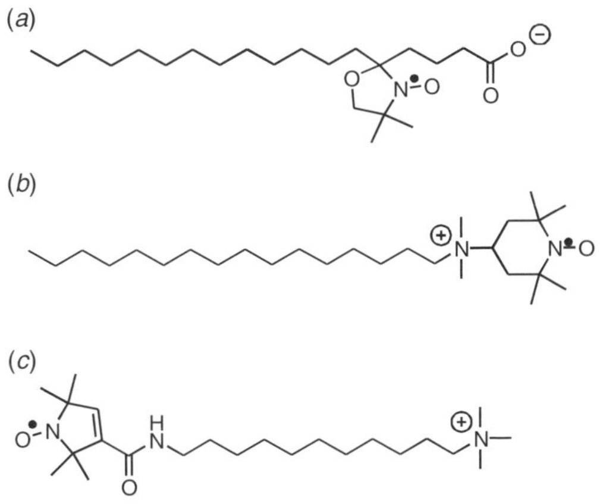

Fig. 3. Structure of spin-labeled surfactants. (*a*) 5-DOXYL-stearate. (*b*) Cat-16. (*c*) Tail-labeled *n*-alkyltrimethylammonium surfactant.

commercially available cationic surfactant Cat-16 (Fig. 3*b*) is somewhat better suited. However, attachment of the label at the site of the cationic anchor group is still expected to change the interaction with the surface compared to the native surfactants trimethylhexadecylammonium and tributylhexadecylphosphonium. Specifically designed surfactants, such as the tail-labeled ammonium surfactant shown in Fig. 3*c*, are thus required to study the properties of the probe that are similar to the properties of the native surfactants.

3. MACROMOLECULES IN SOLUTION

3.1. Choice of Solvent and Temperature

Continuous wave ESR spectra of nitroxide radicals in solution contain information on rotational dynamics and may contain information on the rate of radical–radical collisions if exchange broadening is significant (Chapter 1). If the conformation and behavior of a macromolecule in a given solvent are of interest, indirect information can be obtained by studying how spectra change with solvent variation. Examples are polyelectrolytes whose behavior in aqueous solution is the main interest. Nevertheless, analysis of the differences between aqueous solutions and solutions in mixtures of water with an organic solvent lead to a better understanding of the behavior in pure water.[14–16,18]

In the context of ESR studies on macromolecules, four solvent properties are of main interest. First, the viscosity of the solvent directly influences the rotational correlation time τ_r. A rotational correlation time of ≈ 3.5 ns provides X-band (9.6-GHz) spectra that are most sensitive to details of the rotational motion (restrictions, anisotropy) and to small variations of τ_r. Hence, it may be advantageous to select a solvent in which dynamics is as close as possible to this regime. The situation may be somewhat different for complex structures in which several motions are superimposed. In this case, the viscosity should be adjusted, so that the motional process of interest dominates the spectral line shape or at least makes a significant contribution. Such a case[31] is discussed in more detail in Section 3.4.

The second important solvent property is solvent quality with respect to the macromolecule. Depending on solvent quality, polymers adopt different conformations.[37] In a theta solvent, interactions between monomers are canceled by interactions with the solvent, and the polymer thus behaves as an ideal chain with zero excluded volume. In good and athermal solvents, the polymer behaves as a real chain with positive excluded volume, that is, it adopts a more loosely packed coil conformation. In poor solvents, the excluded volume is negative and the chain is more compact than an ideal chain. Obviously, these differences in conformation may influence the mobility of spin labels and of spin probes interacting with the chain. Both solvent quality and viscosity depend on temperature. For example, many solvent–polymer pairs have a theta temperature at which ideal-chain behavior is adopted.

The remaining important solvent properties are polarity and ability to act as a hydrogen-bond donor. If there is a contrast in these properties between the solvent and the polymer, precise measurements of the hyperfine splitting and the g-value (at high field) can provide information on whether the nitroxide moiety is solvated or surrounded by polymer chains. Such approaches have recently become popular in studies of protein structure[34,35] and appear underutilized in polymer science.

3.2. Frozen Solutions

Measurements of label-to-label distances and label-to-nucleus distances depend on dipole–dipole couplings between spins. As these couplings are averaged in liquid solution, distance measurements have to be performed in the solid state. In many cases, however, one is interested in the structure in solution. As ESR techniques do not require long-range order, they could provide information on solution structure if this structure could be preserved during freezing of the solution. The best approximation of a solution structure by a solid-state structure is obtained by a freeze-quenching, if the solvent forms a glass. Within a few tens Kelvin (K) around the glass transition temperature (T_g), correlation times change by eight orders of magnitude. One may thus assume that on the time scale of ESR measurements there is a dynamic equilibrium of polymer chain conformation significantly above T_g, while the distribution of conformations is static below T_g. This static distribution should then correspond approximately to a snapshot of the equilibrated ensemble at T_g.

Some deviations from this simple approximation should be considered. In a glassy system, motions on different length scales may be frozen at different temperatures, so that there may be no exact correspondence between a freeze-quenched structure and a solution structure at a given temperature. Strain that is built up during vitrification of the matrix may lead to some distortion of local structure. Although these effects may have to be kept in mind, they probably correspond to small conformational changes compared to the diversity of an ensemble of conformations in solution. Generally, deviations from the approximation can be minimized by a rapid freeze-quench. By shooting small droplets into cold isopentane, solutions can be vitrified on a time scale of a few milliseconds.[38,39]

3.3. Polyelectrolyte–Counterion Interactions

Polyelectrolytes are macromolecular substances that are soluble in water or other ionic solvents and dissociate into macromolecular ions that carry multiple charges (polyions) together with an equivalent amount of ions of small charge and opposite sign. The polyion chains can exhibit complex behavior in solution, since entropic effects and electrostatic interactions within the chain and with the counterions influence chain conformation in different ways. Suppose for the moment that a polyelectrolyte with high charge density of the polyion dissociates completely. Repulsion of the like charges on the chain may overcompensate the entropic penalty associated with stretching the chain. Compared to an uncharged macromolecule with similar intrinsic flexibility, the polyion will thus have an enhanced persistence length.

On a sufficiently short length scale, the polyion can thus be considered to be linear (Fig. 4). The radial distribution of the dissociated counterions around such a linear charged chain can be computed by a Poisson–Boltzmann approach.[40] The dielectric permittivity of the solvent as well as temperature can be considered by an electrostatic screening length called the Bjerrum length λ_B. The electrostatic potential of the polyion is then described by the Manning parameter $\xi_M = \lambda_B/b$, where b is the charge spacing (Fig. 4). When computing the radial distribution of counterions of charge z, a singularity appears for $\xi_M \geq 1/|z|$. The physical interpretation of this singularity is that only a fraction $\beta = 1/\xi_M$ of the counterions can be dissociated. The remaining counterions are condensed to the chain, thus reducing the effective spacing of uncompensated charges on the polyion to the Bjerrum length. In this picture of *Manning Counterion Condensation,* three types of counterions can be distinguished: *(1)* condensed or *site-bound* ions, *(2) territorially bound* ions that feel a significant electrostatic interaction with the polyion, and *(3) free* ions that do not interact with the polyion.

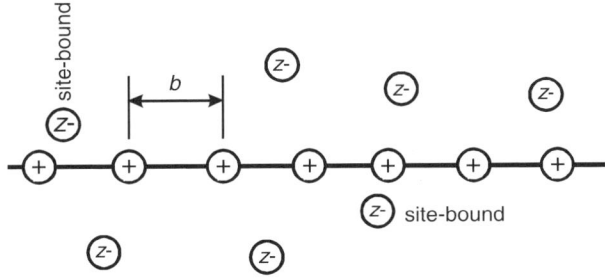

Fig. 4. Model of a stretched positively charged polyelectrolyte chain with charge spacing b and site-bound (labeled) and territorially bound counterions with charge z^-.

This relatively simple description is valid only if the effective linear charge density e/λ_B of the polyion is still sufficient to overcome the entropic tendency of the chain to coil. Otherwise, the relation between chain conformation and counterion distribution becomes rather complex. In the regime close to full stretching of the chain, one can consider a worm-like chain model and apply the linearized Poisson–Boltzmann equation and a simple screened Coulomb potential to compute an electrostatic persistence length that adds to the persistence length of the uncharged polymer.[41,42] Experimental tests of this Odijk–Skolnick–Fixman theory have demonstrated, however, that it predicts a wrong scaling behavior of the persistence length for flexible polyelectrolytes. The concept of an electrostatic persistence length was applied to flexible polyelectrolytes in a combined experimental and theoretical study based on light scattering and numerical computations based on a mean-field approach.[43] Although qualitative agreement was obtained for the scaling of the electrostatic persistence length with ionic strength, persistence lengths were found to be shorter than predicted. Furthermore, the chain-length dependence of the electrostatic persistence length is not successfully predicted by this theoretical approach. To date, the interplay between chain conformation and counterion distribution cannot be described quantitatively by any of the existing theoretical approaches.

While at least some information on chain conformation can be gained from scattering techniques, information on counterion distribution and on molecular details of the interaction of counterions with polyelectrolytes is difficult to obtain with established techniques of polymer characterization. In contrast, ESR spectroscopy of ionic spin probes allows for a direct observation of the counterions, and is sensitive to structure on length scales between a few angstroms and a few nanometers and to dynamics on time scales between 10 ps and 1 μs. For such studies, Fremy's salt dianion (FS^{2-}, Fig. 2a) was found to be particularly well suited.[14–16,18,19] A 0.5 mM aqueous solution of FS^{2-} in the absence of polyelectrolyte exhibits a spectrum with three narrow lines of almost equal amplitude, corresponding to motion in the fast limit with a rotational correlation time of < 10 ps (Fig. 5a). Addition of poly(diallyldimethylammonium chloride), PDADMAC, at a 20-fold excess of polyelectrolyte repeat units over FS^{2-} counterions, causes line broadening (Fig. 5b). From the differences in amplitude between the three lines, it can be concluded that the rotational diffusion of the counterion is slowed down by interaction with the polyelectrolyte. Closer inspection reveals that the relative amplitudes cannot be explained by isotropic tumbling nor by tumbling about the C_2 symmetry axis of the FS^{2-} counterions. Rather, the preferred axis of rotation is almost parallel to one of the probe N–S bonds (Fig. 5c). Motion about this axis is 15 times faster than about axes perpendicular to it. Even larger ratios are found in mixtures of water with ethanol, N-methylpropionamide, or glycerol. This symmetry breaking suggests that electrostatic site attachment is mediated by only one of the two charged groups of FS^{2-}.

Rotational diffusion about the fast axis is on a time scale of a few tens of picoseconds, while rotation about the slow axes is still on a subnanosecond time scale in water, and on a time scale of a few nanoseconds in mixtures of water with organic solvents.[44] The spectra can be simulated nearly perfectly without assuming a distribution of rotational correlation times. This excludes the existence of a fraction of free counterions, as these should have nearly the same rotational correlation time as FS^{2-} anions in pure water. Furthermore, rotational diffusion is not significantly different for site-bound and territorially bound counterions, otherwise a bimodal distribution of rotational correlation times would be expected. As it appears unlikely that in water all FS^{2-} ions are site bound and would still have rotational correlation times in the subnanosecond range about all axes, we have to assume a dynamic equilibrium between site- and territorially-bound counterions. Exchange between the two fractions must be sufficiently fast to average the rotational diffusion tensor. This corresponds to a *dynamic electrostatic attachment* of the counterions with a lifetime of the site-bound state of < 1 ns. The slowdown of rotational dynamics by dynamic electrostatic attachment can be more precisely characterized in a water–glycerol mixture, since due to the higher viscosity of the solvent, the correlation times are closer to the transition between the fast and slow tumbling regimes.[15] In this solvent, one observes a stronger dependence of τ_r on the ratio R between spin probe and polyelectrolyte concentration than in a mixture of water with *N*-methylpropionamide. In both cases, the dependence is linear with negative slope, but the slope is six times larger in the water–glycerol mixture. This suggests that the lifetime of the attached state is longer in more viscous solvents.

Electron spin echo envelope modulation (ESEEM) modulation experiments were performed on shock-frozen solutions of FS^{2-}/PDADMAC in a glycerol–water mixture to determine the geometry of the site-bound state and the fraction of site-bound counterions.[16] The depth and decay of the ^{14}N nuclear modulation depend on the distance of closest approach between the N–O group of the FS^{2-} anion and the quaternary nitrogen in the PDADMAC repeat unit to which the counterion is attached. The depth additionally depends on the fraction of FS^{2-} anions that have a quaternary nitrogen atom in their close vicinity. The best fit is obtained for an average number of 0.2 quaternary nitrogens in a shell of closest approach that is located 0.43 nm from the N–O bond (Fig. 5c). This distance agrees well with the expectation for a contact ion pair (0.4 nm).

The dynamic electrostatic attachment leads to a enhanced concentration of mobile FS^{2-} counterions close to the polyelectrolyte chain. This causes an increased number of collisions between the spin-carrying ions and thus exchange broadening of the lines.[14,16,19] As counterion concentration depends on the radial distance r from the chain, this broadening is expected to be different for counterions close to the chain and counterions far from the chain. Such a distribution of line widths is indeed detected as a deviation of the line shape from the derivative Lorentzian line shape expected for a uniform counterion concentration c (Fig. 6a). Different models for the radial distribution of the counterions can be tested by fitting this line shape.[19] A simple exponential decay $c(r) = c_0 \exp(-kr)$ does not fit the line shape (Fig. 6b). Better fits are obtained when a power law $c(r) = c_0 r^{-2\gamma}$ is assumed for distances $r > 0.4$ nm, as suggested by the charged cylindrical cell model[45] for the distribution of counterions around a linear chain. The best fit in water–ethanol is obtained with $\gamma = 1.0021$, close to the regime of saturated counterion condensation ($\gamma = 1$) in that model. In water–*N*-methylpropionamide $\gamma = 1$ provides the best fit (data not shown).

Hence, the experimental data are consistent with the charged cylindrical cell model. The chain thus appears to be stretched on the length scale of the few nanometers probed by the ESR experiments.

As the water–ethanol mixture has lower permittivity ($\varepsilon \approx 50$) than water ($\varepsilon \approx 80$), while the water–N-methylpropionamide mixture has higher permittivity ($\varepsilon \approx 140$), it is somewhat surprising that ESR spectra for FS^{2-} counterions in an aqueous solution of PDADMAC are *not* consistent with the charged cylindrical cell model. The most likely cause for the failure of the model for pure water are differences in solvation of the polyelectrolyte chain compared to the mixed solvents. The organic components of both solvent mixtures feature ethyl groups that are better suited than water for solvating the hydrophobic parts of the PDADMAC chains. They may thus prevent a local hydrophobic collapse of the chain.

To probe the counterion distribution on a length scale between 1.8 and 5 nm, double electron–electron resonance (DEER) measurements were performed in shock-frozen

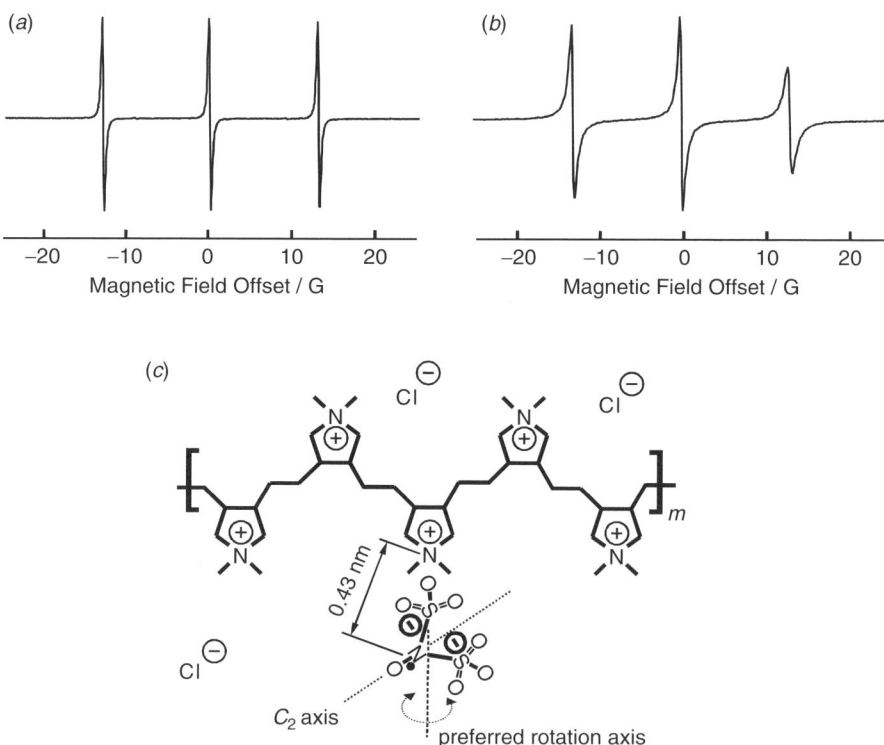

Fig. 5. Dynamic electrostatic attachment of Fremy's salt dianion to PDADMAC. (*a*) The CW ESR spectrum (≈ 9.6 GHz) of a 0.5-mM solution of FS^{2-} in pure water. (*b*) Spectrum after addition of PDADMAC with a concentration of 10-mM repeat units. (*c*) Model of the site-bound state derived from the rotational diffusion tensor and from ^{14}N ESEEM measurements in frozen solution.

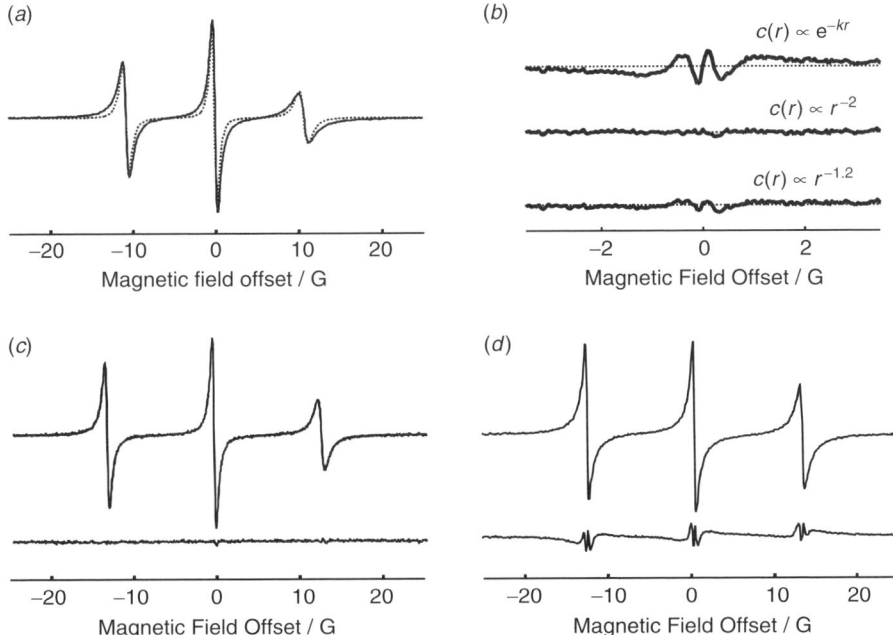

Fig. 6. Radial distribution function of counterions and exchange broadening in CW ESR spectra. (*a*) The experimental line shape of 0.5 mM FS^{2-} / 8 mM PDADMAC in water–glycerol (solid line) cannot be fitted without considering exchange broadening (dotted line). (*b*) Fit residuals for the central line in the spectrum of 0.5 mM FS^{2-}/7.5 mM PDADMAC in water–ethanol assuming different radial distribution functions $c(r)$. (*c*) Spectrum (top trace) and fit residual (bottom trace) for $c(r)$ r^{-2} for 0.5 mM FS^{2-}/ 7.5 mM PDADMAC in water–ethanol. (*d*) Spectrum (top trace) and fit residual (bottom trace) for $c(r)$ r^{-2} for 0.5 mM FS^{2-}/ 7.5 mM PDADMAC in pure water.

solutions of FS^{2-} in a water–glycerol mixture, varying the concentration of added PDADMAC polyelectrolyte.[16] In the absence of polyelectrolyte, data are well fitted by a homogeneous spatial distribution in three dimensions. This is also the case at high polyelectrolyte concentration. In the presence of polyelectrolyte, a higher local concentration of FS^{2-} is found. At low polyelectrolyte concentration, the DEER decay deviates strongly from the exponential decay expected for a homogeneous spatial distribution, and can be fitted by a homogeneous distribution along only one dimension (linear distribution). At intermediate polyelectrolyte concentrations a superposition of a linearly and a spatially distributed fraction fits the data. This behavior suggests an electrostatic persistence length that exceeds the length scale of the DEER experiment. For counterions distributed in a polyelectrolyte coil, one might expect a distribution with dimension 2 (Gaussian coil), but certainly not with dimension 1. The DEER data are thus consistent with the model of a locally stretched chain that was applied in the analysis of exchange broadening in solutions of FS^{2-} and PDADMAC in mixtures of water with ethanol or *N*-methylpropionamide.

Both CW ESR and DEER experiments were also applied to FS^{2-} ions in solutions of weakly charged rigid polyelectrolytes.[18] These polyelectrolytes are coordination polymers consisting of Ru^{2+} ions separated by rigid spacer groups with two coordinating terpyridyl groups. Again, dynamic electrostatic attachment of the counterions is observed. Interestingly, the CW ESR experiments reveal a slightly different type of rotational motion. In this case, where the repeat unit carries two positive charges and the next charged repeat unit is > 1 nm away, the preferential axis of rotation is the C_2 axis of the FS^{2-} molecule. Apparently, in this situation the counterion attaches to the polyelectrolyte with both charged groups. The pair correlation function of the counterions obtained by the DEER experiment exhibits distinct peaks at distances corresponding to the $Ru^{2+}-Ru^{2+}$ separation in the coordination polymer. The full width at half-height of these peaks varies between 0.5 and 1 nm, while molecular dynamics simulations suggest that the $Ru^{2+}-Ru^{2+}$ separation varies by < 0.1 nm. Such smearing of the counterion peaks is expected, since the charge of the Ru^{2+} ions is shielded by the coordinating terpyridyl units and the electrostatic nature of the attachment does not give rise to a fixed ion–ion distance.

No evidence for electrostatic cross-linking of the PDADMAC chains was found with the small FS^{2-} counterion. A different situation was encountered with the triarylmethyl (TAM) trianion radical (Fig. 7a) as a spin carrying counterion.[17] In this case, addition of a small amount of polyelectrolyte (10 repeat units per counterion) to a solution of TAM^{3-} in water–glycerol causes strong line broadening in echo-detected high-field ESR spectra. Further addition of polyelectrolyte leads to a reduction of line broadening, and at a ratio of 350 repeat units per counterion the line shape almost matches the one observed in the absence of polyelectrolyte. The line shape at the lowest polyelectrolyte concentration can be modeled by assuming that 60% of all TAM^{3-} ions have a nearest neighbor at a distance as short as 1.4 nm. This distance coincides remarkably well with the distance of closest approach of two TAM^{3-} anions along their respective C_2 symmetry axes. At higher polyelectrolyte concentration, the distance of closest approach changes only slightly, to $\approx$ 1.5 nm, while the fraction f of densely packed counterions decreases linearly with ratio R between the concentration of spin carrying counterions and concentration of PDADMAC repeat units (Fig. 7b). These findings exclude a homogeneous spatial distribution of counterions as well as a homogeneous one-dimensional distribution along a stretched chain, and suggest that the TAM^{3-} counterions cluster when binding to PDADMAC. Such clustering may be expected as a consequence of cross-linking of the polyelectrolyte chains by TAM^{3-} counterions. Once two locally stretched polyelectrolyte chains are attached to the same TAM^{3-} ion, they offer favorable sites for cross-linking by additional ions. Attachment of further trianions at such sites provides almost the same gain in enthalpy due to optimization of electrostatic interactions as at any other site, but does not require such a large reduction in entropy, as the chains are already close to each other.[17]

3.4. Coconformation and Motional Modes in [2]Catenanes

The quest for functional nanomaterials is one of the most exciting current trends in macromolecular science. Nanomachinery does exist in Nature, where proteins with

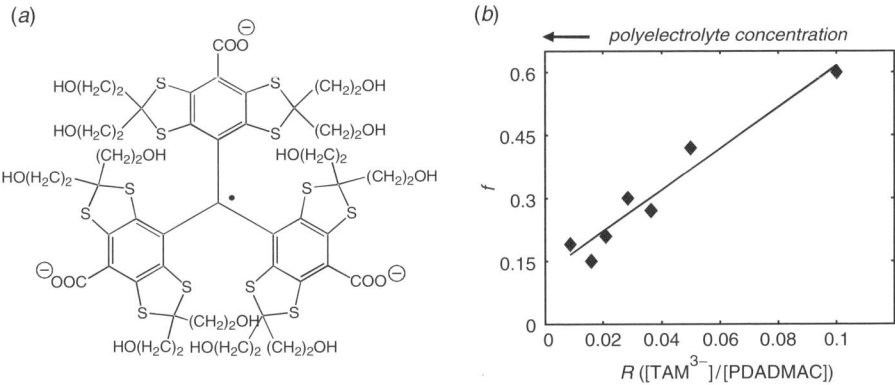

Fig. 7. Physical cross-linking of PDADMAC by large rigid organic counterions TAM^{3-}. (*a*) Structure of the TAM^{3-} ion. (*b*) Fraction of clustered TAM^{3-} counterions as a function of the ratio R of TAM^{3-} concentration to PDADMAC repeat unit concentration.

dimensions between 3 and 10 nm are the devices that perform the basic steps in transport of matter, metabolism, or information processing. In the past few years, the mechanisms of such steps have been elucidated for quite a few proteins. These mechanisms are almost invariably linked to well-defined structural transition of the protein, which is in turn based on the combination of rigid (α-helices, β-sheets) and flexible (loops) building blocks in protein architecture. Such a combination of rigid and flexible building blocks may also be one of the design requirements for synthetic nanodevices. As a basis for rational design of molecules with well-defined structural transitions, we need to understand how the rigid and flexible building blocks of macromolecules move with respect to each other.

This question is encountered, for example, in [2]catenanes, which consist of two concatenated macrocycles (Fig. 8). The motion of the two macrocycles is constrained by the concatenation, that is, the macrocycles cannot separate from each other and may collide during rotational diffusion. Additional effects may constrain the relative motion. For example, the flexible alkyl chains might coil or even collapse. The two macrocycles would then be locked in a compact relative conformation. A similar lock of the coconformation may occur due to π-stacking of the rigid building blocks. To test whether such additional effects constrain the relative motion of the two macrocycles, we performed four-pulse DEER measurements on doubly spin-labeled [2]catenanes.[46] As the distance r between the two labels depends on the relative orientation (coconformation) of the two macrocycles (Fig. 9*a*), different distance distributions are expected for the situation in which the whole coconformational ensemble is sampled, and the situation where the macrocycles are locked. To model the former situation, we consider the simplified geometric model of the [2]catenane shown in Fig. 9*a*. The macrocycles are modeled as circular rings, whose planes include an angle θ, and whose centers are translated by a vector ***T***. The translation is limited by the concatenation constraint, which depends on angle θ. The position of the labels on

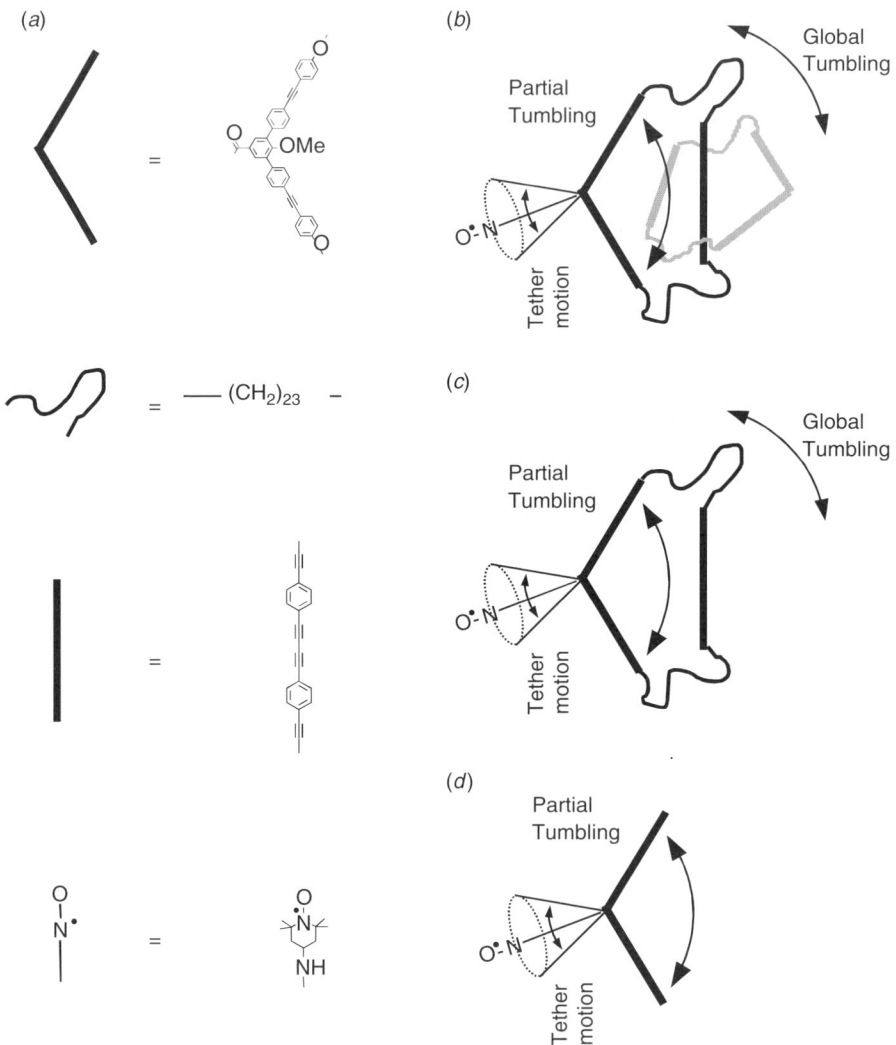

Fig. 8. Building blocks and motional modes in a [2]catenane. (*a*) Rigid and flexible building blocks of a spin-labeled macrocycle. (*b*) Schematic structure of the singly labeled [2]catenane and types of motions. (*c*) Schematic structure of the spin-labeled macrocycle. (*d*) Schematic structure of the spin-labeled rigid part of the macrocycle.

the circular ring are parametrized by angles ξ_1 and ξ_2. To compute the label-to-label distance distribution, averaging over the possible ranges of θ (0−90°), ξ_1 and ξ_2 (0−360°) and over all translation vectors allowed by the concatenation constraint is necessary. The length L of the tether of the nitroxide spin label was estimated as 0.93 nm, from a force-field calculation. Thus, the only free parameter of this model is the effective radius r_{eff} of the macrocycle. By equating the contour length of the

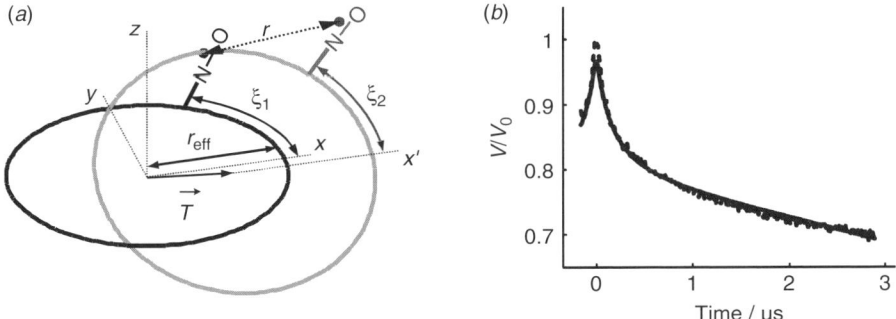

Fig. 9. Characterization of the coconformational distribution of a [2]catenane by DEER measurements. (*a*) Simplified geometrical model of the [2]catenane consisting of two circular macrocycles to which spin labels are attached by a tether with length $L = 0.93$ nm. (*b*) Experimental four-pulse DEER trace (dots) of the [2]catenane in frozen chloroform and fit by the simplified geometrical model (solid line) giving an effective radius $r_{\text{eff}} = 1.50$ nm for the macrocycle.

macrocycle and the circumference of the circular ring, an approximate value of r_{eff} can be predicted.

Despite the considerable simplification of this model with respect to the true molecular geometry, it fits experimental DEER data for [2]catenanes with macrocycles of three different sizes in frozen chloroform. An example for the fit quality is shown in Fig. 9*b*. The best fit for this [2]catenane, corresponding to the structure in Fig. 8, is obtained with $r_{\text{eff}} = 1.50$ nm, while the contour length would suggest $r_{\text{eff}} = 1.75$ nm. For the [2]catenanes with smaller and larger macrocycles, the best fits also give r_{eff} values that are smaller than the predicted values, indicating perhaps that the alkyl chains are not fully extended. However, the trend of the experimental r_{eff} values well reproduces the expected trend. In contrast, experimental DEER data are not well fitted and the variation of experimental r_{eff} values with the size of the macrocycle does not agree well with predictions from contour length for the same series of [2]catenanes in glassy frozen *o*-terphenyl. In particular for the medium-sized and largest macrocycle, we find much too small effective radii. This suggests that the alkyl chains are not expanded in *o*-terphenyl, which is not as good a solvent for aliphatic hydrocarbons as chloroform. The good agreement of the simplified geometric model for [2]catenanes in frozen chloroform indicates that in this solvent the relative movement of the two macrocycles is hindered only by the concatenation constraint.

These measurements in frozen solution can answer the question of whether all coconformations are accessible, but it cannot provide an estimate of the time scale of the relative motion of the macrocycles. To obtain such an estimate, we analyzed CW ESR spectra of the [2]catenanes in terms of the rotational diffusion of the nitroxide spin label. This approach requires interpretation of label motion in terms of the different possible motional processes in the [2]catenane (see Fig. 8*b*). To simplify the interpretation, measurements were performed not only on the singly labeled [2]catenane, but in addition on a singly labeled macrocycle (Fig. 8c), a single-labeled angular rigid building block of the macrocycle (Fig. 8d), and on a doubly labeled [2]catenane. At high or moderate viscosity, the spectra of all singly labeled compounds are virtually indistinguishable; hence we can conclude that tether motion is the dominating contribution. In chloroform at a temperature of 325 K, where viscosity is low, the spectra are distinguishable and can be fitted by assuming an axial rotational diffusion tensor. The tensor component perpendicular to the fast axis decreases only by a factor of 0.8 when going from the angular rigid building block to the macrocycle and by another factor of 0.73 when going from the macrocycle to the [2]catenane. According to the Stokes–Einstein equation, the effective radius of the reorienting unit is proportional to the inverse cubic root of the rotational diffusion coefficient. The effective radius thus increases by only a factor of 1.08 when going from the angular block to the macrocycle, and by another factor of 1.1 when going from the macrocycle to the [2]catenane. This result suggests that motion of the angular rigid building block is only slightly hindered in the macrocycle and [2]catenane. In other words, the flexible alkyl chains appear to allow for an almost independent reorientation of the rigid angular block. The rotational correlation time corresponding to the slow axes is 0.33 ns for the [2]catenane.

In the spectrum of the doubly labeled [2]catenane, we observe significant exchange broadening with respect to the singly labeled [2]catenane (Fig. 10). The exchange frequency of 20 MHz obtained by fitting the spectrum in Fig 10b implies that effective collisions between the two nitroxide labels occur on average every 50 ns. As the line shape is purely Lorentzian, we can exclude that there are different fractions of coconformations with slower and faster exchange frequency. This suggests that the whole coconformational ensemble is sampled at a submicrosecond time scale. In chloroform, the [2]catenane thus consists of expanded macrocycles with high internal flexibility that reorient freely with respect to each other on that short time scale.

4. MATERIALS WITH MESOSCOPIC STRUCTURE: INTERFACES AND NANOSTRUCTURE

4.1. Polymer Latices

The current strong interest in nanomaterials sometimes hides the fact that the length scale between 10 nm and 1 μm has been utilized in materials science for decades. Examples are mesoscopically structured inorganic compounds, such as zeolites and their synthetic counterparts as well as colloids, in particular polymer latices (or latexes). Such latices typically consist of spherical polymer particles with a size between a few tenths of nanometers and a few micrometers that are dispersed in an aqueous serum. The polymerization is usually performed in the presence of surfactants, which also stabilize the latex against coagulation of the polymer particles. Polymer latices are applied as paints, adhesives, textile and paper coatings, printing inks, and binders, since they are easier to handle and less polluting than solutions of

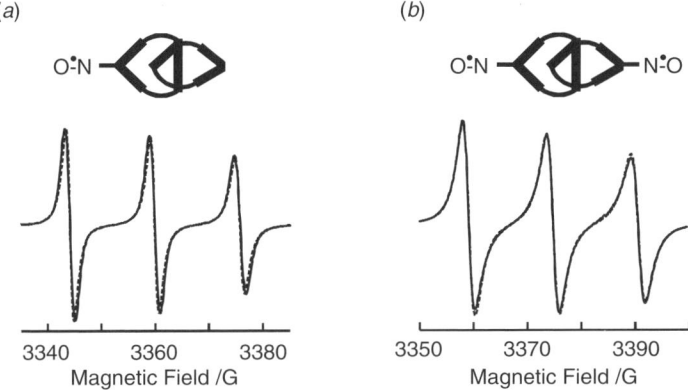

Fig. 10. Experimental (solid lines) and simulated (dashed lines) CW ESR spectra (≈9.6 GHz) of (*a*) the singly labeled [2]catenane and (*b*) the doubly labeled [2]catenane.

polymers in organic solvents, yet can form continuous polymer films of the same high quality. In contrast to polymer solutions in organic solvents, aqueous dispersions contain surfactants and salts. These additional components might potentially compromise the quality of the final polymer film, hence their behavior during film formation is of some interest. In many applications, for example, exterior paintings, Teflon coated frying pans, or car coatings, the final polymer film has to be waterproof. Rewetting behavior is thus another topic of interest.

Polymer dispersions as well as the final films are multicomponent systems with many different sites of interest (Fig.11). Spin probes are particularly well suited to address these sites one by one.[26] In dispersion, the surfactant spin probes insert into the sodium dodecylsulfate surfactant layer that covers the polymer spheres, as can be inferred from the dependence of the rotational correlation time on the T_g of the polymer (Fig. 12). For poly(butylacrylate) with 1% acrylic acid comonomer content [P(BA)^{AA}], the T_g is below ambient temperature and a rotational correlation time of ≈ 2 ns is found for both the head and tail-labeled surfactant. Both spectra can be analyzed with the model of microscopic order–macroscopic disorder.[10] A fairly high degree of order ($\langle P_2 \rangle = 0.56$), comparable to the one of solutes in liquid crystals, is found for the head-labeled surfactant. The higher mobility perpendicular to the chain axis at the end of the alkyl chain may explain the smaller degree of order for the tail-labeled surfactant ($\langle P_2 \rangle = 0.21$). Poly(butylmethacrylate) (PBA) and a copolymer of PBA with 28% methylmethacrylate and 1% acrylic acid (P(BMA)Co2) have T_g above ambient temperature. This results in longer rotational correlation times of the head-labeled surfactant of ≈ 8–9 ns, and a lower degree of order of the surfactant layer ($\langle P_2 \rangle = 0.3...0.4$). The tail-labeled surfactant exhibits bimodal rotational dynamics, with the slow component corresponding to the spectrum of the same probe in the dry polymer film. The dynamics of the fast component is similar to that in PBA, suggesting that a fraction of the surfactant does not insert with the alkyl chain end into the polymer sphere, but only associates with the surfactant layer.

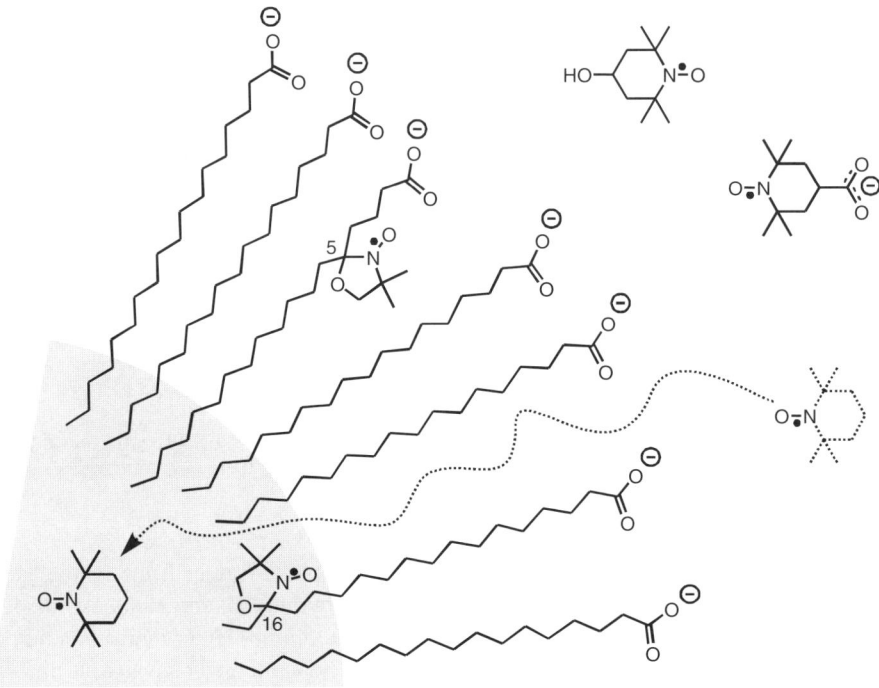

Fig. 11. Site selection by different spin probes in an aqueous polymer dispersion (latex). Spin-labeled surfactants insert into the surfactant layer and may probe insertion of the surfactant tail into the polymer (16-DOXYL-stearate) or order of the surfactant layer (5-DOXYL-stearate). The weakly polar probe TEMPO diffuses (dotted arrow) from the aqueous serum into the polymer sphere (gray). More polar probes such as 4-hydroxy-TEMPO and TEMPO-4-carboxylate reside in the aqueous serum.

The CW ESR spectra of the ionic spin probe TEMPO-4-carboxylate and the hydrophilic spin probe 4-hydroxy-TEMPO in polymer dispersions are virtually indistinguishable from spectra in pure water. These probes do not interact significantly with the dispersed polymer particles. In contrast, spectra of the weakly polar probe TEMPO are bimodal, with one component corresponding to the spectrum in pure water, and the other corresponding to the spectrum in a dry polymer film. The ratio between the two components depends on the time that has passed after admixing the spin probe to the dispersion. This behavior can be interpreted as diffusion of the probe TEMPO from the aqueous serum into the polymer particles. By fitting the time dependence with an appropriate diffusion equation we can calculate diffusion coefficients D for TEMPO in the polymer particles.[28] These diffusion coefficients also depend on T_g; for $T_g = 320$ K, $D \approx 1.2 \times 10^{-15}$ cm^2 s^{-1}, and for $T_g = 338$ K, $D \approx 0.5 \times 10^{-15}$ cm^2 s^{-1}. The diffusion coefficient depends somewhat on particle size, which may be either due to different ageing of the polymer in particles of different size, or to the inhomogeneity of the particle surface.

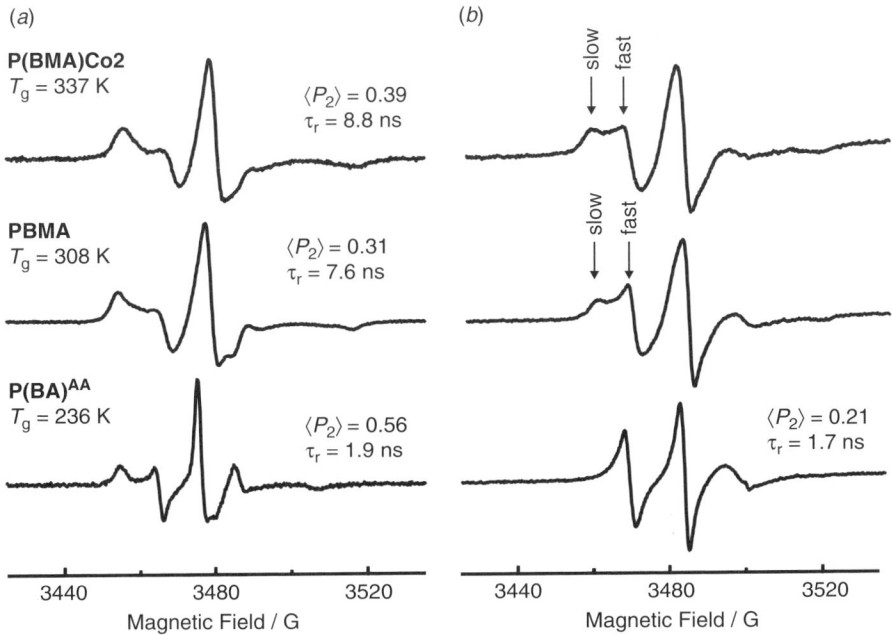

Fig. 12. The CW ESR spectra (≈ 9.6 GHz) of surfactant spin probes in aqueous polymer latices. (a) Surfactant labeled near the ionic head group (5-DOXYL-stearate). (b) Surfactant labeled near the hydrophobic tail end (16-DOXYL-stearate).

Site selectivity of the spin probes is also observed in the polymer films obtained by drying the dispersions.[25,26] For example, high-field CW ESR spectra differ strongly for TEMPO and TEMPO-4-carboxylate probes, although both probes are of approximately the same size (Fig. 13). The spectrum of TEMPO in P(BA)^{AA} can be simulated by assuming an isotropic rotational diffusion tensor with a rotational correlation time $\tau_r = 0.26$ ns. In contrast, the spectrum of TEMPO-4-carboxylate in a poly(fluoroalkyl acrylate) can only be simulated assuming an axial diffusion tensor with the fast axis parallel to the bisector of the carboxylate group. About the fast axis, the rotational correlation time is also 0.26 ns, but about axes perpendicular to it, the correlation time is 3.6 ns. This can be interpreted as immobilization of the probe via the carboxylate group. The site of attachment are most likely ionic clusters, associated either with the charged acrylic acid groups of the polymer, or with inverse surfactant micelles.

During drying of the dispersions[27] all spin probes go through a regime where the CW ESR spectrum is a superposition of the spectrum in dispersion (mobile fraction) and the spectrum in the final film (immobilized fraction). The dependence of the immobilized fraction on the water content is different for the different spin probes (Fig. 14a). As expected, the weakly polar probe TEMPO is immobilized first, the amphiphilic spin probes 16-DOXYL-stearate and 5-DOXYL-stearate at somewhat lower water contents, and the hydrophilic spin probes 4-hydroxy-TEMPO and

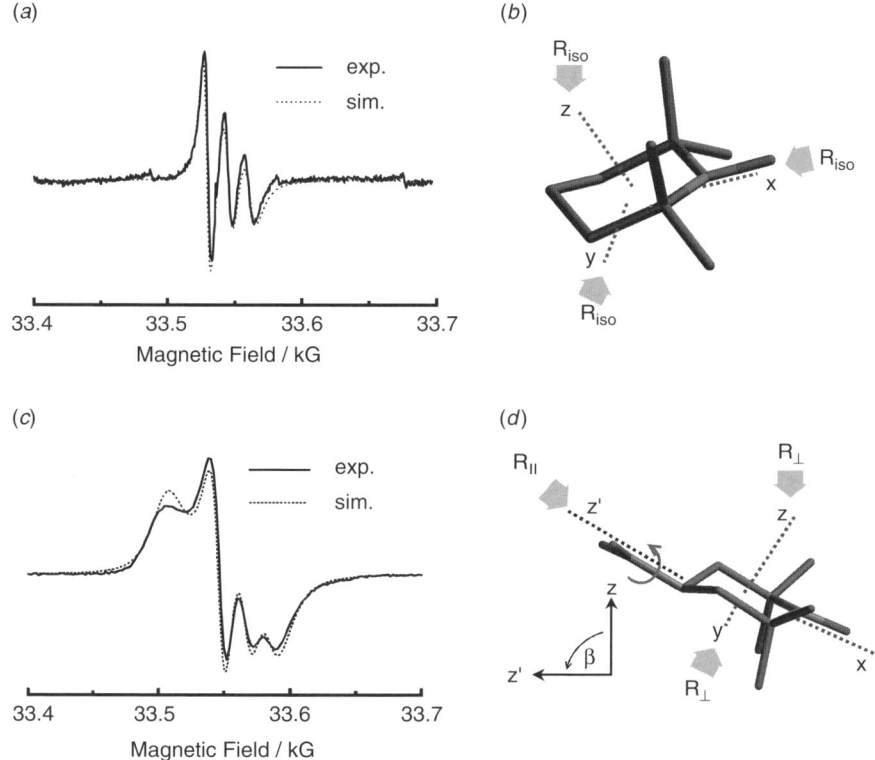

Fig. 13. High-field CW ESR spectra ($\approx$94.2 GHz) of different spin probes in polymer films. (*a*) TEMPO in poly(butylmethacrylate) with 1% acrylic acid comonomer (PBA^{AA}). (*b*) Motional model for TEMPO in PBA^{AA}. (*c*) TEMPO-4-carboxylate in poly(fluoroalkylacrylate). (*d*) Motional model for TEMPO-4-carboxylate in poly(fluoroalkyl)acrylate.

TEMPO-4-carboxylate at the lowest water contents. Even at a water content as low as 15%, as much as 80% of the surfactant head groups are still mobile.

The spin probes that remain in the polymer film can also be used to study rewetting behavior.[27] In films that were formed and kept below the T_g, polar probes can be completely remobilized (see, e.g., filled circles in Fig. 14*b*), suggesting that in such samples there exists a percolating surfactant network in which these probes are localized. Somewhat surprisingly, no significant remobilization is observed for the surfactant probes. Probably the alkyl chains are strongly entangled with the polymer and loading of the head group region with water is not sufficient to mobilize the whole molecule. In films that were annealed above the T_g, TEMPO-4-carboxylate probes can still be remobilized completely or almost completely.[27] In contrast, the majority of 4-hydroxy-TEMPO probes is not remobilized in such samples even after keeping them fully immersed in water for 6 h (Fig. 14*b*).

If the water used for rewetting contains the weakly polar probe TEMPO, one observes its uptake by the polymer by an *increase* of the immobilized fraction with

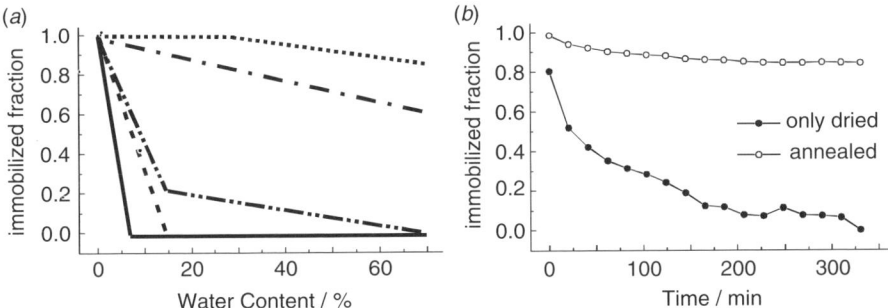

Fig. 14. Behavior of spin probes during film drying and rewetting. (*a*) Schematic plot of immobilization of different spin probes as a function of water content during drying of PBMA dispersion. Solid line: TEMPO-4-carboxylate. Dashed line: 4-hydroxy-TEMPO. Dash–dot–dot line: 5-DOXYL-stearate. Dash–dot line: 16-DOXYL-stearate. Dotted line: TEMPO. (*b*) Remobilization of TEMPOL in rewetted PBMA films (1% acrylic acid comonomer) as a function of time after immersing the film in water. The film samples were obtained by mere drying of the dispersion at ambient temperature (filled circles) or by 4 h of annealing of the dried dispersion at $T/T_g = 1.13$ (open circles).

time.[27] The increase is much faster in films formed and kept below T_g than in films annealed above T_g, thus confirming the hypothesis of a percolating surfactant network, and a large inner surface of the former films.

4.2. Ionomers

Established techniques for polymer characterization, such as small-angle scattering, have strongly contributed to our understanding of self-organized structures in soft materials. With increasing complexity of the materials, however, the possibilities of contrast variation in these techniques may be too limited to provide a complete picture of the structure. This situation was encountered with ionically end-capped polymers (ionomers), in which the distance between ion clusters can be estimated from small-angle X-ray scattering (SAXS) data as long as they were based on homopolymers, but not when they were based on diblock copolymers.[47] In the latter case, the SAXS profiles were dominated by block copolymer peaks. With ESR, the ionic clusters can be selectively addressed by ionic spin probes, such as TEMPO-4-carboxylate, as is clearly revealed by comparing line shapes obtained in ionomers doped with this spin probe and with TEMPO.[22,47,48] The distance between ion clusters can thus be measured by applying the four-pulse DEER experiment[49] to these spin probes.

Time-domain DEER data (see Chapter 2) of the weakly polar probe TEMPO in ionomers can be fitted by an exponential decay, while data for TEMPO-4-carboxylate exhibit two additional features.[20] This result suggests that TEMPO is homogeneously distributed in the ionomers, but that the spatial distribution of TEMPO-4-carboxylate features two characteristic distances within the distance range accessible to DEER experiments (1.75–8 nm). By fitting the data with a distance distribution consisting of two Gaussian peaks, we found that the shorter characteristic distance r_1 does not vary significantly (dotted line in Fig. 16d), and ranges between 1.8 and 2.2 nm, with a width of the Gaussian peak (standard deviation) that varies somewhat more strongly between 0.3 and 0.7 nm.[21] Approximately two TEMPO-4-carboxylate molecules per ionic cluster were doped into the ionomers, so that the chances of finding the nearest-neighbor spin on the same or on a different cluster are roughly the same. Furthermore, molecular modeling suggests typical distances of 1.8 nm between probes located on the surface of the same cluster. The peak at r_1 can thus be ascribed to intracluster distances.[20] The long characteristic distance r_2 varies between 4.8 and 7.7 nm, showing a systematic dependence on the type of polymer and on the location and type of the charged groups (dashed and solid lines in Fig. 15d). For the homopolymers, this distance is between 1 and 2 nm shorter than the distance corresponding to the ionomer peak in SAXS data. Since experiments with surfactant spin probes suggest that the ionic group of the probe is located on the cluster surface,[20] this difference may be explained by the fact that DEER measures surface-to-surface distances between ionic clusters, while SAXS measures center-to-center distances. The long characteristic distance can thus be assigned to the typical intercluster distance between direct neighbor clusters.

By studying this intercluster distance r_2 as a function of the type of ionomer and on the molecular weight M, information can be obtained on the distribution of the ionic clusters in ionically functionalized microphase-separated block copolymers.[21] For monoionic diblock copolymers, r_2 increases with molecular weight (circles in Fig. 15d). In fact, data for a monoionic homopolymer (diamonds) appear to lie on the same curve. The data can be fitted by a scaling law $r_2 = 2.09\, M^{0.334}$ (solid line in Fig. 15d). Assuming that the density of the polyisoprene microphase does not depend on M, a power law $r_2 = A\, M^{1/3}$ would be expected if the ionic clusters were distributed homogeneously throughout the whole microphase. The surprisingly close agreement with this expectation suggests that the clusters are neither repelled nor attracted by the interface between the microphases.

In contrast, r_2 does not significantly depend on M for the α,ω-zwitterionic diblock copolymer, in which low molecular mass counterions have been removed by dialysis. As the cations and anions are located at the chain ends of the two different blocks, the clusters may be expected to reside in or near the interface between the microphases (Fig. 15b). As the interface area depends on the number of chains, but not their length, a constant r_2 is expected if the average number of ions per cluster is independent of M. Deviations from this simplified picture might occur due to the fuzziness of the interface between the microphases, but we failed to detect such deviations.

The mesoscopic structure of ionomers based on diblock copolymers is determined by two effects: microphase separation of the diblock copolymer and clustering of the ions. For the α,ω-zwitterionic diblock copolymers, this leads to the confinement of the ion clusters to the interface between the two microphases. Within the two microphases, polymer dynamics is different. If the fuzziness of this interface extends to lengths significantly beyond the diameter of the ion clusters, a broad distribution of rotational correlation times is expected for a spin probe attached to the cluster surface. In contrast, if the interface were infinitely sharp, a bimodal distribution is expected, with the two sharply defined mean correlation times corresponding

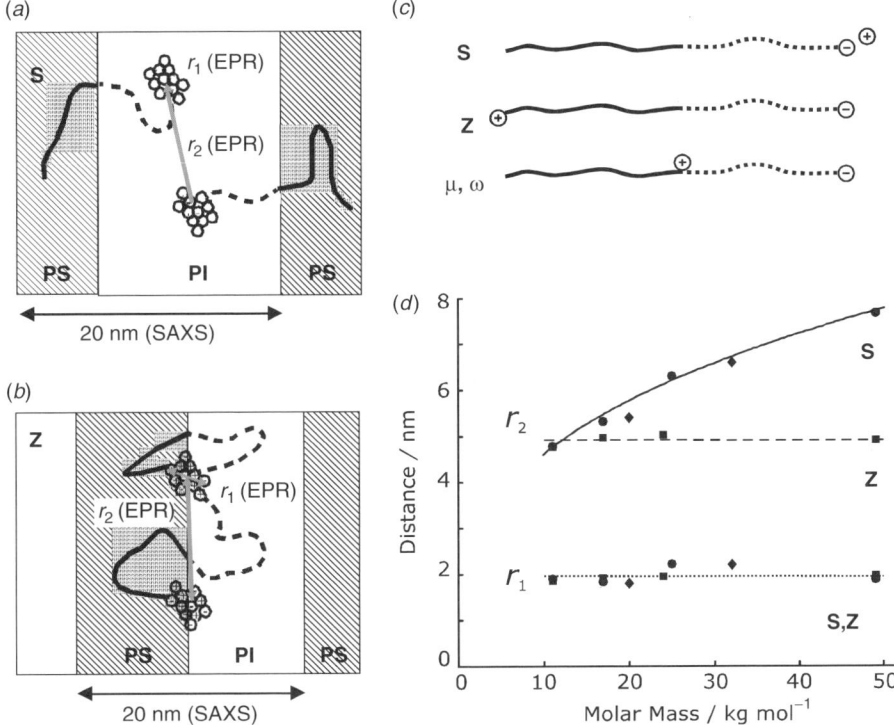

Fig. 15. Pulsed ELDOR (DEER) distance measurements on the ionic spin-probe TEMPO-4-carboxylate attached to ionic clusters in ionically modified diblock copolymers. (*a*) Schematic structure of a monoionic polystyrene–polyisoprene diblock copolymer modified by sulfonate end groups on the polyisoprene block (sample series S). (*b*) Schematic structure of an α,ω-zwitterionic polystyrene–polyisoprene diblock copolymer modified by a quaternary ammonium end group on the polystyrene block and a sulfonate end group on the polyisoprene block (sample series Z). (*c*) Schematic structures of the polymer chains. The solid line corresponds to the harder block polystyrene, the dotted line to the softer block polyisoprene. (*d*) Dependence of ionic cluster size (r_1) and intercluster distance (r_2) on molecular weight. Squares correspond to sample series Z, circles to sample series S, and diamonds to monionic homopolymers (polystyrene modified with quaternary ammonium end groups). The dotted and dashed lines are fits of a constant function. The solid line is the best-fit scaling law $r_2 = 2.09 \, M^{0.334}$.

roughly to clusters in the polystyrene and polyisoprene phase. In earlier work, it was found that the real situation is closer to the limit of a sharp interface. The spectra of α,ω-zwitterionic diblock copolymers could be fitted well by a superposition of three experimental spectra of corresponding monoionic polymers, two of which were dominating.[48] For the subspectrum with faster dynamics, a temperature shift of −20 to −30 K with respect to monoionic poly(isoprene) gave the best fits, suggesting that the polyisoprene chain is somewhat immobilized in the vicinity of the interface.

Ionic clusters in α,ω zwitterionic diblock copolymers have a preference for the diblock copolymer interface, since they are attached to both polystyrene and polyisoprene chains. However, this preference may compete with other effects on the cluster. It is therefore interesting to compare these systems with systems in which the ionic clusters are *strictly* confined to the interface. This can be achieved by locating the charge at the junction point of the two blocks. Indeed, the dynamics of TEMPO-4-carboxylate in both monoionic and μ,ω-zwitterionic diblock copolymers of this type is somewhat simpler than in the α,ω-zwitterionic diblock copolymers.[21] In the monionic and μ,ω-zwitterionic diblock copolymers the spectra can be fitted by a superposition of only two spectra without assuming a variable temperature shift, so that the only fitting parameter is the relative weight of the less mobile component. In the monionic ionomer with a quaternary ammonium group at the block junction point, this weight is 0.7 ± 0.05 throughout the temperature range from 300 to 370 K. This indicates a higher preference of the probe for the more rigid polystyrene microphase, and thus probably an exposure of a larger part of the surface of the ionic cluster to that microphase. As for the α,ω-zwitterionic diblock copolymer, the morphology is lamellar in this monoionic ionomer (see Fig. 15). In contrast, a cubic bicontinuous morphology with a curved interface between the microphases was found by SAXS measurements for the dialyzed μ,ω-zwitterionic diblock copolymer. Here, the positive charge of the quaternary ammonium group is located at the junction point, and the negative charge of the sulfonate group at the polyisoprene chain end. A back folding of this chain end to the block junction point leads to a situation where for each polystyrene chain segment that enters the ionic cluster, two polyisoprene chain segments must enter the cluster. In this case, the weight of the less mobile component is only 0.28 ± 0.04 throughout the temperature range 300–370 K. This result is in line with the expectation that the cluster interface is more strongly exposed to the polyisoprene microphase.

This analysis of CW ESR line shapes is well suited for studying large-amplitude rotational dynamics of spin probes, that is, to characterize dynamics in the vicinity of or above the T_g of a polymer. For the TEMPO-4-carboxylate spin probe attached to ionic clusters, rigid-limit spectra are observed at temperatures of 300 K and below. The dynamics of this probe below the T_g transition temperature is limited to small-angle librations that do not influence the CW ESR line shape significantly. Thus, for the glassy state of the ionomers no information can be obtained from line shape analysis. However, the small-angle librations influence several processes that contribute to spin relaxation.[23] A systematic study of relaxation times can thus provide information on dynamics in the glassy state. Such a study benefits from the enhanced orientation selection that can be obtained in high-field ESR (see Chapter 1), since the orientation dependence of the relaxation time provides an additional test for motional models. To simplify matters, we restricted ourselves to monoionic ionomers based on homopolymers or diblock copolymers. These materials exhibit a monomodal, narrow distribution of rotational correlation times at higher temperatures. We can thus make the working assumption that each motional process can be described by a single rotational diffusion tensor at lower temperatures as well.

The relaxation times measured by different electron spin echo experiments are influenced by motion on different time scales.[23] As the longitudinal relaxation time T_1 obtained by saturation recovery experiments is governed by spectral density at the electron Zeeman frequency of ≈ 95 GHz, it is sensitive to fast libration on a picosecond time scale. In contrast, the transverse relaxation time obtained from decay of a two-pulse echo is governed by motional processes on a microsecond time scale. An intermediate time scale of some hundreds of nanoseconds can be accessed by measuring the decay of the stimulated echo with respect to the second interpulse delay, which is influenced by spectral diffusion.

All these experiments were performed on TEMPO-4-carboxylate in polyisoprene homopolymer with a sulfonate end group, a molar mass of 10 kg mol^{-1}, and T_g = 281 K. Since analysis of the CW ESR line shape showed that the reorientation angle of the probe on a time scale of 50 ns is <10°, analysis of the relaxation times obtained by the saturation recovery and stimulated echo decay experiments could be restricted to small-angle motion. For the two-pulse echo experiment, the highly significant orientation dependence of T_2 excludes large-angle reorientation. All motional processes detected by the relaxation experiments can thus be considered as small-angle jumps of the probe in a cage formed by the surrounding polymer chains and the ionic cluster.

In saturation recovery experiments, the decay is nearly exponential and can be observed up to times of 150–250 μs at 180 K. Data obtained at the x and y positions in the spectrum virtually coincide, whereas T_1 is two times as long at the z position. Furthermore, T_1 along z is virtually independent of temperature between 180 and 295 K, while T_1 along x and y decreases with increasing temperature above T_g. This suggests that the cage opens up above T_g. The relaxation times observed at 295 K can be fitted by assuming wobbling in a cone with maximum polar angle of 9° and a rotational diffusion coefficient of 3.3 Gs^{-1}. Fits with larger cone angles are possible, but are inconsistent with the CW ESR line shape.

The two-pulse echo decay is exponential within experimental accuracy at all temperatures and orientations. The transverse relaxation times T_2 range between 270 and 650 ns at temperatures between 200 and 295 K. Below 200 K, T_2 is increasingly influenced by freezing of the methyl group rotation of the probe molecule. The effect of this process on T_2 is negligible at and above 260 K; between 90 and 180 K it leads to such a fast transverse relaxation that echo experiments are virtually impossible. In contrast to the orientation dependence of T_1, the one of T_2 does not exhibit axial symmetry. The T_2 relaxation time is maximum along z, but the difference of the T_2 value along z and at the x or y orientations is not as remarkable as for T_1, and is generally comparable to the difference between T_2 values at the x and y orientations. By assuming anisotropic free diffusion with an *axial* diffusion tensor, the data at 295 and 260 K can be fitted quite well. The nonaxiality of T_2 stems from the nonaxiality of the g-tensor and thus of the resonance frequency change for a given reorientation angle. Along the z axis of the molecular frame, the rotational diffusion coefficient is 3200 s^{-1} at 260 K and 9600 s^{-1} at 295 K, whereas along the x and y axes, diffusion is slower (1350 s^{-1} at 260 K and 1960 s^{-1} at 295 K). The experimental data could also be fitted by assuming restricted diffusion in a cone, which leads to higher diffusion coefficients.

However, this model is inconsistent with both the CW ESR spectra and data from stimulated echo decay.

Stimulated echo decay probes a longer time scale than two-pulse echo decay, since in the former experiment spins loose their memory with the longer longitudinal relaxation time T_1, while in the latter experiment they loose memory with the shorter transverse relaxation time T_2. In the ionomers, we observed clearly bimodal decay behavior, with the fast component having a similar time scale as T_2. The slow component has a time scale well separated from T_2. The contribution of spin–lattice relaxation can be removed numerically from the data, as T_1 is known from the saturation recovery experiments. The remaining slow component of the decay can thus be interpreted solely in terms of spectral diffusion. The dynamical process responsible for this spectral diffusion must be different from the one contributing to the decay of the two-pulse echo, since the temperature dependence is different. Indeed, theoretical considerations show[23] that a single jump of the probe orientation is sufficient to cancel its contribution to the stimulated echo. In contrast to CW ESR, saturation recovery, and two-pulse echo decay experiments, the stimulated echo is thus rather insensitive to the magnitude of the reorientation angle.

Taken together, the relaxation data demonstrate that there is a hierarchy of dynamic processes. The spin probe undergoes fast intramolecular librations on the time scale of a few picoseconds, experiences local rearrangement of the cage on the time scale of about hundreds of nanoseconds, and performs cooperative reorientation over time scales comparable to or longer than several microseconds in the vicinity of the glass transition.

These findings can be used to interpret differences in the relaxation behavior of TEMPO-4-carboxylate between a polyisoprene homopolymer with sulfonate end groups and a polystyrene–polyisoprene diblock copolymer with sulfonate end groups on the styrene chain. Although the ionic clusters in both cases are situated in a polyisoprene environment, spin–lattice relaxation[24] and transverse relaxation[22] are slightly, but significantly, slower in the diblock copolymer. Such an effect is also observed for the decay of the stimulated echo along the x and y axes of the molecular frame.[22] This implies that the dynamical constraints imposed on the polyisoprene chains by attachment to the more rigid polystyrene block propagate over the whole chain consisting of ≈ 170 monomer units.

5. CONCLUSIONS

Electron spin resonance spectroscopy on spin probes and labels can provide a wealth of information on macromolecular materials. Liquid solutions, soft matter, and glassy systems are accessible to this approach. The particular strength of such methods is their ability to selectively address sites of interest in complex materials. To distinguish between properties of the probe and of the investigated material and thus to avoid overinterpretation of the results, it is necessary to understand in some detail which processes influence probe or label dynamics, and which driving forces are responsible for probe attachment. This requires a reasonably large set of data acquired under different conditions or in materials with well-defined changes in composition. Because of its sensitivity and relative simplicity, CW ESR at X-band frequencies is the method of choice to obtain these data. In CW ESR, the main effort is usually not measuring the spectra, but analyzing them in a way that provides maximal reliable information.

After acquisition, analysis, and interpretation of the X-band CW ESR data there are usually open questions that cannot be answered with this technique. In some cases, the enhanced sensitivity of high-field CW ESR to small changes in dynamics can provide the answers. In many cases, application of pulsed ESR techniques for measuring relaxation times, detecting proximity to certain types of nuclei in the vicinity of the spin probe or label, and for measuring distance distributions are required to obtain a more complete picture. Such a combined approach can provide a detailed understanding of the site of interest on length scales between 0.2 (van der Waals contact) and 8 nm, and on time scales between a few picoseconds and several microseconds.

ACKNOWLEDGMENTS

Many of the ideas and most of the experimental work were contributed by my former and present Ph. D. students M. Pannier, S. E. Cramer, E. Narr, D. Hinderberger, S. Schleidt, and Q. Mao. The macromolecular materials were due to A. Godt, S. Veit, U. Wiesner, V. Schädler, M. Schöps, X. X. Zhu, M. Krause, M. Rehahn, and O. Schmelz. Long-standing collaborations with H. W.Spiess and D. Leporini contributed insights into polymer dynamics. A Dozentenstipendium by Fond der Chemischen Industrie is gratefully acknowledged.

REFERENCES

1. Hoffmann, A.K.; Henderson, A.T. *J. Am. Chem. Soc.* **1961**, *83*, 4671.
2. Rozantsev, E.G.; Neiman, M.B. *Tetrahedron* **1964**, *20*, 131.
3. Berliner, L.J., Ed. *Spin Labeling: Theory and Applications*, Academic Press: New York, 1976.
4. Berliner, L.J., Ed. *Spin Labeling II: Theory and Applications*, Academic Press: New York, 1979.
5. Berliner, L.J., Ed. *Biological Magnetic Resonance, Vol. 14: Spin Labeling: The Next Millenium*, Plenum Press: New York, 1998.
6. Boyer, R.F., Keineth, S.E., Eds. *Molecular Motions in Polymers by E.S.R.*, Symposium Series Vol. 1; MMI Press: Harwood, Chur, 1980.
7. Cameron, G.G.; Davidson, I.G. In *Polymer Spectroscopy*, Fawcett, A.H., Ed., John Wiley d Sons, Inc.: Chichester, UK, 1996, Chapt. 9, pp. 231–252.
8. Wassermann A.M.; Khazanovich, T.N. In *Polymer Yearbook 12*, Pethrich, R.A., Ed. Harwood, New York, 1995, p. 153–185.
9. Jeschke, G. *Macromol. Rapid. Commun.* **2002**, *23*, 227.

10. Schneider, D.J.; Freed, J.H. In *Biological Magnetic Resonance*, Vol. 8, Berliner, J.H.; Reuben J., Eds.; Plenum Press: New York, 1989, pp. 1–76.
11. Molin, Yu.N.; Salikohov, K.M; Zamaraev, K.I. *Spin Exchange. Principles and Applications in Chemistry and Biology*, Springer: Berlin, 1980.
12. Marsh, D.; Kurad, D; Livshits, V.A. *Chem. Phys. Lipids* **2002**, *116*, 93.
13. Schweiger, A.; Jeschke, G. *Principles of Pulse Electron Paramagnetic Resonance*, Oxford University Press: Oxford, 2001.
14. Hinderberger, D.; Jeschke, G.; Spiess, H.W. *Macromolecules* **2002**, *35*, 9698.
15. Hinderberger, D.; Spiess, H.W.; Jeschke, G. *Macromol. Symp.* **2004**, *211*, 71.
16. Hinderberger, D.; Spiess, H.W.; Jeschke, G. *J. Phys. Chem. B* **2004**, *108*, 3698.
17. Hinderberger, D.; Jeschke, G.; Spiess, H.W. *Colloid Polym. Sci.* **2004**, *282*, 901.
18. Hinderberger, D.; Schmelz, O.; Rehahn, M.; Jeschke, G. *Angew. Chem. Int. Ed.* **2004**, *43*, 4616.
19. Hinderberger, D.; Spiess, H.W.; Jeschke, G. *Europhys. Lett.* **2005**, *70*, 102.
20. Pannier, M.; Schädler, V.; Schöps, M.; Wiesner, U.; Jeschke, G.; Spiess, H.W. *Macromolecules* **2000**, *33*, 7812.
21. Pannier, M.; Schöps, M.; Schädler, V.; Wiesner, U.; Jeschke, G.; Spiess, H.W. *Macromolecules* **2001**, *34*, 5555.
22. Leporini, D.; Schädler, V.; Wiesner, U.; Spiess, H.W.; Jeschke, G. *J. Non-Cryst. Solids* **2002**, *307,* 510.
23. Leporini, D.; Schädler, V.; Wiesner, U.; Spiess, H.W.; Jeschke, G. *J. Chem. Phys.* **2003**, *119*, 11829.
24. Leporini, D.; Jeschke, G. *Philos. Mag.* **2004**, *84*, 1567.
25. Cramer, S.E.; Bauer, C.; Jeschke, G.; Spiess, H.W. *Appl. Magn. Reson.* **2001**, *21*, 495.
26. Cramer, S.E.; Jeschke, G.; Spiess, H.W. *Macromol. Chem. Phys.* **2002**, *203*, 182.
27. Cramer, S.E.; Jeschke, G.; Spiess, H.W. *Macromol. Chem. Phys.* **2002**, *203*, 192.
28. Cramer, S.E.; Jeschke, G.; Spiess, H.W. *Colloid Polym. Sci.* **2002**, *280*, 569.
29. Leporini, D.; Zhu, X.X.; Krause, M.; Jeschke, G.; Spiess, H.W. *Macromolecules* **2002**, *35*, 3977.
30. Jeschke, G.; Panek, G.; Schleidt, S.; Jonas, U. *Polym. Eng. Sci.* **2004**, *44*, 1112.
31. Godt, A.; Jeschke, G. *Magn. Reson. Chem.*, **2005**, *43*, S110.
32. Hawker, C.J.; Bosman, A.W.; Harth, E. *Chem. Rev.* **2001**, *101*, 3661.
33. Keana, J.W.F. In: *Spin Labeling II: Theory and Applications*, Berliner, L.H., Ed. Academic Press: New York, 1979, pp. 115–172.
34. Owenius, R.; Engstrom, M.; Lindgren, M.; Huber, M. *J. Phys. Chem. A* **2001**, *105*, 10967.
35. Plato, M.; Steinhoff, H.J.; Wegener, C.; Torring, J.T.; Savitsky, A.; Mobius, K. *Mol. Phys.* **2002**, *100*, 3711.
36. Faetti, M.; Giordano, M.; Leporini, D.; Pardi, L. *Macromolecules* **1999**, *32*, 1876.
37. Rubinstein, M.; Colby, R.H. *Polymer Physics*, Oxford University Press: Oxford, 2003.
38. Bray, R.C. *Biochem. J.* **1961**, *81*, 189.
39. Bray, R.C.; Peterson, R. *Biochem. J.* **1961**, *81*, 194.
40. Manning, G. *Acc. Chem. Res.* **1979**, *12*, 443.
41. Odijk, T. *J. Polym. Sci., Polym. Phys.* **1977**, *15*, 477.

42. Skolnick, J.; Fixman, M. *Macromolecules* **1977**, *10*, 944.
43. Förster, S.; Schmidt, M.; Antonietti, M. *J. Phys. Chem.* **1992**, *96*, 4008.
44. Hinderberger, D., Ph. D. thesis, University of Mainz, 2004.
45. Deshkovski, A.; Obukhov, S.; Rubinstein M. *Phys. Rev. Lett.* **2001**, *86*, 2341.
46. Jeschke, G.; Godt, A. *ChemPhysChem* **2003**, *4*, 1328.
47. Schädler, V.; Franck, A.; Wiesner, U.; Spiess, H.W. *Macromolecules* **1997**, 30, 3832.
48. Schädler, V.; Kniese, V.; Thurn-Albrecht, T.; Wiesner, U.; Spiess, H.W. *Macromolecules* **1998**, *31*, 4828.
49. Pannier, M.; Veit, S.; Godt, A.; Jeschke, G.; Spiess, H.W. *J. Magn. Reson.* **2000**, *142*, 331.

8

ESR METHODS FOR ASSESSING THE STABILITY OF POLYMER MEMBRANES USED IN FUEL CELLS

EMIL RODUNER
University of Stuttgart, Stuttgart, Germany

SHULAMITH SCHLICK
University of Detroit Mercy, Detroit, Michigan

Contents
1. Introduction and Motivation — 198
2. Fuel Cell Processes and Their Effects on Membrane Stability — 201
3. Direct ESR and Spin Trapping as Tools for the Detection of Radical Intermediates — 204
4. Study of Membrane Stability by ESR — 206
   4.1. Reaction of Oxygen Radicals with Model Compounds in Solution — 206
   4.2. Radical Species in Nafion Membranes Treated in Fenton Media — 210
      4.2.1. Oxygen-Centered Radicals — 211
      4.2.2. Fluorinated Alkyl Radicals — 214
   4.3. Membrane-Derived Fluorinated Radicals in UV-Irradiated Nafion and Dow Ionomers: Effect of Counterions and H_2O_2 — 215
      4.3.1. Fluorinated Radical Fragments — 217
   4.4. *In Situ* ESR Experiments — 222
5. Conclusions and Prospects — 225
Acknowledgments — 226
References — 226

Advanced ESR Methods in Polymer Research, edited by Shulamith Schlick.
Copyright © 2006 John Wiley & Sons, Inc.

1. INTRODUCTION AND MOTIVATION

"I believe that water will one day be used as a fuel, that the hydrogen and oxygen of which it is constituted will be used, simultaneously or in isolation, to furnish an inexhaustible source of heat and light, more powerful than coal can ever be."

Jules Verne, 1874[1]

In recent years, the science and technology of fuel cells have captivated the attention of scientists and engineers, because of their potential as an alternative source of energy. An important moment in the development of fuel cell (FC) technology was the Ballard bus program, which culminated in 1998 when three buses powered by H_2-fueled FC stacks rolled on Chicago streets and became part of the public transport system.[2] As an antidote to general skepticism and a strong signal to the automotive, insurance, and oil companies, this achievement showed that FCs may power our future. The US government expressed its commitment at the 2003 US State of the Union address by establishing the Hydrogen Fuel Initiative and FreedomCAR Partnership, with the intention to develop "clean, hydrogen-powered automobiles." At the present time, most major automobile manufacturers in the world are working on the development of fuel cell vehicles.[3] More exotic ideas on hydrogen production and storage are proposed and tried, for example, the development of hydrogen-producing organisms.[4] Although the verdict in the Introduction to a series of articles in *Science* on the topic of the hydrogen economy was "Not so simple,"[5] the problem is too urgent, and the solution too promising to be abandoned. Massive introduction of the new technology is expected to transform the way we live as individuals and as a community. Major advantages are less dependence on dwindling fossil energy sources, especially oil reserves, with ramifications to security and the local economies; and a cleaner environment, because of zero emission of soot and hydrocarbons, and near zero emission of carbon monoxide (CO) and nitrogen oxides, NO_x.[2,3,5] The modern FC and the drive to a hydrogen economy are anchored on the 1839 discoveries of Schönbein[6] and Grove[7] that oxygen and hydrogen can react to produce electricity. Figure 1 illustrates Grove's experiment and the use of the energy produced in the reaction to decompose water into its elements.

A modern FC used in transportation and other applications is shown in Fig. 2.[8] Its key elements are the electrodes, the catalyst, and the proton exchange membrane (PEM); the cell is fueled by hydrogen or methanol at the anode and oxygen or air at the cathode. The membrane electrode assembly (MEA) that is the heart of FCs includes the proton exchange membrane, a polymer modified to include ions, typically sulfonic groups: an *ionomer*.[9] In the presence of water, ionomers self-assemble into microphase separated domains that allow the movement of H^+ in one direction only, from the anode to the cathode. The membrane performance was first demonstrated by Nafion, the ionomer made by DuPont, which consists of a perfluorinated backbone and pendant chains terminated by sulfonic groups, $-SO_3^-$. Nafion was the major component in the PEMFC developed by General Electric for

INTRODUCTION AND MOTIVATION

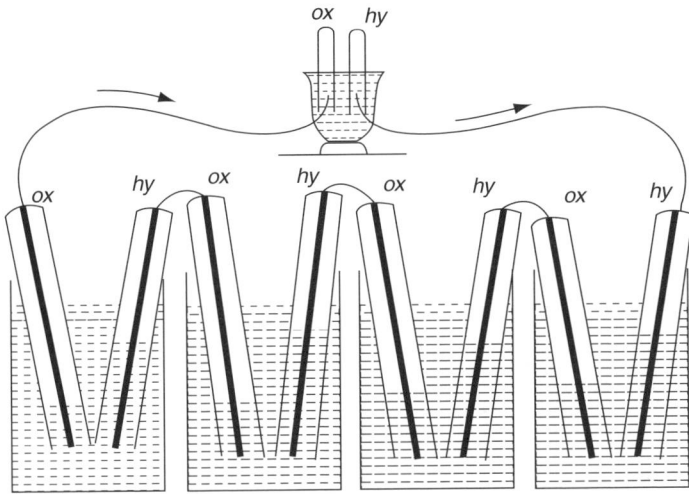

Fig. 1. Grove's experimental "gas voltaic battery" (1842). Oxygen and hydrogen are in contact with a Pt filament in the lower reservoirs and react in the sulfuric acid solution to form water. The electrical current is used to electrolyze water to oxygen and hydrogen in the upper tube[7b] (website http://fuelcells.si.edu/origins/orig1.htm, accessed 15 February 2005).

the NASA Gemini mission in the 1960s, to provide both energy and drinking water for space vehicles;[10] its conductive properties, and those of the similar Dow membrane shown below, depend on the equivalent weight (EW), which is the mass that contains 1 mol of sulfonic groups. The original FC was built around Pt as the catalyst and Nafion as the PEM, and operated at $\approx$ 80°C. In many fuel cell applications, Nafion is the membrane of choice. However, if function is the most important consideration in the space program, cost is a decisive factor in the development of fuel cells as an alternative to the internal combustion engine. At the present Nafion price, there is great motivation for the development of alternative membranes.[11] Additional objectives are improving the catalytic step, especially at the cathode; identifying nonprecious metal catalysts as an alternative to Pt; and reducing "poisoning" of the anode due to CO during processing of methanol and water into H_2, CO, and CO_2 at 280°C.

Perfluorinated Membranes, in Acid Form

$$-(CF_2CF_2)_m CF_2CF- \\ | \\ OCF_2CFOCF_2CF_2SO_3H \\ | \\ CF_3$$

$$-(CF_2CF_2)_n CF_2CF- \\ | \\ OCF_2CF_2SO_3H$$

Nafion, EW 1100 (m = 6.5)   **Dow, EW 800 (n = 5.2)**

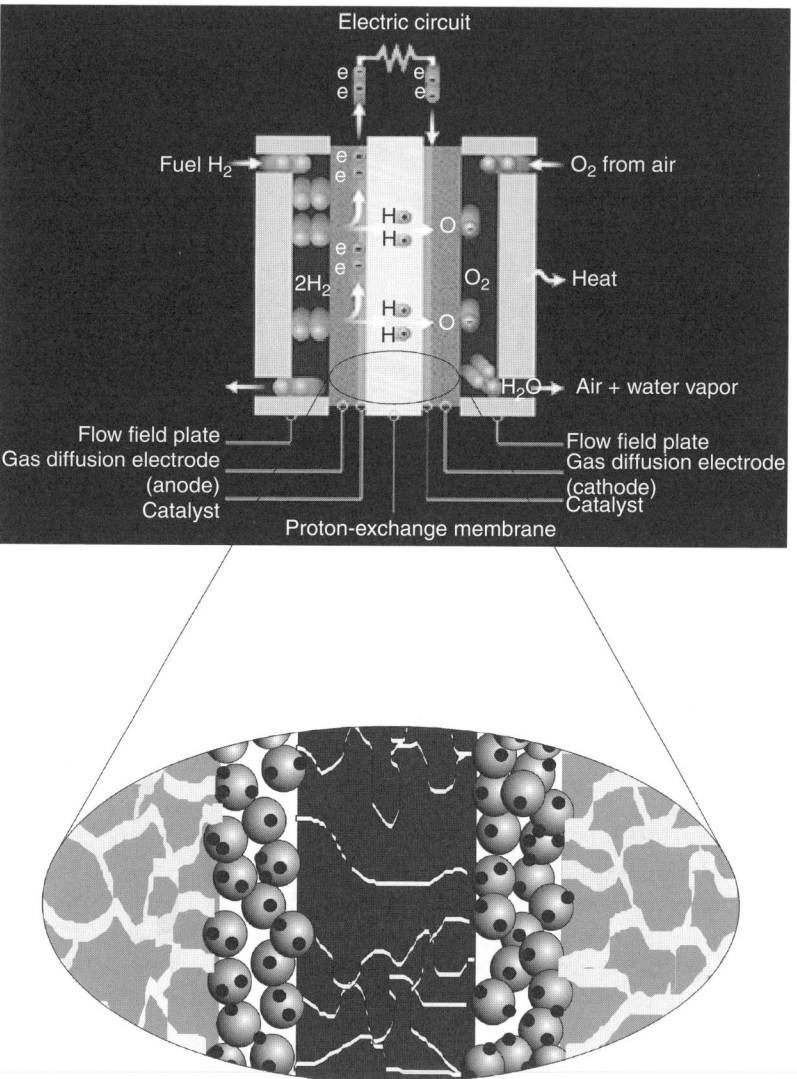

Fig. 2. Typical membrane-electrode assembly in a proton exchange membrane fuel cell (PEMFC).[8] The enlarged part shows that on a microscopic scale the gas diffusion layers, the electrodes and the membrane are inhomogeneous, leading to restrictions for the transport of gases, electrons, and protons. Under heavy loads these conditions cause large inhomogeneities in temperature, pH, and electrochemical potential, which are highly significant for membrane degradation processes.

Important advantages in terms of water management and CO tolerance in fuel cells can be gained by operating at higher temperatures, typically ≈120°C; this operating temperature imposes additional stringent requirements in terms of membrane stability in a highly oxidative environment. Even at an operating temperature of 50°C, evidence for the deterioration of Nafion membranes was detected by X-ray powder diffraction (XRD) and X-ray photoelectron spectroscopy (XPS); moreover, it was reported that the type of structural damage is different at the anode and cathode sides.[12] Currently, membranes do not have the lifetime required for commercial automotive applications; thus, their durability has become a major issue and obstacle in the wide adoption of fuel cell-powered vehicles. Work in various laboratories has demonstrated the involvement of reactive oxygen species, such as hydrogen peroxide and short-lived oxygen radicals in the degradation of Nafion and other perfluorinated membranes, leading ultimately to membrane thinning and hole formation. These reactive intermediates can be produced in the laboratory, and testing protocols have been developed. Currently, however, lifetimes of PEMs deduced from laboratory tests are longer compared to those of the same membranes in actual FC conditions, but present protocols are suitable for a relative ranking of membranes in terms of their stability.

Current studies in our laboratories are focused on the chemical degradation of membranes, which is assumed to occur by an oxidative process involving free radicals. Major objectives are to determine the presence and identify the structure of radicals produced as a result of specific membrane treatments, to develop laboratory protocols for the study of membrane degradation that correlate well with the extent of degradation produced during automotive operating conditions, and finally to understand the degradation mechanism. Electron spin resonance (ESR) methods are particularly suitable for this investigation, because of their sensitivity and specificity for the detection of radical species.

This chapter describes the reactions that occur in a FC environment; assesses various methods used to produce oxygen-centered radicals that decrease the membrane stability; and summarizes progress made in our laboratories in the detection (directly by ESR or via spin trapping) of oxygen radicals and membrane-derived fragments.

2. FUEL CELL PROCESSES AND THEIR EFFECTS ON MEMBRANE STABILITY

Hydrogen in a PEMFC and methanol in a direct methanol fuel cell (DMFC) are oxidized at the anode. The protons diffuse through the ionomer membrane, and the electrons circulate through the load where their electrical energy is used. The net reactions for the four-electron reduction of oxygen are $2H_2 \rightarrow 4H^+ + 4e^-$, and $O_2 + 4H^+ + 4e^- \rightarrow 2H_2O$. In practice, the process is more complicated, and the catalyst is involved in the formation of numerous adsorbed oxygen species. The following mechanism has been suggested.[13]

$$Pt + O_2 \rightarrow Pt-O_2 \qquad (1)$$

$$Pt-O_2 + H^+(aq) + e^- \rightarrow Pt-OOH \qquad (2)$$

$$Pt-OOH + H^+(aq) + e^- \rightarrow Pt-O + H_2O \qquad (3)$$

$$Pt-O + H^+(aq) + e^- \rightarrow Pt-OH \qquad (4)$$

$$Pt-OOH + H^+(aq) + e^- \rightarrow Pt-(OHOH) \qquad (5)$$

$$Pt-(OHOH) + H^+(aq) + e^- \rightarrow Pt-OH + H_2O \tag{6}$$

$$Pt-OH + H^+(aq) + e^- \rightarrow Pt-OH_2 \tag{7}$$

Adsorbed hydroperoxyl radicals HOO• (Reaction 2), hydroxyl radicals HO• (reactions 4 and 6), and hydrogen peroxide (Reaction 5) are expected intermediates of oxygen reduction. If these reactants are desorbed from the catalyst surface before full reduction to water, they reach, and react with, the membrane. Under vacuum conditions, HO• radicals are predicted to be adsorbed strongly, with an adsorption energy of 2.23 eV on Pt(111), and even higher on other crystal faces.[14] Coadsorbed water provides additional stabilization for surface-adsorbed species, but small clusters of bulk water also stabilize the desorbed radicals. The water balance near the electrode is therefore crucial for the adsorption equilibrium of reaction intermediates. Hydrogen peroxide is expected to be unstable at dry Pt surfaces and to dissociate into hydroxyl radicals. On the catalyst surface, the driving force is provided by their large binding energies, while the homolytic dissociation of free H_2O_2 into two HO• radicals is endothermic with $\Delta H = 214$ kJ mol^{-1}. The only HOO• configuration predicted to be stable is the bridge position between two Pt atoms on a (111) surface.[14]

A reaction that interferes with the above redox system is the oxidation of water molecules,

$$Pt-OH_2 \rightarrow Pt-OH + H^+ + e^- \tag{8}$$

which provides an additional source of adsorbed hydroxyl radicals.[15] The above mechanism explains the formation of reactive oxidative species at the cathode. In agreement with this prediction, hydrogen peroxide has been detected in FC product water.[16,17]

Other experiments, however, gave evidence that the main membrane degradation is associated with loss of sulfonic acid groups and occurs at the anode.[18] The mechanism involves oxygen diffusion through the membrane (crossover), as follows:

$$H_2 \rightarrow 2H\bullet \text{ (anode Pt catalyst)} \tag{9}$$

$$H\bullet + O_2 \text{ (from crossover)} \rightarrow HOO\bullet \tag{10}$$

$$HO_2\bullet + H\bullet \rightarrow H_2O_2 \tag{11}$$

$$H_2O_2 + Me^{n+} \rightarrow Me^{(n+1)+} + HO\bullet + HO^- \tag{12}$$

$$HO\bullet + H_2O_2 \rightarrow H_2O + HOO\bullet \tag{13}$$

The Me^{n+} transition metal ion can be oxidized further, for example, Fe(II), Cu(I), or Ti(III), and can activate oxidative cleavage of hydrogen peroxide, as in the Fenton mechanism. It was suggested that the hydroperoxyl radical, HOO•, is the species that attacks the membrane.[18] As will be clearly seen below, some of the transition metal ions may react directly with the sulfonic groups of the membrane, thus providing an additional degradation pathway.

There is general agreement that high temperature, drying out of the membrane, and the presence of transition metal ions represent unfavorable conditions, which facilitate membrane degradation and induce early failure. Similarly, start–stop cycles are known to result in notable degradation;[19] little attention has been paid, however, to the electrochemical running condition of the cell and to the consequences of current–voltage cycling, even though it is well known in electrochemistry that binding energies and redox properties of adsorbates at the catalyst surface are strong functions of the electrochemical potential. This effect can be used to advantage by applying negative voltage pulses to oxidize adsorbed CO, which poisons the anode catalyst surface.[20] The effect has been studied further in Fourier transform infrared (FTIR) experiments of adsorbed CO,[21] and theoretically for oxygen species.[22,23] When a current is drawn from the fuel cell, the anode potential becomes more positive and the cathode potential more negative, resulting in strengthening of some bonds and weakening of others.[22] In particular, HO• radicals could desorb from the catalyst surface and damage the membrane more easily under conditions of heavy loads.

Furthermore, local inhomogeneities in current density, voltage, temperature, and pH may be greatly enhanced under conditions of high currents. These conditions lead to significant local stress that can be fatal for membrane performance, as will be justified briefly below with reference to the close-up in Fig. 2.

The fuel cell reaction that converts 1 mol of H_2 to liquid H_2O gives rise to a reaction enthalpy $\Delta H_R = -285.8$ kJ mol^{-1} and a Gibbs free energy change at standard temperature and gas pressures of $\Delta G_R = -237.2$ kJ mol^{-1}. The ΔG_R parameter is the maximum energy that is convertible to electrical work, and corresponds to a theoretical equilibrium voltage of $\Delta G_R/2F = 1.23$ V for a single cell at 298 K (F is the Faraday constant). Under practical conditions, due mostly to the high overpotential of the oxygen reduction reaction, only ≈ 0.9 V are obtained even at a very low current density. When operated at typically 0.7 V and a current density of 0.4 A cm^{-2}, the generated power density amounts to 0.28 W cm^{-2}, which is only 47% of ΔH_R; the rest (0.31 Wcm^{-2} or 152 kJ mol^{-1} of oxidized H_2) is dissipated as heat. More than 80% of the waste heat is generated at the cathode, and the rest is due to ohmic resistance of proton conduction within the membrane. The useful voltage and thus the fraction of electrically used energy drops rapidly when the current density reaches a level under which it becomes limited by hindered diffusion of the fuel gases through the porous electrode structures (above ≈ 1.6 A cm^{-2}), in particular when the pores become partially flooded by liquid water.

The above picture gives the normal macroscopic view in which the membrane is a homogeneous proton conducting continuum and both electrodes are ideally homogeneous with finely distributed catalyst grains, all in electrical contact with the electrode and located at the phase boundary to the wet membrane. In reality, the membrane is a microheterogeneous structure with hydrophobic and hydrophilic regimes, through which the protons have to find a continuous conduction path of percolated hydrophilic and hydrated domains in the membrane. A fraction of the catalyst particles is not located directly near a conduction path; moreover, catalyst particles may experience electrical contact resistance to the carbon electrode, so that their electrochemical potential is different compared to the macroscopically measured potential. The various size pockets and pores in the gas diffusion layer represent locally different diffusion resistances, and water condensation will set in first in the narrower pores. As a result, not all catalyst grains are used with the same efficiency, and the pH, the electrochemical potentials, and the actual chemistry are all local properties. As will be shown below, change of pH can result in a dramatically different chemistry of HO• radicals.

Part of the waste heat is generated initially in localized "hot spots" that can lead to transient evaporation of water, and may be the starting point for the formation of pinholes. An increased local temperature will also support desorption of adsorbed reactive intermediates from catalyst grains. In general, deviations from equilibrium are expected to be larger when higher power is drawn from the cell.

3. DIRECT ESR AND SPIN TRAPPING AS TOOLS FOR THE DETECTION OF RADICAL INTERMEDIATES

Nafion membranes have provided the standard material for low-temperature fuel cells and their stability in the presence of reactive oxygen species, such as H_2O_2 and HO•, has been investigated as a reference. The membrane has considerable stability: The C–F bond is strong, but shows a significant dependence on the local structure. The bond dissociation energy was calculated as 478, 433, and 386 kJ mol^{-1} for primary, secondary, and tertiary fluorine atoms, respectively,[24] thus providing the material with its remarkable inertness. Nevertheless, Nafion is prone to electrochemical degradation, but the mechanism is not fully understood. Product water analysis showed the presence of fluoride ions, low-molecular-weight perfluorocarbon sulfonic acid, and CO_2.[18,25] The susceptibility toward oxidative degradation of Nafion has been attributed to traces of –CHF_2 groups that are hard to avoid during synthesis, or to –$CF=CF_2$ groups formed in the termination step of polymerization. Furthermore, the tertiary fluorine atom, which is further activated by the inductive effect of the ether link of the –$OCF_2CF_2SO_3H$ side chain, is a potential site of F abstraction by HO• radicals. A radical defect on the chain can, in a domino-type process, lead to continuous disintegration by bond cleavage and/or oxidation.

In the laboratory, oxygen radicals can be produced by the Fenton reaction, where the major step is H_2O_2 + Fe(II) → Fe(III) + HO• + HO$^-$.[26] The formation of radical species from the attacked substrate, for example, RH, can occur by hydrogen-abstraction, HO• + RH → H_2O + R•; polymer-derived R• radicals can initiate a degradation cascade. Fenton reagents based on Ti(III) instead of Fe(II) have also been used.[26] Currently, the stability of FC membranes to oxidation is based on the Fenton test: The polymer is soaked in an aqueous solution of H_2O_2 (3–30%) and $FeSO_4$ (molar ratio [Fe(II)]/[H_2O_2] ≈ 1/500). Membrane stability is measured in terms of weight loss at a given treatment temperature (ambient to ≈ 80°C) as a function of immersion time, or in terms of the time required for the membrane to break or to begin dissolving.[27,28] This test provides no details on reactive intermediates, degradation mechanism, or factors that may affect the extent of structural damage.

An additional method for producing oxygen radicals in the laboratory is photolysis of aqueous solutions of H_2O_2,[29] and the main expected reactions are

$$H_2O_2 \rightarrow 2HO\bullet \quad (14)$$

$$HO\bullet + H_2O_2 \rightarrow HOO\bullet + H_2O \quad (15)$$

$$HOO\bullet + H_2O \rightarrow O_2^{\bullet-} + H_3O^+ \quad (16)$$

DOO• and DO• radicals are expected in ultraviolet (UV)-irradiated H_2O_2/D_2O solutions.

The objectives of our recent studies on membrane stability are to achieve a better understanding of the degradation mechanism in PEMs by developing specific treatment protocols in the laboratory. In our laboratories, ESR spectroscopy is used to identify reactive oxygen intermediates as well as membrane fragments formed during the degradation process. The major difficulties in this approach are the high reactivity and short lifetimes of the radicals; low temperatures, and/or spin trapping techniques can be used to improve the Boltzmann factor, increase the lifetimes, and capture radicals as they form.

The nature of unstable intermediates in Nafion and in self-assembled systems in general was interrogated in early work in our laboratory. The formation of the superoxide radical anion, $O_2^{\bullet-}$, was detected in dry Nafion fully exchanged by Ti(III) ions and exposed to oxygen, and was examined as a function of exposure time to oxygen.[30] The radicals are stable at and < 300 K, and decrease in intensity > 310 K. Interestingly, the intensity of $O_2^{\bullet-}$ radicals increased when the membrane protons were exchanged by a mixture of Al(III) and Ti(III) cations. This result was an early indication that the formation of reactive intermediates is sensitive to the specific nature of counterions in the membranes.[30] In addition, the presence of oxygen radicals in UV-irradiated aqueous H_2O_2 was confirmed in our laboratory by direct ESR, using a combination of UV-irradiation and ESR measurements at low temperature, typically 77 K, followed by gradual annealing of irradiated samples for short intervals (≈ 3 min) > 77 K; in this way ESR signals from radicals HO•, DO•, HOO•, DOO•, and $O_2^{\bullet-}$ have been detected.[31] The radicals mentioned above were identified by their magnetic parameters (g-values, hyperfine interactions, line widths, and line shapes).

In some experiments, the reactive intermediates are too short lived and cannot be detected directly. In such cases, spin trapping can be used to transform the unstable intermediates into long-lived radicals that can be studied by ESR.[32] The method is based on the reaction of a diamagnetic molecule (the "spin trap") with a short-lived, reactive free radical R• to form a stable radical, the "spin adduct". In most cases, the adduct is a nitroxide radical, as shown below for nitroso (a) and nitrone (b) spin traps.

Spin Trapping by Nitroso and Nitrone Spin Traps

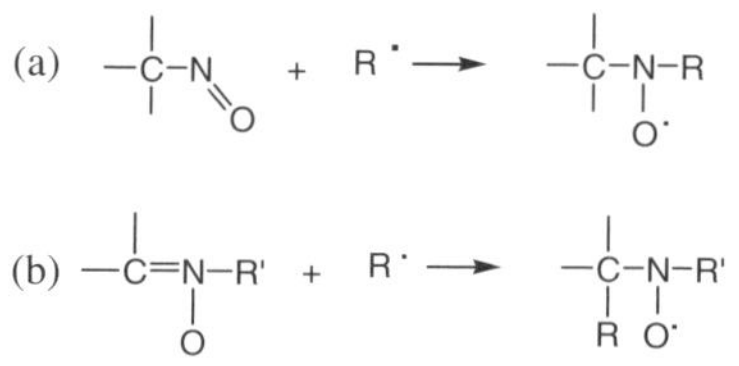

(a) $-\underset{|}{\overset{|}{C}}-N\diagdown_O \; + \; R^\bullet \; \longrightarrow \; -\underset{|}{\overset{|}{C}}-\underset{\overset{|}{O^\bullet}}{N}-R$

(b) $-\underset{\overset{|}{O}}{\overset{|}{C}}=N-R' \; + \; R^\bullet \; \longrightarrow \; -\underset{\overset{|}{R}}{\overset{|}{C}}-\underset{\overset{|}{O^\bullet}}{N}-R'$

Two common spin traps, α-phenyl-*tert*-butylnitrone (PBN), and 5,5-dimethylpyrroline-*N*-oxide (DMPO), are shown below.

Spin Traps α-Phenyl-*tert*-butylnitrone, and 5,5-Dimethylpyrroline-*N*-oxide

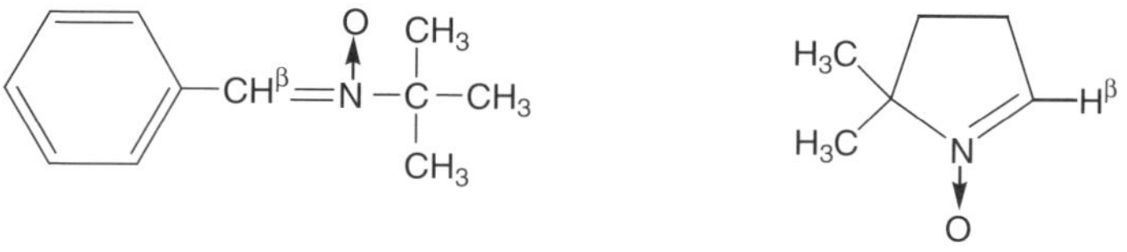

PBN DMPO

The corresponding nitroxide spin adducts exhibit hyperfine splittings from the ^{14}N nucleus and the H_β proton. It is usually easy to decide if a short-lived radical is present, and more of a challenge to identify the radical. Alkoxy, alkyl, and oxygen radicals, such as HO•, HOO•, and $O_2^{\bullet-}$ have been detected by this approach.[32–36] The spin trapping database is a useful guide.[36]

Recent results obtained by direct ESR spectroscopy and by spin trapping will be described in Section 4.

4. STUDY OF MEMBRANE STABILITY BY ESR

This section describes: (*1*) The effect of HO• radicals, formed by UV irradiation of H_2O_2, on low molecular weight model sulfonated compounds at ambient temperature as a function of pH in a flow system; (*2*) the detection and identification of radicals in Nafion membranes exposed to the Fenton reagent based on Ti(III) ($TiCl_3$ + H_2O_2); (*3*) the identification of membrane-derived radical fragments in UV-irradiated Nafion exchanged by Cu(II), Fe(II), and Fe(III), including membranes exposed to H_2O_2 solutions prior to UV irradiation; and (*4*) spin trapping experiments in a fuel cell that was inserted in the ESR resonator.

4.1. Reaction of Oxygen Radicals with Model Compounds in Solution

Some of the membranes considered as alternatives to Nafion contain aromatic building blocks, in particular sulfonated phenyl groups or grafted polystyrene side chains. For this reason, toluene sulfonic acid was chosen as a model compound to investigate its reactions with HO• radicals as a function of pH. The HO• radicals were generated by direct photolysis of liquid aqueous solutions of H_2O_2 in the presence of p-toluene sulfonic acid in a flow cell in the resonator of the ESR spectrometer.[37] Additional species, such as HOO•, $O_2^{•-}$ at higher pH, and $O^{•-}$, also play important roles; moreover, small amounts of •SO_3^- derived from the acid group were also detected. The reaction products were detectable and the results are presented in Fig. 3. The chemistry depends strongly on pH, a fact that is highly significant in view of the possible pH inhomogeneities in fuel cells at high current. The intermediate pH regime is dominated by the cyclohexadienyl radical as a product of HO• addition to the ring. The benzyl radical is thought to arise at low pH by acid-catalyzed H_2O elimination from the cyclohexadienyl radical, a reaction that requires a hydrogen atom at the α-carbon in the aromatic ring. At high pH, the elimination of water may be base catalyzed; alternatively, the benzyl radical may be formed via direct methyl hydrogen abstraction by $O^{•-}$.

It is remarkable that the formation of phenoxyl radicals was observed only in the pH range close to the pK_a of HO• radicals (pK_a 11.9) and of H_2O_2 (pK_a 11.7). As shown in Fig. 4, this behavior reflects the relative rate of formation of the superoxide radical anion, $O_2^{•-}$, which is enhanced by the facile hydrogen abstraction from HO_2^- by HO•, but suppressed at higher pH when HO• is deprotonated.[37] Furthermore, singlet oxygen is formed from $O_2^{•-}$ by reaction with HO• or with $O^{•-}$; the latter is another very reactive species that can lead to severe membrane damage via various routes, provided these extreme pH conditions are reached.

Another instructive example of oxidative degradation in substituted aromatic compounds is given in Fig. 5,[37,38] which demonstrates that semiquinone radicals are obtained not only from toluene sulfonic acid, but also from hydroxytoluene sulfonic acid:

Formation of Semiquinone Radicals by Reaction with HO• Radicals

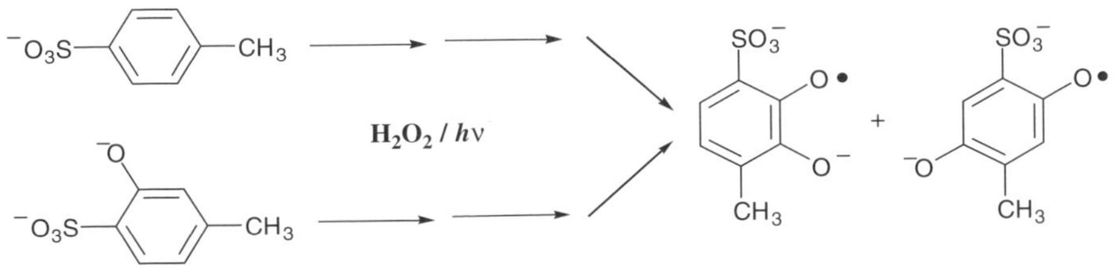

The result implies that in this case oxygenation occurs in increments of atomic rather than molecular oxygen.

Aromatic groups in fuel cell membranes are often linked via ether bridges. Therefore, it is significant that for pH > 10 phenoxy-type radicals are observed, which do not show any splitting attributable to the –OCH_3 group. Obviously, the ether group is not stable under these conditions and becomes substituted by HO•.

The main conclusion drawn from these experiments is that HO• radicals are extremely reactive, and expected to react rapidly with aromatic hydrocarbons. The positions susceptible to attack are shown below for various fuel cell membranes. The preferential reaction is by addition to the aromatic ring, in particular in the ortho positions to activating substituents, such as alkyl or alkoxy groups. The para position

(a)

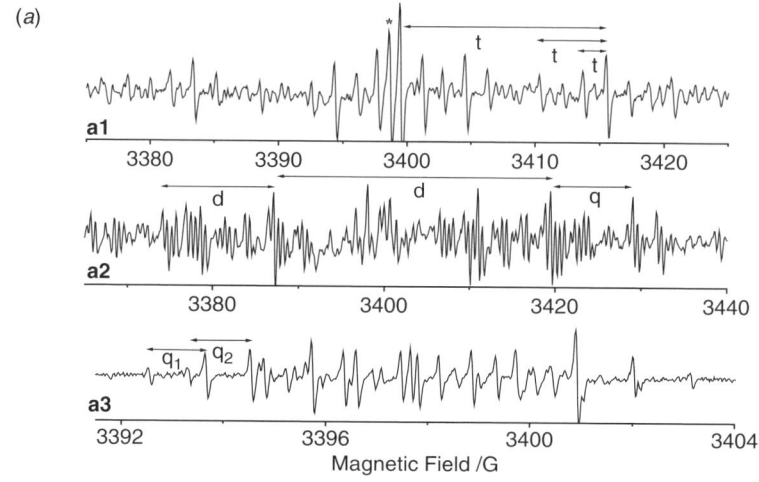

(b)

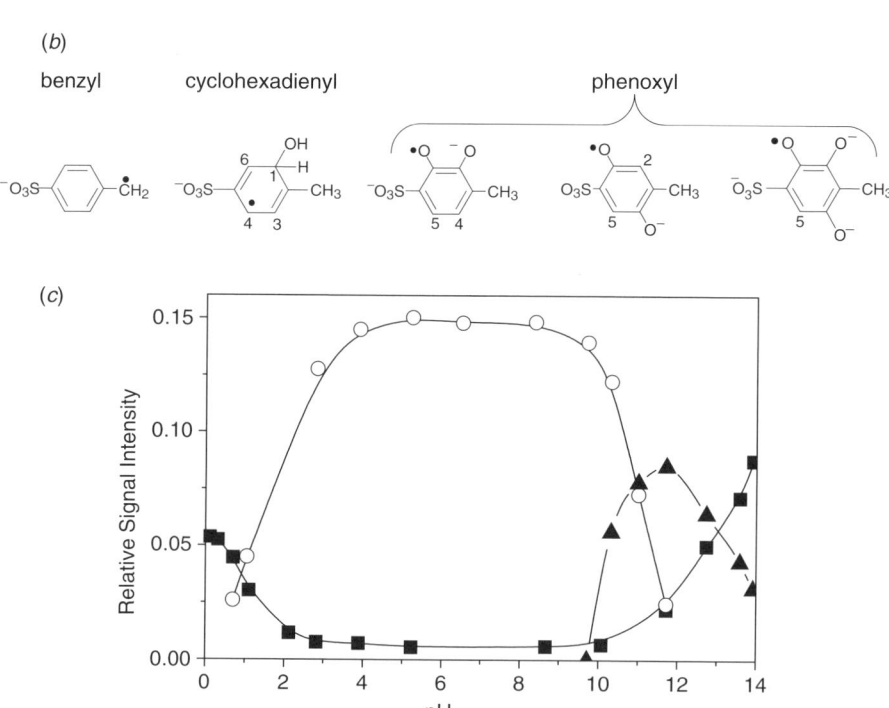

Fig. 3. (a) The ESR spectra obtained upon UV irradiation of aqueous H_2O_2 in the presence of p-toluene sulfonic acid at pH 0.4 (a1, identified as the benzyl radical), pH 4.8 (a2, cyclohexadienyl radical), and pH 11.5 (a3, a superposition of spectra from three phenoxyl type radicals). Note the different scales. Arrows marked d, t, and q indicate doublet, triplet, and quartet splitttings, respectively. The corresponding radical structures are shown in (b). The pH dependence of total signal amplitudes (obtained by integration of an isolated line and then scaled to represent the entire spectrum) are given in (c) for the benzyl radical (■), the cyclohexadienyl radical (○), and the sum of the three phenoxyl radicals (▲). (From Ref. 37 with permission.)

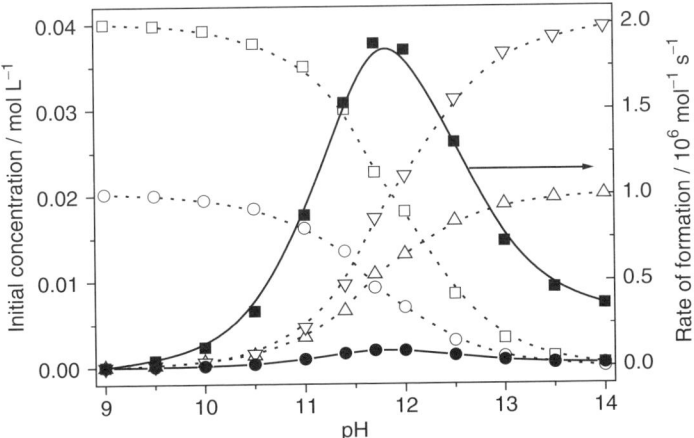

Fig. 4. Simulation of the relative rates of formation of the superoxide radical anion ($O_2^{\bullet-}$, ■), and singlet oxygen ($^1O_2$, •), derived from the pH-dependent concentrations of HO• (□), $O^{\bullet-}$ (▽), H_2O_2 (○), and HO_2^- (△). (From Ref. 37.) Open symbols relate to the left vertical scale, closed ones to the right.

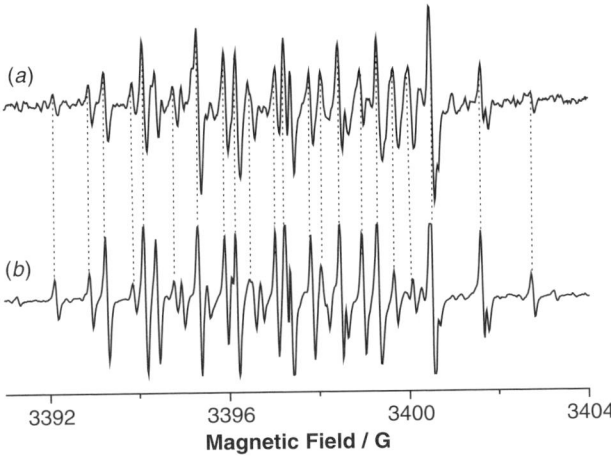

Fig. 5. The ESR spectra of semiquinone radicals observed for *p*-toluene sulfonic acid at pH 11.5 (*a*), and for 3-hydroxytoluene-4-sulfonic acid at pH 13.5 (*b*). The two spectra have identical line positions and represent a superposition of the same two radicals. (See Fig. 3 legend, first two phenoxyl radical entries) with slightly different relative intensities. (From Refs. 37 and 38 with permission.)

is also susceptible, but it is often already blocked by a substituent. The α-hydrogens are a weak point for direct abstraction or in context with H_2O elimination from cyclohexadienyl radicals. So far, no reaction of HO• with perfluoroalkanes has been detected.

Expected Attack Sites of Hydroxyl Radicals on PEMs: SPEEK, PSU, FEP-*g*-PSA, and SPTFS of the Ballard BAM3G series

Sulfonated Polyether(ether)ketone (SPEEK)

Polyethersulfone (PSU)

Poly(tetrafluoroethylene-co-hexafluoropropylene) grafted with Polystyrene Sulfonic Acid (FEP-g-PSA)

Sulfonated Polytrifluorostyrene (SPTFS of the Ballard BAM3G series)

4.2. Radical Species in Nafion Membranes Treated in Fenton Media

Nafion membranes exposed to the Fenton reagent based on Ti(III) were studied by ESR; this cation can be detected by ESR and its disappearance during reaction can be monitored.[30] As will be clearly demonstrated below, the results indicated that several types of oxygen radicals are formed, which lead to the formation of membrane-derived radicals.

In this section, evidence will be provided for the formation of HOO•, $O_2^{•-}$, TiOO•, and fluorinated alkyl radicals in Nafion/Ti(III)/H_2O_2. Separate signals from these radicals species were obtained by variation of sample preparation methods, temperature used for spectra acquisition, and annealing conditions. The formation and stability of these radicals were compared to those of radicals present in the Fenton reagent (TiCl$_3$ + H_2O_2).[39]

Upon addition of the H_2O_2 solution, the exchanged membranes became yellow-red, indicating the presence of diamagnetic Ti(IV) complexes; the color was detected also in the Fenton reagent (no Nafion). The UV spectra of the Fenton reagent consist of two major signals, at 200 ± 2 nm and in the range 338–354 nm. These signals are similar to those assigned recently to ligand-to-metal charge transfer (LMCT) in diamagnetic dimeric peroxo Ti(IV) complexes,[40] and have the same metal-ion electronic configuration as the yellow-red complexes of V(V) catalysts reacted with excess of H_2O_2 solutions.[41] In the Nafion system, a band at 198 nm and a broad band in the vicinity of 300 nm were detected, and represent corresponding diamagnetic Ti(IV) species.

4.2.1. Oxygen-Centered Radicals.
X-band ESR spectra at 200 K of Nafion membranes fully exchanged by Ti(III) and in contact with an aqueous solution containing 30% H_2O_2 are shown in Fig. 6, as a function of contact time. The molar ratio Ti(III)/H_2O_2 was ≈1:300. The broad high-field signal is attributed to Ti(III),[30] whose spectrum is prominent for exposure times of 20 and 50 min. The weak anisotropic signal near $g \approx 2$ seen after exposure time of 20 min increased in intensity with contact time and dominated the spectrum after 100 min of contact. This signal, with $g_1 = 2.0261$, $g_2 = 2.0088$, $g_3 = 2.0031$ ($g_{iso} = 2.0127$), and $a_1 = 5$ G, was attributed to HOO• radicals; g_1 is in the direction of the O–O bond, g_2 is in the molecular plane, and g_3 is perpendicular to the molecular plane. The HOO• radicals are produced in a Fenton reagent from HO• radicals in the presence of excess H_2O_2, as seen in Section 3. Their ESR line shape, Fig. 6, is similar to that assigned to HOO• radicals detected at 77 K in UV-irradiated aqueous H_2O_2 solutions and H_2O/H_2O_2 solutions adsorbed on silica gel,[31] with $g_{iso} = 2.0138$, and $a_1 = 12.3$ G. HOO• radicals were also obtained in Nafion/Ti(III) for lower H_2O_2 concentrations. The intensities of these signals were slightly lower and a_1 slightly higher: a_1 was 6 G for the Nafion system containing 20% H_2O_2 (molar ratio of Ti(III)/H_2O_2 ≈ 1:200), and 7 G for 3% H_2O_2 (molar ratio of Ti(III)/H_2O_2 ≈1:25).

The effect of temperature on the signal from HOO• radicals is shown in Fig. 7, for fully Ti(III)-exchanged Nafion swollen by H_2O_2/H_2O (30% w/v H_2O_2): spectra at 300

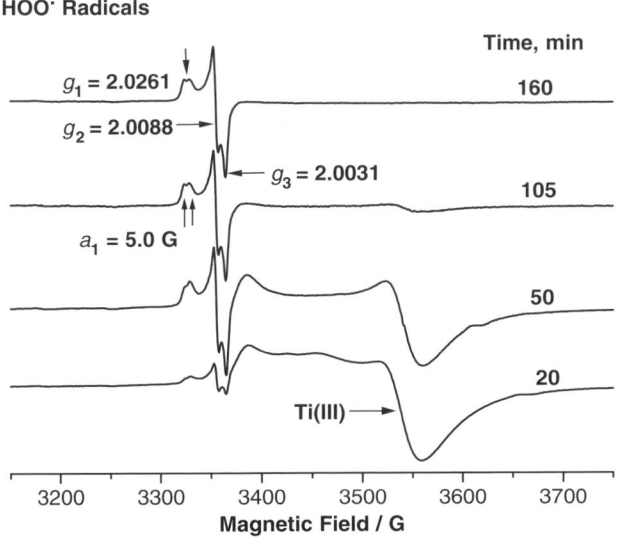

Fig. 6. The ESR spectra at 200 K of dry Nafion fully equilibrated by Ti(III) and exposed to the H_2O_2 solution (30% w/v), as a function of contact time. The g-tensor components and the proton hyperfine splitting for HOO• are indicated. (From Ref. 39 with permission.)

K and above are assigned to peroxy radicals TiOO•. The spectrum measured at 300 K has clear signals corresponding to $g_1 = 2.0228$, $g_2 = 2.0092$, and $g_3 = 2.0046$ ($g_{iso} = 2.0122$). The motionally averaged signal with $g_{iso} \approx 2.0122$ is evidence for a dynamical effect. The principal values of the g-tensor are in the range of values determined for peroxy radicals in various systems.[42] The dynamic effect was demonstrated by the reproducibility of the spectra as the temperature was cycled in the range 290–320 K. The formation of these species is described in the following two reactions:

$$HOO^{\bullet} + H_2O \leftrightarrow O_2^{\bullet -} + H_3O^+ \qquad (17)$$

$$O_2^{\bullet -} + Ti(IV) \rightarrow TiOO\bullet + H_2O \qquad (18)$$

As expected, HOO• was more stable in Nafion samples that were less than fully exchanged by Ti(III), for example, 80%; the membrane acidity is of course higher at lower degrees of exchange by cations.

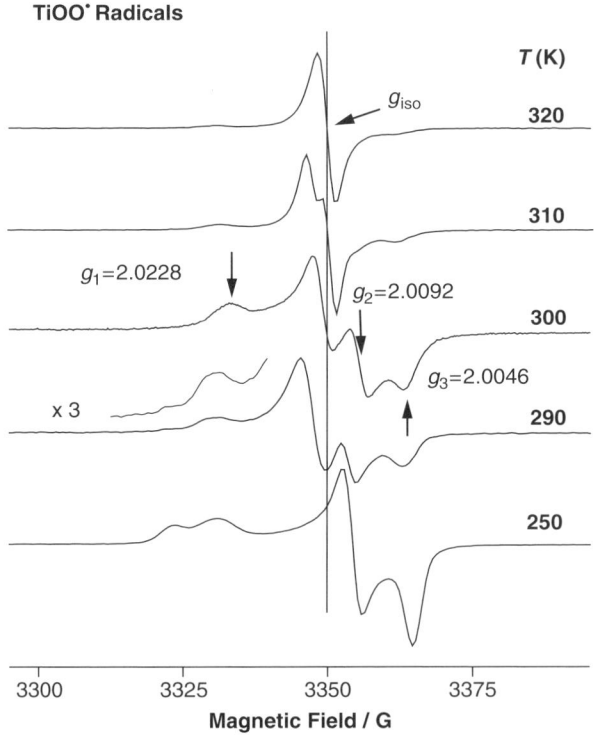

Fig. 7. The ESR spectra at 200 K of dry Nafion fully equilibrated by Ti(III) and exposed to the H_2O_2 solution (30% w/v), as a function of temperature. The g-tensor components for TiOO• are indicated; the vertical line is drawn at a magnetic field corresponding to g_{iso}. (From Ref. 39 with permission.)

The ESR spectra in Nafion membranes fully exchanged by Ti(III), exposed to H_2O_2 solutions (30% w/v), and dried in vacuum are presented as a function of temperature in Fig. 8. In accord with literature data, the signal was assigned to $O_2^{\bullet-}$, with $g_1 = 2.0190$, $g_2 = 2.0092$, and $g_3 = 2.0036$ ($g_{iso} = 2.0106$) at 300 K. The g_1-value, along the O–O bond, is sensitive to the local electric field from the surrounding electrical charges and is used to identify the local environment of the radical in different media. The g-tensor components are stable in the temperature range 160–360 K, but the signal intensity decreased on heating > 360 K. In samples kept at room temperature, the signal disappeared after ≈14 day.

The results presented in Figs. 7 and 8 indicate that two radicals, TiOO• and $O_2^{\bullet-}$, are formed from the initial HOO• radicals in Nafion treated in Fenton media. The TiOO• radical was detected in swollen Nafion and characterized by the principal values of the g-tensor and by the dynamical process that led to averaging of the tensor components. The $O_2^{\bullet-}$ radicals were detected only in dry Nafion.

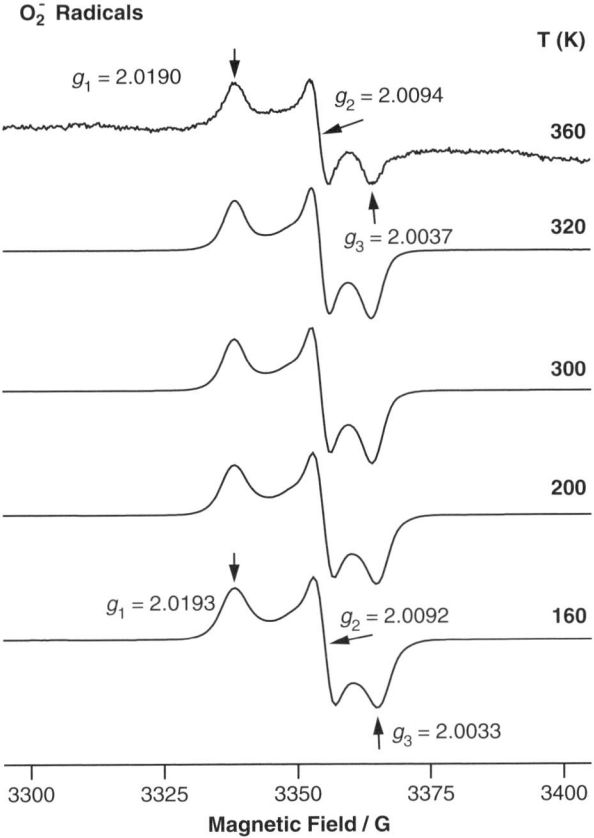

Fig. 8. The ESR spectra of $O_2^{\bullet-}$ in the dry Nafion/Ti(III)/H_2O_2 sample as a function of temperature. The g-tensor components at 160 and 360 K are indicated. (From Ref. 39 with permission.)

4.2.2. Fluorinated Alkyl Radicals. The evolution of ESR spectra at 300 K in Nafion/Ti(III)/ H_2O_2 is shown in Fig. 9 as a function of storage time at ambient temperature. The bottom spectrum is a mixture of TiOO• and $O_2^{\bullet-}$ radicals; the signal from $O_2^{\bullet-}$ dominates, but the shoulder at the low-field side is at g ≈ 2.023, indicating the presence of TiOO• as well. After 14 day, the central narrow signals are weaker and appear superimposed on a broad signal with ≈ 2.0023, typical for carbon-centered radicals. The broad signal became stronger after 92 days (top spectrum); the arrows in Fig. 9 are separated by ≈ 84 G. This signal was assigned to fluorinated alkyl radicals, because hyperfine splittings from fluorinated radicals are large, in the range 60–150 G for α-fluorine, and 20–70 G for β-fluorine nuclei.[43] At this stage, the exact nature of the fluorinated alkyl radicals responsible for the broad signal was not determined, but the width and g-value of the signal were in the range expected for fluorinated radicals. The fluorinated radicals were formed after the $O_2^{\bullet-}$ radicals were expected to disappear; it is possible therefore that Nafion chains were attacked by the peroxide radicals, TiOO•.

Two radicals, HOO• and TiOO•, were identified also in the Fenton reagent in the absence of Nafion. The stability of HOO• radicals is lower than in the Nafion system, and at 200 K only TiOO• signals were detected; the isotropic signal due to dynamical averaging was detected at a lower temperature, 250 K. The same spectra were obtained when the molar ratio H_2O_2/ $TiCl_3$ was increased to 300:1, to mimic the conditions in the Nafion system. No signals were detected when the concentration of H_2O_2 was lower, 20 or 3% w/v. The results emphasize the effect of the environment on the stability of radical species obtained in the Fenton reaction. The main effects of the Nafion matrix are increased stability of HOO• radicals, decreased rate of dynamical averaging

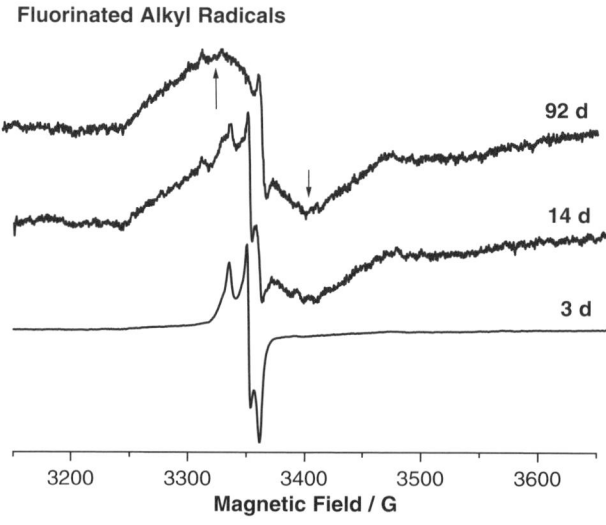

Fig. 9. The ESR spectra at 300 K of the swollen Nafion/Ti(III)/H_2O_2 sample, as a function of contact time with H_2O_2. Arrows indicate the approximate line width of the broad signal (84 G), which was assigned to fluorinated alkyl radicals. (From Ref. 39 with permission.)

for the TiOO• radicals, and formation of $O_2^{•-}$ radicals. Radical stabilization by the matrix is also reflected in the higher concentration of H_2O_2 needed to form the radical species in the Fenton reagent (no Nafion).

4.3. Membrane-Derived Fluorinated Radicals in UV-Irradiated Nafion and Dow Ionomers: Effect of Counterions and H_2O_2

Radicals formed in UV-irradiated Nafion membranes partially or fully exchanged by Cu(II), Fe(II), and Fe(III) were analyzed by ESR spectroscopy; some samples were exposed to H_2O_2 solutions prior to the UV irradiation. Selected experiments were performed with the Dow membranes. The major goals were to identify the radicals and to assess the effects of H_2O_2 and of specific cations. Cu(II), Fe(II), and Fe(III) are paramagnetic and can be detected by ESR. In addition, their intensity during the formation of membrane-derived radicals can be tracked. These counterions were selected because of their known effect on polymer durability: A dramatic decrease of thermal stability in poly(acrylic acid) containing $CuNO_3$ was reported;[44] moreover, perfluorinated membranes were less stable when stainless steel end plates were used in FCs, but no deleterious effects were observed with aluminum end plates.[45] The results presented here are evidence for the formation of specific membrane-derived radical fragments, and suggest that a possible point of attack is at or near the pendant chain of the ionomer.[46]

The ESR signals at 200 K of Cu(II) in Nafion membranes (degree of exchange 2%, by $CuCl_2$) are presented in Fig. 10. The gradual formation of the "quintet" signal is seen with increasing irradiation time. The five lines are approximately in the ratio 1:4:6:4:1, as expected for hyperfine splittings (hfs) from four equivalent nuclei with $I = 1/2$, and isotropic coupling $a_{iso} = 22.5$ G. Annealing above 200 K led to the formation of the "quartet" signal, as seen in Fig. 11; the new signals can be assigned to hfs from three equivalent nuclei with $I = 1/2$ and $a_{iso} \approx 14$ G. The quartet was also detected in Nafion/Fe(II) and the additional small splittings (≈ 5 G) of each line were assigned to spin flip satellites from neighboring protons (see below). Arguments for the assignments of the hfs for the quintet and quartet signals to splittings from ^{19}F nuclei will be presented below.

In Nafion exchanged by Fe(II) as $FeSO_4$, weak ESR signals from Nafion-derived radicals were detected at 77 K even in the absence of H_2O_2. The degree of exchange was important in determining the signal intensity; in the range 2–40%, signals were strongest for 10% neutralization. A weak signal was seen even in non-UV-irradiated samples. This result suggested that the role of UV irradiation is to accelerate the formation of a signal that is produced even in the dark, and possibly during fuel cell operation.

The ESR intensity was significantly higher when the ionomer was treated with aqueous H_2O_2 prior to evacuation and UV irradiation, as seen in Fig. 12. By comparison with published data on fluorinated radicals obtained by γ-irradiation or photolysis of fluorinated compounds, including polytetrafluoroethylene (PTFE, Teflon),[43,47–49] the signal shown in Fig. 12 was assigned to a chain-end perfluorinated radical $RCF_2CF_2^{•}$. The presence of Fe(III) is seen in Fig. 12, at $g = 4.31$ and 2.00, for tetrahedral and octahedral coordination, respectively; the weak signal from a forbidden transition of Fe(III) in a distorted tetrahedral or octahedral coordination

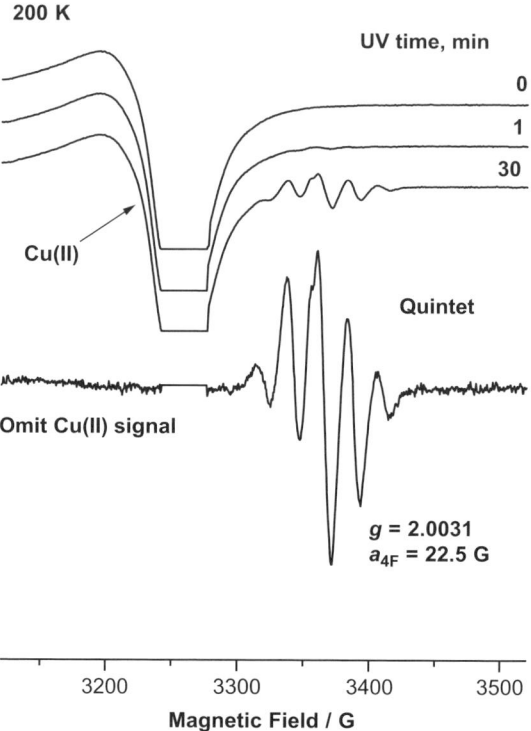

Fig. 10. Evolution of the "quintet" ESR spectrum at 200 K as a function of UV-irradiation time in Nafion 10% exchanged by $CuCO_3$ and evacuated for 1 day prior to irradiation. (From Ref. 46 with permission.)

at $g \approx 3.4$ is also present. The intensity of signals attributed to $RCF_2CF_2^\bullet$ increased with irradiation time, while that of Fe(III) decreased, indicating that the presence of Fe(III) is crucial for radical formation. In order to prove this point, Nafion membranes that were partially exchanged by Fe(III) were examined; in these membranes strong signals from $RCF_2CF_2^\bullet$ radicals were detected upon UV irradiation in the absence H_2O_2, thus providing additional support for the conclusion that Fe(III) reacts with the membranes to form the radicals. Identical ESR signals from the $RCF_2CF_2^\bullet$ radicals were also detected in the UV-irradiated Dow membranes partially exchanged by Fe(III). Figure 13a shows X-band ESR spectra at 77 K in UV-irradiated Nafion/Fe(II)/H_2O_2 and Nafion/Fe(III), together with the simulated spectrum for the $RCF_2CF_2^\bullet$ fragment. Comparison of ESR spectra measured at 133 K at X- and Q-bands, Fig. 13b, indicated that the spectrum is dominated by hyperfine splitting from ^{19}F and not by g-anisotropy.

For some combinations of temperature, degree of neutralization by Fe(II), and $RCF_2CF_2^\bullet$, and (low) H_2O_2 concentrations, the signal from the chain-end radical is transformed into the "quartet", as seen in Fig. 14; this signal is identical to that detected in Nafion/Cu(II), Fig. 13. The additional small splitting in the ESR spectrum

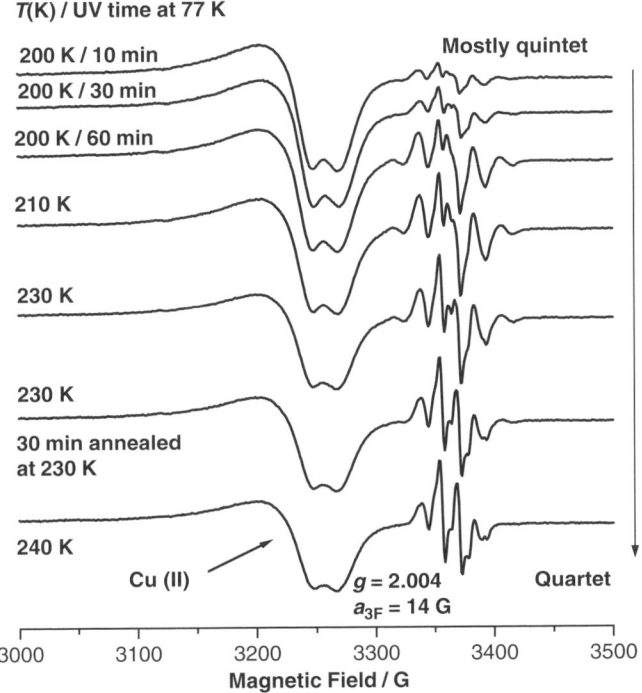

Fig. 11. Transformation of the "quintet" detected in the sample described in Fig. 9 into the "quartet" ESR spectrum upon annealing above 200 K. (From Ref. 46 with permission.)

of the quartet disappeared when the radical was formed in Nafion/Fe(II)/H_2O_2/D_2O. Simulation of the spectra allowed a more accurate determination of the main splittings. The hyperfine splitting from protons, $a_H \approx 5$ G, are spin-flip satellites: small splitting from neighboring protons.[50]

4.3.1. Fluorinated Radical Fragments. The ESR spectrum of the chain-end radical was simulated (Fig. 13a) based on hfs from four ^{19}F nuclei, two in α and two in β positions, and by assuming a planar structure around the carbon atom bearing the unpaired electron. The magnetic parameters used for the simulation are shown in Table 1. The principal values of the g- and ^{19}F hyperfine tensors for the chain-end radical $RCF_2CF_2^{\bullet}$ obtained by simulation of the ESR spectrum (Table 1), were compared with literature values.[43,47–49] The principal values of the hyperfine tensor, the F_α nuclei in the $RCF_2CF_2^{\bullet}$ radical, are 222, 18, and 18 G, close to the corresponding values of 225, 17, and 17 G determined for the propagating chain-end radical in PTFE.[48,49] The isotropic splitting, $a_{iso} = 86$ G, is identical in both systems, and close to the value of 87.3 measured for the fragment $RCF_2CF_2^{\bullet}$ detected in the high-temperature photolysis of perfluorinated polyethers.[51]

The principal values of the hyperfine tensor for the F_β nuclei in the $RCF_2CF_2^{\bullet}$ radical in Nafion are 30, 38, and 38 ($a_{iso} = 35$ G), different compared to the corresponding

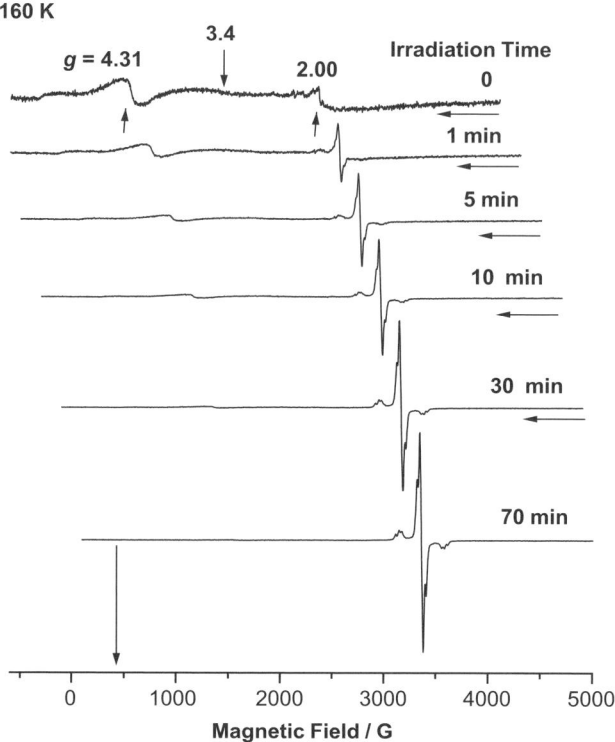

Fig. 12. Effect of UV-irradiation time on the ESR intensity of the chain-end radical $RCF_2CF_2^\bullet$ measured at 160 K in Nafion/Fe(II)/H_2O_2. Degree of neutralization: 40%. Note the progressive horizontal displacement of the spectra to the left, to better visualize the signal from the perfluorinated fragment. The g-values for Fe(III) are indicated. (From Ref. 46 with permission.)

TABLE 1. Magnetic Parameters of the Radical $RCF_2CF_2^\bullet$ in Nafion[a]

$g_x = 2.0023$	$g_y = 2.0023$	$g_z = 2.0030$	$g_{iso} = 2.0025$
$n(F_\alpha) = 2$		$n(F_\beta) = 2$	
$A_{xx}(F_\alpha) = A_{yy}(F_\alpha) = 18$ G		$A_{xx}(F_\beta) = A_{yy}(F_\beta) = 38$ G	
$A_{zz}(F_\alpha) = 222$ G		$A_{zz}(F_\beta) = 30$ G	
$a_{iso}(F_\alpha) = 86$ G		$a_{iso}(F_\beta) = 35$ G	

[a] $n(F_\alpha)$ and $n(F_\beta)$ are the number of α and β interacting ^{19}F nuclei, respectively. The simulation is based on Gaussian line shapes, line widths 38 (along x,y), and 18 G (along z), $\theta = 100$ and $\varphi = 1$.

values of 10, 17, and 17 ($a_{iso} = 15$ G) determined for the propagating chain-end radical in PTFE. Isotropic hyperfine splittings from β fluorine nuclei are sensitive to the radical conformation, and usually follow the expression $a_{iso}(F_\beta) = b_1 + b_2 \cos^2\theta$, where θ is the dihedral angle, between the direction of the unpaired electron and the projection of the C_β–F_β bond on a plane perpendicular to the direction of the C_α–C_β

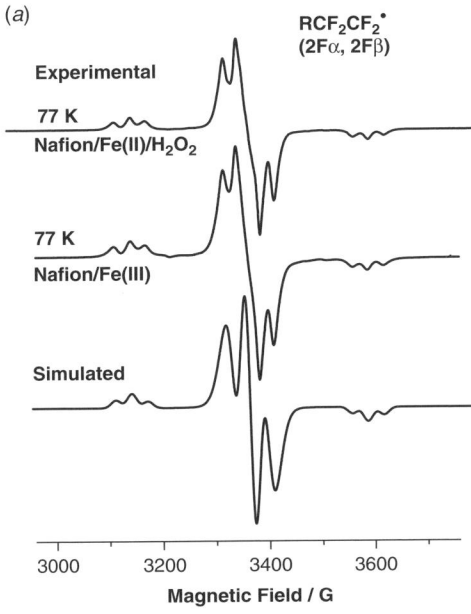

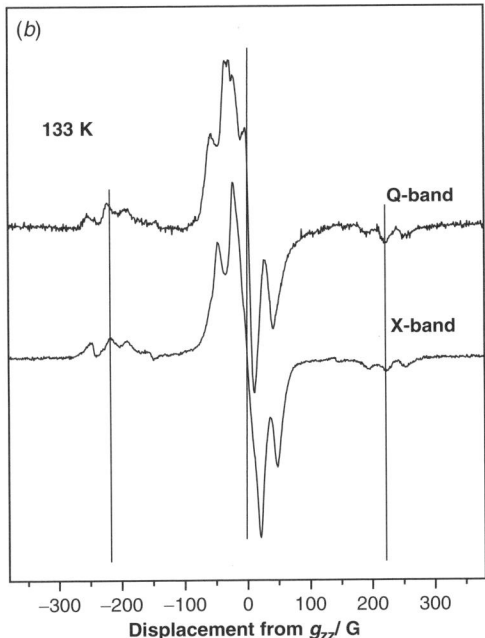

Fig. 13. (a) The ESR spectrum at 77 K of the chain-end radical RCF$_2$CF$_2^{\bullet}$ in UV-irradiated Nafion/Fe(II)/H$_2$O$_2$/H$_2$O (60 min, degree of neutralization 40%), in UV-irradiated Nafion/Fe(III) (55 min, degree of neutralization 10%), and the corresponding simulated spectrum. The magnetic parameters used in the simulated spectrum are given in Table 1. (b) Comparison of ESR spectra at X- and Q-band (34 GHz), $T = 133$ K. (From Ref. 46 with permission.)

direction. For ^{19}F nuclei, $b_1 \approx 0$, and b_2 is known to vary in the range 50–80 G, depending on the particular system.[43] With $b_2 \approx 65$ G, the value $\theta = 43°$ is obtained for the chain-end radical in Nafion, compared to 61° in the propagating chain-end radical in PTFE. The large difference in the dihedral angle for these two systems may indicate that the radical in Nafion is not part of the backbone, but part of the side chain, as shown in Fig. 15. A possible route for the radical formation is UV scission of the C–S bond in the pendant chain of Nafion. The weak signal detected in Nafion/Fe(II) in the absence of H_2O_2 and the strong signal in the presence of H_2O_2 (Fig. 14) suggest that Fe(III) facilitates such scission. The proposed process is $-OCF_2CF_2SO_3^- + Fe(III) \rightarrow -OCF_2CF_2SO_3^\bullet + Fe(II)$.[52] Recombination of $-OCF_2CF_2SO_3^\bullet$ radicals can lead to SO_2, O_2, and to the chain-end radical $-OCF_2CF_2^\bullet$.

The ESR spectrum of the chain-end radical $RCF_2CF_2^\bullet$ was recently simulated by an automatic fitting procedure, using as input the hyperfine coupling tensors of the two Fα and two Fβ nuclei as well as the corresponding directions of the principal values from density functional theory (DFT) calculations. An accurate fit was obtained only for different orientations of the coupling tensors for the two Fα nuclei, indicating a nonplanar structure about the Cα radical center. The isotropic hyperfine splittings for the two Fβ nuclei in the Nafion radical are slightly different (24.9 and 27.5 G), and significantly larger than for the chain-end radical in Teflon (15 G), implying different radical conformations in the two systems.[53] These studies are in progress.

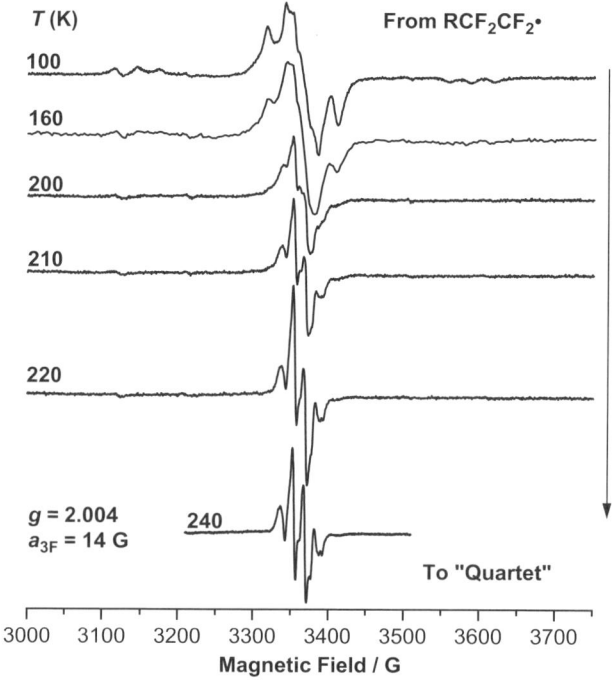

Fig. 14. Transformation of the chain-end radical $RCF_2CF_2^\bullet$ in Nafion/Fe(II)/H_2O_2/H_2O into the "quartet", with hyperfine splittings from three ^{19}F nuclei. Degree of neutralization: 10%. (From Ref. 46 with permission.)

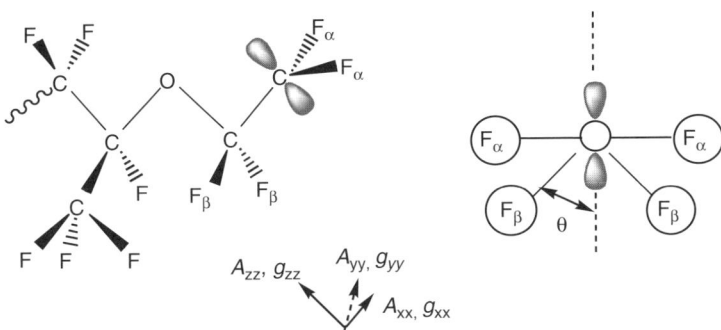

Fig. 15. Proposed structure and conformation of the RCF$_2$CF$_2^\bullet$ radical fragment detected in Nafion, and the directions of the principal values of the ^{19}F hyperfine-H and g-tensors. The $2p\,\pi$ orbital of the unpaired electron is shown; θ is the dihedral angle, equal to $\approx 43°$.

It was reasonable to assume that the quartet signal, which is formed by the disappearance of the perfluorinated chain-end radical in Nafion/Fe(II)/ H$_2$O$_2$ and was detected in Nafion/Cu(II), is also a perfluorinated radical. Isotropic hyperfine splittings from three fluorine atoms, with $a_{iso} = 14.2$ and $g_{iso} = 2.004$, were deduced by simulation. The g_{iso}-value suggests a carbon centered radical. The isotropic hfs in both quartet and quintet radicals are too low to arise from radicals where the unpaired electron is centered on the carbon in a C–F fragment; fragments with no α fluorine nuclei are more likely candidates. Note that a radical showing interactions with three β fluorine nuclei and $a_{iso} = 11.5$ G was assigned to the radical CF$_3$C$^\bullet$O, a value similar to that in the quartet radical in Nafion.[54]

The quartet is formed in Nafion/Cu(II) by the transformation of the quintet radical, with $a_{iso} = 22.5$ G. A similar quintet ($a_{iso} = 24$ G) was detected during the photolysis of perfluoroketones and assigned to radical **1** below.[54] By analogy, radical **2** was proposed as responsible for the quintet detected in Nafion. Structure **2** implies that the unpaired electron is located on the carbon atom in the Nafion backbone that is linked to the pendant chain by loss of a fluorine atom. This structure is in agreement with the detection of fluorine anions during fuel cell operation.[25]

Tentative Assignment of the Quintet Radical in Nafion

—CF$_2$CF$_2$C$^\bullet$CF$_2$CF$_3$
          |
       OCF$_2$CF$_3$

(1) $a_{iso}(4F_\beta) = 24$ G (Ref. 54)

—CF$_2$CF$_2$CF$_2$C$^\bullet$CF$_2$CF$_2$—
                     |
              OCF$_2$CFOCF$_2$CF$_2$SO$_3$H
                     |
                    CF$_3$

(2) Quintet in Nafion, $a_{iso}(4F_\beta) = 22.5$ G

The proton spin flip satellites detected for the quartet (Fig. 14) also suggest the proximity of the pendant side chain to the radical; the proton could come from water protons, or from nonexchanged sulfonic groups.

The identification of these Nafion fragments suggests the possibility that the point of attack is at, or near, the ionomer side chain.

4.4. In situ ESR Experiments

Conventional *in situ* analysis of fuel cells is limited in most cases to static and dynamic electrochemical methods, in particular to current–voltage characteristics and impedance spectroscopy or cyclovoltammetry. These techniques provide an integral type of information and cannot easily differentiate between cathode and anode. Aiming at complementary and more selective information, a fuel cell capable of operating in the resonator of an X-band spectrometer was developed (Fig. 16), which permits direct observation and monitoring of radical formation.[55] This study had the objective to identify fuel cell operating conditions that lead to oxidative membrane degradation.

Carbon-supported Pt catalysts with carbon-based gas diffusion electrode backings gave an intrinsic ESR signal of Dysonian line shape with a humidity dependent amplitude.[56] An alternative solution was advantageous: a catalyst-coated membrane (CCM) in which dry Pt powder of $\approx$ 30-nm grain size was blown onto the membrane and fixed by rolling it through a calender. These CCMs, which were prepared with a Pt loading of $\approx$ 0.77 mg Pt cm^{-2} on both Nafion and fluorine-free membranes, had no intrinsic ESR signal. As the concentration of free radicals produced in a fuel cell is low and their lifetime is short, direct ESR detection was not possible.

For sufficiently high concentrations, radical defects in a polymer are easily detectable, as demonstrated for electron-irradiated Teflon FEP (Fig. 17).[38,57] The peroxy radical was observed, with its characteristic axial signal at low temperature (Fig. 17*a*). The signal narrows and becomes more symmetric due to partial averaging of dipolar contributions to the line width when rapid chain motion sets in above the glass transition (Fig. 17*b*). The perfluorinated peroxy radical loses oxygen (O_2) at higher temperature, and the signal from the alkyl radical, in the present case a

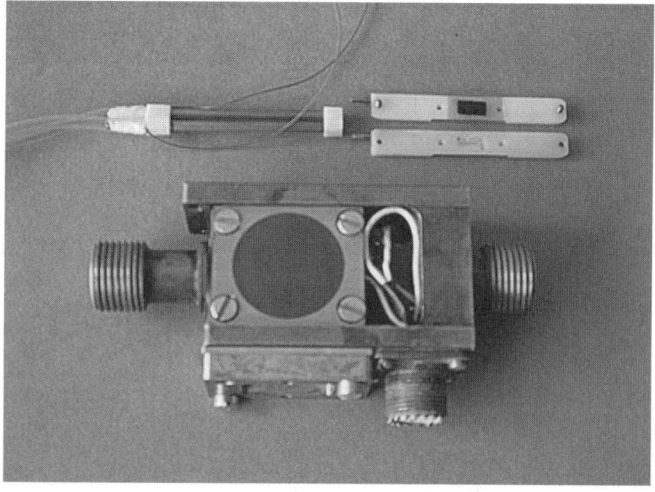

Fig. 16. Fuel cell and X-band microwave resonator for *in situ* ESR experiments with membrane-electrode assembly (black rectangle) and electrical contacts and gas feeds (from the left).

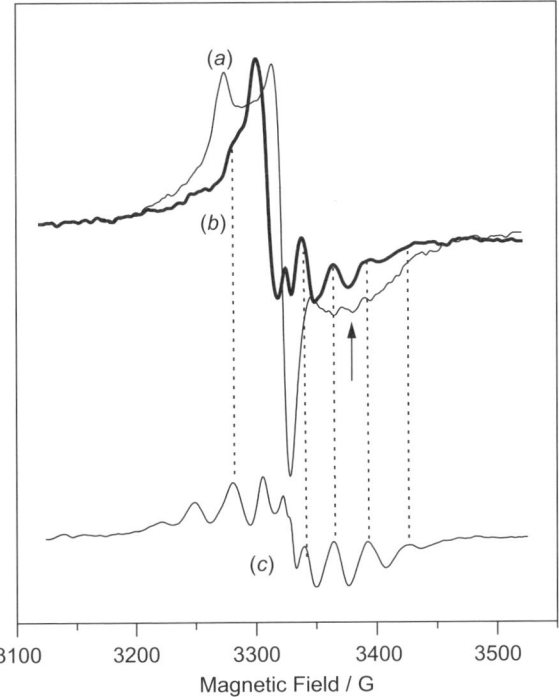

Fig. 17. The ESR spectra obtained by electron irradiation of Teflon FEP in air: (*a*) FEP-peroxy radical at 163 K, (*b*) same sample 333 K, showing a superposition of peroxy radicals and alkyl radicals (arrow and dotted lines); (*c*) perfluoroalkyl radical obtained under argon at 333 K. (From Ref. 57 with permission.)

superposition of a secondary and a tertiary radical, dominates (Fig. 17*c*). The detection limit depends strongly on line width and splitting multiplicity. For the present examples, this limit was ≈ 2 × 10^{-8} mol spins/g polymer. This means that 1 out of 10^6 CF_2 groups was damaged, a rather high value. For lower concentrations of defects, spin trapping technique transforms short-lived radicals into longer lived species, so that a detectable concentration can be accumulated, as will be shown below.

Figure 18 displays ESR spectra obtained when 10 μL of a 1 M aqueous solution of α-(4-pyridyl-1-oxide)-*N-tert*-butylnitrone (POBN) was deposited at the anode. After 5 min of closed-circuit operation, the POBN–H• adduct was detected, with $g =$ 2.0056, $a_N = 15.1$ G (1:1:1 triplet), and $a_H = 9.8$ G (1:2:1 triplet), independently of the type of membrane. The same spectrum was detected when the cell was operated with either H_2 or D_2 gas. However, when D_2O water was used as a solvent, the observed spectrum was predominantly that of the D adduct. Thus, we conclude that either the spin trap is first reduced to its anion and then protonated by the solvent water as shown below, or that the surface-adsorbed hydrogen exchanges rapidly in the aqueous environment.

Two-Step Spin Trapping of H (D) Atoms Using POBN

[POBN reaction scheme showing spin trapping with H₂O and D₂O]

POBN

Of interest in view of polymer membrane degradation are the observations at the cathode. Under the same experimental conditions as above, the detected signal was a nitrogen triplet with $a_N = 15.1$ G split by a hydrogen doublet with $a_H = 2.8$ G (Fig. 19). A similar spectrum was observed in the presence of ethanol, when the ethylol radical (CH₃C•HOH) was trapped;[55] in the absence of ethanol the signal was much broader. This demonstrates that the additional line width is not due to spin exchange with O_2, and must be assigned to incomplete averaging of the dipolar anisotropy. We conclude that the only explanation for these spectra is a radical defect located on the membrane surface. This is key evidence of membrane degradation, and in agreement with the expectation that the signal observed with the more inert Nafion membrane is an order of magnitude weaker compared to that obtained with a fluorine-free membrane.

Another important observation was obtained with DMPO as the spin trap. When DMPO was applied to the cathode of a Nafion-based CCM, a four line spectrum was obtained, with hfs characteristic for the HO• adduct: $a_H = a_N = 14.8$ G (below, and Fig. 20). The lines are narrow, as expected for a radical that can tumble rapidly. It is significant that this radical was observed mostly with the Nafion

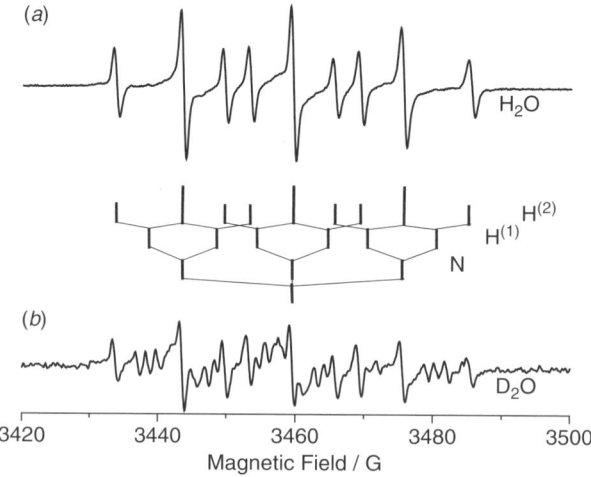

Fig. 18. The ESR spectra (*a*) obtained with an aqueous POBN solution at the anode side of the *in situ* fuel cell, with H₂O ($g = 2.0056$, $a_N = 15.10$ G, $a_H = 9.82$ for two H atoms) and D₂O (*b*) as solvents for the spin trap. Spectrum *b* is a superposition of the deuterium-adduct ($a_D = 1.50$ G for two D atoms) and the hydrogen-adduct from residual protons. (From Ref. 55 with permission.)

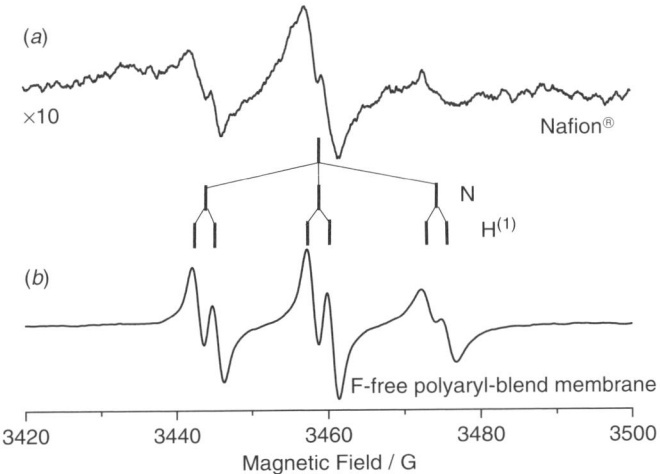

Fig. 19. The EPR spectra obtained with aqueous POBN solution at the cathode side of the *in situ* fuel cell. Catalyst-coated membrane based on Nafion membrane (*a*) and on a fluorine-free membrane (*b*, a_N = 15.10 G, a_H = 2.68 G). (From Ref. 55 with permission.)

membrane: Nafion does not consume the HO• radicals, and these are then trapped by DMPO.

5. CONCLUSIONS AND PROSPECTS

This chapter described ESR methods for the detection of oxygen radicals; demonstrated that the chemistry of membrane degradation is strongly pH dependent; detected the involvement of transition metal ions in reactions with PEMs that lead to the formation of membrane fragments even in the stable Nafion and Dow membranes; and proved the presence of spin-trapped HO• radicals or membrane radicals in Nafion or less inert membranes, respectively. Taken together, these results have suggested that the process of degradation is complex and intimately related to the membrane structure. In the perfluorinated PEMs, for example, the process of membrane collapse can start not only at the backbone as suggested so far, but also at the pendant chains. Much remains to be done for a full and quantitative understanding of the kinetics and mechanism. We single out the need to determine FC materials and operating conditions that lead to the formation of membrane fragments.

The durability of polymer membranes in fuel cells depends, however, on numerous parameters, of which the chemical stability of the membrane is only one factor. Fuel cell materials, the nature of the catalyst, mechanical and thermal stresses, and

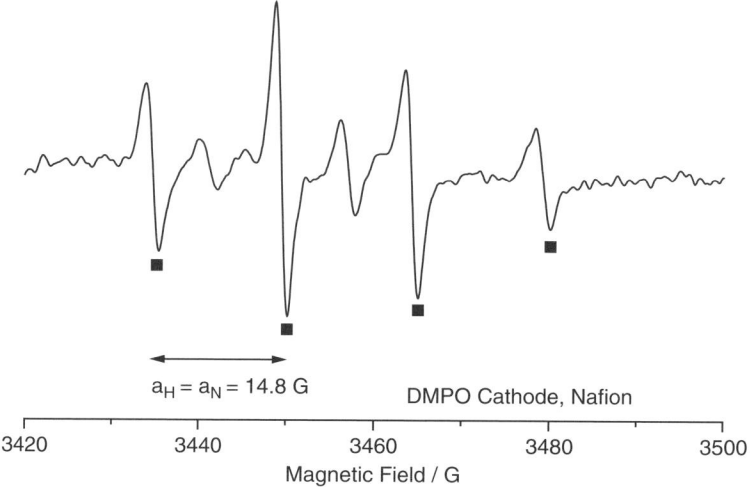

Fig. 20. The ESR spectrum obtained with an aqueous DMPO solution at the cathode side of a Nafion-based catalyst-coated membrane. The lines marked with squares are from the DMPO-HO• adduct. (From Ref. 55 with permission.)

macroscopic inhomogeneities, such as pressure or humidity gradients along over fuel gas flow lines as well as local microscopic inhomogeneities near catalyst grains and proton conduction paths, must also be considered. For these reasons, the full picture of membrane durability requires, and is expected to evolve through, combined efforts and collaboration of an interdisciplinary team with expertise in materials, electrochemistry, and, of course, spectroscopy.

ACKNOWLEDGMENTS

Support in the US laboratory was provided by the General Motors Fuel Cell Activities Program and by the Polymers Program of the National Science Foundation. Experiments at the National Biomedical EPR Center, Medical College of Wisconsin in Milwaukee, Wisconsin were supported by grant EB001980. S. Schlick is grateful to John Healy, Charlene Hayden, Tao Xie and Hubert Gasteiger (GM) and Mariana Pinteala for illuminating discussion on radical formation in fuel cell membranes. E. Roduner appreciates the fruitful collaborations with Günter Scherer and Hans-Peter Brack at the Paul Scherrer Institute, Jochen Kerres and his team from the University of Stuttgart, and Till Kaz at the German Aerospace Center in Stuttgart. Parts of this chapter would not exist without key contributions from Gerold Hübner and Alexander Panchenko performed during their Ph.D. work.

REFERENCES

1. Cyrus Smith (the "engineer") In Verne, J. *The Mysterious Island*, 1874. New translation by Jordan Stump, Random House: New York, 2001, p. 327.
2. Koppel, T. *Powering The Future: The Ballard Fuel Cell and the Race to Change the World*; John Wiley & Sons, Inc.: Toronto, 2001.

3. Fuel Cells — Green Power, Los Alamos National Laboratory. See also the following website: www.education.lanl.gov/resources/fuelcells.
4. Trager, R.S. *Science* **2002**, *298*, 947.
5. Coontz, R.; Hanson, B. *Science* **2004**, *305*, 957. This Introduction is the opening part of seven articles under the general topic "Toward a Hydrogen Economy", in the same issue of *Science*, pp. 958–976.
6. Schoenbein, C.F. *Philos. Mag.* **1839**, *14*, 43.
7. Grove, W.R. (a) *Philos. Mag.* **1839**, *14*, 127. (b) *Philos. Mag.* **1842**, *21*, 417. See also website http://chem.ch.huji.ac.il/ ~eugeniik/history/grove.htm (accessed 16 Feb. 2005).
8. *Chem. Eng. News* **2005**, *83*, 20 (31 Jan. 2005 issue).
9. *Ionomers: Characterization, Theory, and Applications*, Schlick, S., Ed.; CRC Press: Boca Raton, FL, 1996, 300 pp.
10. Risen, W.M., Jr., In *Ionomers: Characterization, Theory, and Applications*, Schlick, S., Ed.; CRC Press: Boca Raton, FL, 1996; Chapt. 12, pp. 281–300.
11. Kerres, J.A. *J. Membr. Sci.* **2000**, *4711*, 1.
12. Huang, C.; Tan, K.S.; Lin, J.; Tan, K.L. *Chem. Phys. Lett.* **2003**, *371*, 80.
13. Anderson, A.B.; Albu, T.V. *J. Electrochem. Soc.* **2000**, *147*, 4229.
14. Panchenko, A.; Koper, M.T.M.; Shubina, T.E.; Mitchel, S.J.; Roduner, E. *J. Electrochem. Soc.* **2004**, *151*, A2016.
15. Markovic, N.M.; Schmidt, T.J.; Grgur, B.N.; Gasteiger, H.A.; Behm, R.J.; Ross, P.N.; *J. Phys. Chem.* **1999**, *103*, 8568.
16. Scherer, G.G. *Ber. Bunsen. Phys. Chem.* **1990**, *94*, 145.
17. Lu, C.; Rice, C.; Masel, R.I.; Babu, P.K.; Waszczuk, P.; Kim, H.S.; Oldfield, E.; Wieckowski. A. *J. Phys. Chem. B* **2002**, *106*, 9581.
18. LaConti, A.B.; Hamdan, M.; McDonald, R.C. In *Handbook of Fuel Cells — Fundamentals, Technology and Applications,* Vielstich, W., Gasteiger, H.A., Lamm, A., Eds.; John Wiley & Sons, Inc.: New York, 2003; Vol. 3.
19. Gubler, L.; Kuhn, H.; Schmidt, T.J.; Scherer, G.G.; Brack, H.-P.; Simbeck, K. *Fuel Cells* **2004**, *4*, 196.
20. Carrette, L.P.L.; Friedrich, K.A.; Huber, M.; Stimming, U. *Phys. Chem. Chem. Phys.*, **2001**, *3*, 320.
21. Tkach, I.; Panchenko, A.; Kaz, T.; Gogel, V.; Friedrich, K.A.; Roduner, E. *Phys. Chem. Chem. Phys.* **2004**, *6*, 5419.
22. Panchenko, A.; Koper, M.T.M.; Shubina, T.E.; Mitchel, S.J.; Roduner, E. *J. Electrochem. Soc.* **2004**, *151*, A2016.
23. Nørskov, J.K.; Rossmeisl, J.; Logadottir, A.; Lindquist, L.; Kitchin, J.R.; Bligaard, T.; Jónsson, H. *J. Phys. Chem. B* **2004**, *108*, 17886.
24. Mitov, S.; Panchenko, A.; Roduner, E. *Chem. Phys. Lett.* **2005**, *402*, 485.
25. Healy, J.; Hayden, C.; Xie, T.; Olson, K.; Waldo, R.; Brundage, M.; Gasteiger, H.; Abbott, J. *Fuel Cells* 2005, *5(2)*, 302.
26. Walling, C. *Acc. Chem. Res.* **1975**, *8*, 125.
27. Miyatake, K.; Oyaizu, K.; Tsuchida, E.; Hay, A.S. *Macromolecules* **2001**, *34*, 2065.
28. Fang, J.; Guo, X.; Harada, S.; Watari, T.; Tanaka, K.; Kita, H.; Okamoto, K. *Macromolecules* **2002**, *35*, 9022.
29. Kaczmarek, H.; Linden, L.A.; Rabek, J.F. *Polym. Degrad. Stab.* **1995**, *47, 175.*

30. Alonso-Amigo, M.G.; Schlick, S. *J. Phys. Chem.* **1989**, *93*, 7526.
31. Bednarek, J.; Schlick, S. *J. Phys. Chem.* **1991**, *95*, 9940.
32. Janzen, E.G. *Acc. Chem. Res.* **1971**, *4*, 31.
33. Villamena, F.A.; Hadad, C.M.; Zweier, J.L. *J. Phys. Chem. A* **2003**, *107*, 4407.
34. Brezova, V.; Valko, M.; Breza, M.; Morris, H.; Telser, J.; Dvoranova, D.; Kaiserova, K.; Varecka, L.; Mazur, M.; Leibfritz, D. *J. Phys. Chem. A* **2003**, *107*, 2415.
35. Rosen, G.M.; Beselman, A.; Tsai, P.; Mailer, C.; Ichiawa, K.; Robinson, B.H.; Halpern, H.J.; MacKerrel, Jr., A.D. *J. Org. Chem.* **2004**, *69*, 1321.
36. Spin Trapping Database: http://epr.niehs.nih.gov.
37. Hübner, G.; Roduner, E. *J. Mater. Chem.* **1999**, *9*, 409.
38. Hübner, G.; Brack, H.P.; Scherer, G.G.; Roduner, E. *Proceedings of the 17th Annual Membrane Technology/Separations Planning Conference;* Business Communication: 304 (1999).
39. Bosnjakovic, A.; Schlick, S. *J. Phys. Chem. B* **2004**, *108*, 4332.
40. Dakanali, M.; Kefalas, E.T.; Raptopoulou, C.P.; Terzis, A.; Voyiatsis, G.; Kyrikou, L.; Mavromoustakos, T.; Salifoglou, A. *J. Inorg. Chem.* **2003**, *42*, 4632.
41. Lee, C-H.; Lin, T-S.; Mou, C-H. *J. Phys. Chem. A* **2003**, *107*, 2543.
42. Kevan, L.; Schlick, S *J. Phys. Chem.* **1986**, *90*, 1998, and references cited therein.
43. (a) Kispert, L.D.; Rogers, M.T. *J. Chem. Phys.* **1971**, *54*, 3326. (b) Bogan, C.M.; Kispert, L.D. *J. Phys. Chem.* **1973**, *77*, 1491. (c) Kispert, L.D. In *Fluorine-Containing Free Radicals Kinetics Kinetics and Dynamics of Reactions;* Root, J.W., Ed.; ACS Symposium Series No. 66, 1978; Chapt. 13, pp. 349–385.
44. Dubinsky, S.; Grader, G.S.; Shter, G.E.; Silverstein, M.S. *Polym. Degrad. Stab.* **2004**, *86*, 171.
45. Pozio, A.; Silva, R.F.; De Francisco, M.; Giorgi, L. *Electrochim. Acta* **2003**, *48,* 1543.
46. Kadirov, M.K.; Bosnjakovic, A.; Schlick, S. *J. Phys. Chem. B* **2005**, *109*, 7664.
47. Rogers, M. T.; Whiffen, D. H. *J. Chem. Phys.* **1964**, *40,* 2662.
48. Toriyama, K.; Iwasaki, M. (a) *J. Chem. Phys.* **1969**, *73*, 2919. (b) *J. Chem. Phys.* **1969**, *73,* 2663.
49. Hara, S; Yamamoto, K.; Shimada, S.; Nishi, H. (a) *Macromolecules* **2003**, *36,* 5661. (b) *J. Polym. Sci. Polym. Phys. Ed.* **2004**, *42*, 1539.
50. Schlick, S.; Kevan, L. (a) *J. Magn. Reson.* **1976**, *21,* 129. (b) *J. Magn. Reson.* **1976**, *22,* 171, and references cited therein.
51. Faucitano, A.; Buttafava, A.; Martinotti, F.; Marchionni, G.; De Pasquale, R.J. *Tetrahedron Lett.* **1990**, *31*, 7055, and references cited therein.
52. Allcock, H.R.; Lampe, F.W.; Mark, J.E. *Contemporary Polymer Chemistry*, 3rd ed.; Pearson/Prentice Hall: Upper Saddle River, NJ, 2003; p. 65.
53. Macomber, L.; Stevens, J.; Lund, A.; Schlick, S., unpublished results.
54. Krusic, P.J.; Chen, K.S.; Meakin, P.; Kochi, J.K. *J. Phys. Chem.* **1974**, *78,* 2036.
55. Panchenko, A.; Dilger, H.; Kerres, J.; Hein, M.; Ullrich, A.; Kaz, T.; Roduner, E. *Phys. Chem. Chem. Phys.* **2004**, *6*, 2891.
56. Panchenko, A.; Dilger, H.; Möller, E.; Sixt, T.; Roduner, E. *J. Power Sources*, **2004**, *127*, 325.
57. Hübner, G. Ph.D. *Dissertation*, University of Stuttgart, 1999.

9

SPATIALLY RESOLVED DEGRADATION IN HETEROPHASIC POLYMERS FROM 1D AND 2D SPECTRAL–SPATIAL ESR IMAGING EXPERIMENTS

SHULAMITH SCHLICK
University of Detroit Mercy, Detroit, Michigan

KRZYSZTOF KRUCZALA
Jagiellonian University, Cracow, Poland, and University of Detroit Mercy, Detroit, Michigan

Contents

1. ESR Imaging (ESRI) as a Nondestructive Method for Spatially Resolved Degradation Studies   230
2. Experimental Details   233
3. ESR Spectra of Nitroxides Derived from Hindered Amine Stabilizers in Aged Heterophasic Polymers   235
4. 1D and 2D Spectral-Spatial ESRI in Aged Heterophasic Polymers   238
    4.1. Poly(Acrylonitrile–Butadiene–Styrene) Systems   238
        4.1.1. Thermal Aging of ABS   238
        4.1.2. Photodegradation of ABS   241
    4.2. Heterophasic Propylene–Ethylene Copolymers   243
        4.2.1. Thermal Aging of HPEC   243
        4.2.2. Photodegradation of HPEC   244
    4.3. Comparison of ESRI and FTIR Methods   248
    4.4. Sensitivity of Crystalline Polymers to Microtoming   250
5. Conclusions   251
Acknowledgments   252
References   252

Advanced ESR Methods in Polymer Research, edited by Shulamith Schlick.
Copyright © 2006 John Wiley & Sons, Inc.

1. ESR IMAGING (ESRI) AS A NONDESTRUCTIVE METHOD FOR SPATIALLY RESOLVED DEGRADATION STUDIES

The lifetime of polymeric materials used in various applications is limited. When exposed to heat, mechanical stress, and ionizing or ultraviolet (UV) irradiation in the presence of oxygen, polymers undergo oxidative degradation due to the formation of reactive intermediates, such as free radicals R• and ROO•, and hydroperoxides ROOH.[1-4] Exposure to environmental factors leads to profound changes in polymer properties, on both the molecular and the macroscopic levels. The chemical structure is modified due, for example, to chain scission, cross-linking, formation of double bonds, and increase or decrease of the molecular mass. These modifications result in changes of the elastic properties and of the degree of crystallinity.[1] The degradation chemistry is complicated because even small amounts of chromophores, free radicals, and metallic residues from polymerization reactions can introduce additional reaction pathways that enhance the rate of degradation. The deleterious effects are often not detected immediately, but develop over periods of months or even years. The gradual modification of the polymer and the ultimately grave results are due to trapped radicals that react slowly, to peroxy radicals that decompose in time with formation of reactive radicals and gas molecules, and to trapped gases that lead to local stresses and cracking. While the time scale of these changes may vary, the final results are dramatic: degradation of the structure and collapse of the mechanical properties.

Electron spin resonance (ESR) methods have been used extensively and are extremely useful for detecting and identifying the radicals formed, elucidating the degradation mechanism, and simulating the variation of the ESR spectra with temperature based on specific motional models. Model-specific simulations of the line shapes have been developed for the study of oxidative degradation of polymers due to ionizing radiation.[5] Irradiation in vacuum has allowed the study of the type and mobility of alkyl radicals R• derived from the polymer. Such studies were initially performed on polytetrafluoroethylene (Teflon) and other perfluorinated polymers, because perfluoroalkyl radicals can be stabilized even at ambient temperature; in these polymers mid-chain alkyl radicals and radicals formed by chain scission have been detected. Admission of oxygen led to the formation of the corresponding peroxy radicals, ROO•. Variation of the temperature in the range 77 K to and above ambient temperature has been used to advantage to stabilize some of the more reactive intermediates, and to elucidate the dynamics. The motional mechanism of the peroxy radicals was deduced by simulation of the temperature dependence of the spectra; in this way a correlation between the radical mobility and reactivity has been clearly established.

This approach has also been extended to some protiated polymers. In γ-irradiated polypropylene (PP), the more mobile peroxy radicals, where the main motion was a rotation about the polymer axis, were also more reactive and short lived, compared to peroxy radicals that were rigid on the time scale of the ESR experiment (10^{-8}–10^{-11}s) and involved only in limited local motions.[6]

Recent research on the effects of radiation and thermal treatment of polymeric materials is focused on two important goals: understanding of the degradation mechanism and prediction of polymer lifetime, and the development of protective additives.[7-11] Hindered amine stabilizers (HAS) rank among the most important recent developments for stabilization of polymeric materials. Nitroxides and hydroxylamines are major products of reactions involving HAS. The HAS-derived nitroxides (HAS–NO) are thermally stable, but can scavenge free radicals to yield diamagnetic species; the hydroxylamines can regenerate the original amine, thus resulting in an efficient protective effect. Some of these events are shown in Fig. 1, where >NH denotes the amine, –NO• the nitroxide, and R•, ROO•, and ROOH are the reactive intermediates derived from polymer chains exposed to oxygen and irradiation or heat.

The crucial role of nitroxides (added as such, or derived from HAS) has been documented recently in low-molecular-mass model systems;[12] this study has suggested that the key step that explains the antioxidant role of HAS is the formation, by one-electron oxidation, of the radical cation $-NH^+$ that deprotonates to give the aminyl radical –N•. An important intermediate radical, the N-peroxy radical –NOO•, has been detected by ESR.[10] The most stable radical is –NO•. Though stable, however, nitroxide radicals can react efficiently with alkyl and alkoxyl radicals derived from polymeric precursors.

In polymers, the situation is considerably more complicated compared to low-molecular-mass model systems. The nature and concentration of the various intermediates, including the nitroxide radicals, varies with the nature of the polymer matrix, thermal history, humidity, and type of HAS. In spite of numerous studies, important details on the degradation steps and kinetics are still incomplete or missing altogether.[7-12]

The concept of diffusion-limited oxidation (DLO) developed by Clough and co-workers has greatly contributed to the understanding of the mechanism for polymer degradation: If oxygen diffusion is slow compared to the rate of degradation, as in accelerated degradation in the laboratory, only thin surface layers in contact with air are degraded, while the sample interior is little, if at all, affected; this is the DLO regime.[13] This concept implies that lifetimes of polymeric materials deduced from the study of *average* properties of samples involved in accelerated degradation cannot be used to estimate the durability of polymers in normal exposure. For this reason, methods for measuring the spatial distribution of polymer properties due to degradation are needed, and have been developed: Density profiling measures the change in density, which is expected to increase in aged samples, along the irradiation depth;[14] and modulus profiling measures the tensile modulus, which decreases during degradation.[14,15] The spatial variation of these properties is an excellent

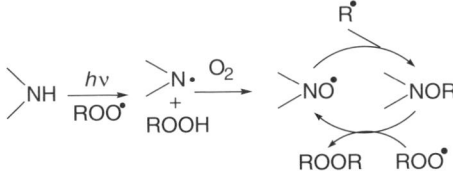

Fig. 1. The chemistry of HAS.

indicator of degradation, especially at advanced stages in the degradation process. Both profiling methods are destructive, in the sense that the sample is cut into sections and each section is studied separately. As described in Section 4, cutting a crystalline polymer can induce morphological changes; for this reason, nondestructive methods are preferable.

Electron spin resonance imaging (ESRI) can be performed in stabilized polymers containing HAS–NO as the imaging contrast by encoding spatial information in the ESR spectra via magnetic field gradients, as described in Chapter 4. This approach was originally suggested by Ohno, who presented two-dimensional (2D) spectral–spatial ESRI images of radicals in polypropylene (PP) containing two different stabilizers, but no detailed analysis.[16]

Lucarini and co-workers have determined[17–19] the distribution of the nitroxide radicals in UV-irradiated PP containing a hindered amine stabilizer by one-dimensional (1D) ESRI. The spatial variation of nitroxide intensity in the thicker samples irradiated for longer times was explained by the diffusion-limited oxidation (DLO) concept. The DLO regime and high oxidation rates lead to narrow penetration depth of oxygen. Recently, Marek et al.[20] reported 1D ESRI experiments on polystyrene and polypropylene plaques exposed to thermal aging and UV irradiation.

Our laboratory developed 1D and 2D spectra–spatial ESRI for the study of heterophasic systems, such as poly(acrylonitrile–butadiene–styrene) (ABS)[21–30] and heterophasic propylene–ethylene copolymers (HPEC) containing bis(2,2,6,6-tetramethyl-4-piperidinyl) sebacate (Tinuvin 770) as the HAS, and exposed to thermal treatment and UV irradiation.[31–33] The major objectives were to examine polymer degradation under different conditions; to assess the effect of rubber phase (polybutadiene in ABS and ethylene–propylene rubber in HPEC) on the extent of degradation; and to evaluate the extent of stabilization by HAS. The repeat units in ABS and the formula of Tinuvin 770 are shown in Fig. 2.

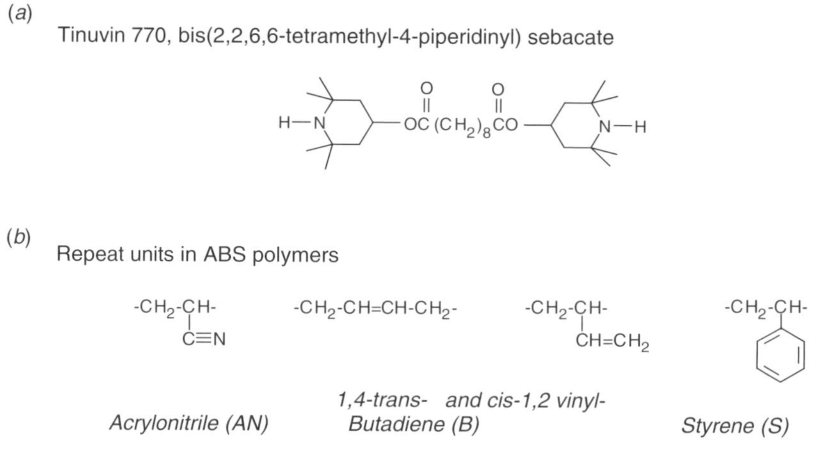

(a) Tinuvin 770, bis(2,2,6,6-tetramethyl-4-piperidinyl) sebacate

$$H-N\text{-piperidinyl}-OC(CH_2)_8CO-\text{piperidinyl}-N-H$$

(b) Repeat units in ABS polymers

$-CH_2-CH(C\equiv N)-$ $-CH_2-CH=CH-CH_2-$ $-CH_2-CH(CH=CH_2)-$ $-CH_2-CH(C_6H_5)-$

Acrylonitrile (AN)   1,4-trans- and cis-1,2 vinyl- Butadiene (B)   Styrene (S)

Fig. 2. Tinuvin 770 and repeat units in ABS.

As described below, the HAS-derived nitroxides in heterophasic polymer systems perform a triple role. First, they provide the contrast needed in the imaging experiments. Second, they enable the visualization of polymer morphology, based on the detection of two dynamically different components detected in the ESR spectra of the nitroxides; in ABS, for example, the two sites, fast (F) and slow (S), have been assigned to location of nitroxides in butadiene-rich and styrene–acrylonitrile (SAN)-rich domains, respectively. Third, the spatial variation of the ESR spectra of nitroxides (in terms of intensity and line shapes) with treatment time, t, provides detailed information on the extent of degradation in the different microdomains. These experiments made possible the determination of the concentration profiles of the nitroxides from 1D ESRI, and also of the spectral profiles from 2D spectral–spatial ESRI, both in a nondestructive way. In these studies the nitroxides, which are the contrast agents, are part of the system; therefore these studies represent the evolution of ESRI techniques *beyond phantoms*.

This chapter is organized as follows. Section 2 describes selected experimental details on sample preparation and treatment, as well as on the determination of the nitroxide profile (1D ESRI) and on 2D spectral–spatial ESRI. The ESR spectra of HAS–NO in the heterophasic polymers are described in Section 3. Both 1D and 2D ESRI experiments are described in Section 4, which includes results for the ABS and HPEC systems, a comparison of ESRI and Fourier transform (FTIR) methods, and our experience with the effect of microtoming on crystalline polymers. Conclusions are presented in Section 5.

2. EXPERIMENTAL DETAILS

The ABS polymers are usually prepared in two steps. In the first step, polybutadiene (PB) with the required degree of cross-linking is prepared. In the second step, PB reacts with acrylonitrile and styrene monomers; during this step two processes take place: copolymerization of acrylonitrile–styrene monomers and grafting of the copolymer to polybutadiene. The resulting complex polymeric materials are phase separated and consist of a continuous acrylonitrile–styrene copolymer matrix phase in which PB particles are dispersed.[34,35] The properties of ABS can be modified by variation of the preparation method, size, and size distribution of the rubber particles, cross-link density, the amount of each repeat unit, and the molecular weight of the free SAN. The size of the rubber particles is in the range 0.1–1 μm for emulsion polymerization, and 0.5–5 μm (that can contain occluded SAN) in mass polymerization. Due to the wide range of variables, the properties of ABS polymers can be modified to suit specific applications. Our experiments were performed on ABS containing 10% PB prepared by mass polymerization, and ABS containing 25% PB prepared by emulsion polymerization.

Heterophasic propylene–ethylene copolymers (HPEC) consist of crystalline polypropylene (PP) modified by an elastomeric component, typically ethylene–propylene rubber (EPR),[36] and are prepared by polymerization of propylene (P) in the presence of catalysts, and sequential polymerization of a propylene–ethylene mixture with the same catalysts.[37] The resulting polymeric materials are heterophasic, but the specific morphology depends on the preparation method and monomer ratio. In scanning electron microscopy (SEM) studies of HPEC, evidence for the presence of "two phases" has been reported: the continuous PP phase, and the dispersed EPR phase (Fig. 1b).[38] The size of the dispersed elastomer particles was found to depend on the polymerization details; for example, in HPEC prepared by sequential polymerization and containing 18 %wt EPR, the particle size measured by SEM was ≤1 μm.[38] More recent papers have recognized the presence of *four* phases in HPEC: crystalline PP, amorphous PP, crystalline EPR (predominantly polyethylene, PE), and amorphous EPR.[39] Our experiments were performed on HPEC with ethylene (E) content of 25% (notation HPEC1), and HPEC with E content of 10% (notation HPEC2).

For the ESR and ESRI experiments, the polymers containing 1–2 % Tinuvin were prepared as 10 × 10 × 0.4-cm plaques, which were obtained by injection molding at 483 K. Thermal treatment of the plaques was performed in convection ovens, at 333, 353, and 393 K for ABS samples, and at 393 and 433 K for HPEC samples. The UV irradiation of the plaques was done in weathering chambers at 338 K with a Xe arc that mimics sunlight, or at 318 K with UVA (320–370 nm) or UVB (290–320nm) sources. For the ESR and ESRI experiments, cylindrical samples 4 mm in diameter were cut through the plaque thickness at selected time intervals, transferred to a 5 mm diameter ESR sample tube, and placed in the ESR resonator with the symmetry axis along the long (vertical) axis of the resonator and parallel to the direction of the magnetic field gradient. The irradiance as a function of wavelength for UVA-340 and UVB-FS20 fluorescent lamps versus sunlight, and the sample collection from the plaque, are shown in Fig. 3.

The intensity profile of the nitroxide radicals was deduced from 1D ESRI experiments. To this end, two spectra are needed: the usual ESR spectrum, and the ESR

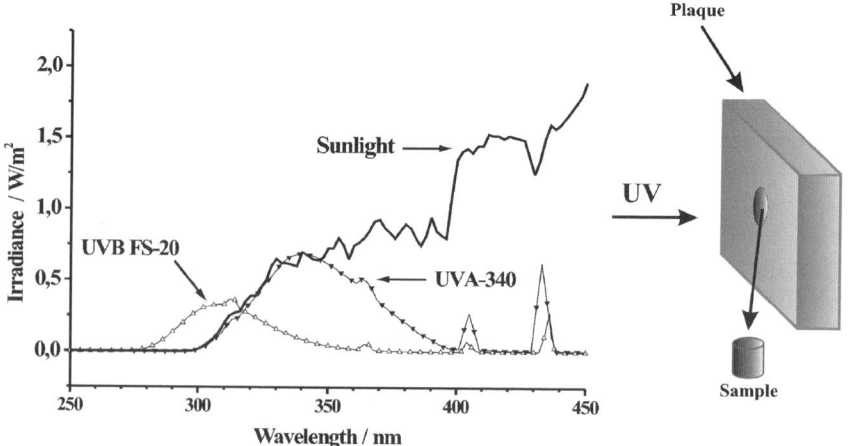

Fig. 3. Irradiance as a function of wavelength for UVA-340 and UVB-FS20 fluorescent lamps and for sunlight (left), and sample selection from the irradiated plaque.

spectrum measured in the presence of the magnetic field gradient (*1D image*). The 1D image is a convolution of the ESR spectrum in the absence of the gradient with the distribution of the paramagnetic centers along the gradient direction (*the profile*). The convolution is correct only if the ESR line shape has no spatial dependence. The two spectra were measured either at 240 K close to the rigid limit of both spectral components, or at 340 K in order to reach the motional narrowing regime of both spectral components; in this way the spatial dependence of the ESR signal was avoided. In most cases, the 1D images were obtained with a field gradient of 200 G cm^{-1}.

The intensity profiles were fitted by analytical functions and convoluted with the ESR spectrum measured in the absence of the field gradient in order to simulate the 1D image. In the most recent work, the genetic algorithm (GA) was used for minimization of the difference between simulated and experimental 1D images; this procedure allowed the best fit to be chosen automatically.[31–33] In some cases the experimental 1D profiles were corrected for the sensitivity profile of the resonator; the difference will be illustrated below for the 1D profiles of HPEC.

The 2D spectral–spatial ESR images were reconstructed from a complete set of projections, typically 128–256, collected as a function of the magnetic field gradient, using a convoluted back-projection algorithm. In the first reconstruction stage, the projections at the missing angles were assumed to be identical to the projection measured at the largest available angle. In the second stage, the projections at the missing angles were obtained by the projection slice algorithm (PSA) with 2–10 iterations.[23–33] Two-dimensional images were saved as 128 × 128 or 256 × 256 matrices.

3. ESR SPECTRA OF NITROXIDES DERIVED FROM HINDERED AMINE STABILIZERS IN AGED HETEROPHASIC POLYMERS

Selected X-band ESR spectra at 300 K of HAS–NO in ABS, for the indicated irradiation times with the Xe source in the weathering chamber, are shown in Fig. 4. All spectra, except that corresponding to the longest irradiation time in Fig. 4, consist of a superposition of two components, from nitroxides differing in their mobility: a "fast" component (F) with a total width of ≈32 G, and a "slow" component (S) with a spectral width of ≈ 64 G. The corresponding rotational correlation times, τ_c, are 4×10^{-9} s rad^{-1} and 5×10^{-8} s rad^{-1}, respectively, deduced by simulations of the spectra.[21,23] The spectra indicate the presence of nitroxides in two different environments. It is reasonable to assign the fast and slow components to nitroxides located, respectively, in low-T_g domains dominated by polybutadiene sequences ($T_g \approx$ 200 K), and in high-T_g domains dominated by polystyrene ($T_g \approx$ 370 K) or polyacrylonitrile sequences ($T_g \approx$ 360 K). This assignment was verified by spin probe studies of polystyrene, polyacrylonitrile, and polybutadience and ABS.[40] As seen in Fig. 4, the relative intensity of the F component as a function of irradiation time increases to a maximum (25%), decreases, and becomes negligible at the longest irradiation time. The decrease of the relative intensity of F with irradiation time is due to the consumption of the HAS-derived nitroxide radicals located in the butadiene-rich domains of the polymer, as butadiene is expected to be more vulnerable to degradation compared to the other repeat units in ABS.

The ESR spectrum corresponding to the F component was isolated by subtracting the spectrum of the slow component presented at the top of Fig. 4 from one of the composite spectra. After this step it was easy to superimpose the two components, and to reproduce all composite spectra; the relative concentration of each spectral component was then obtained by double integration. After examination of numerous samples, we became convinced that the same F and S components are detected in different samples, and the corresponding composite spectra differ only in the relative intensity of the two components. The percentage of the fast component, %F, calculated as described above, is given for all spectra shown in Fig. 4.

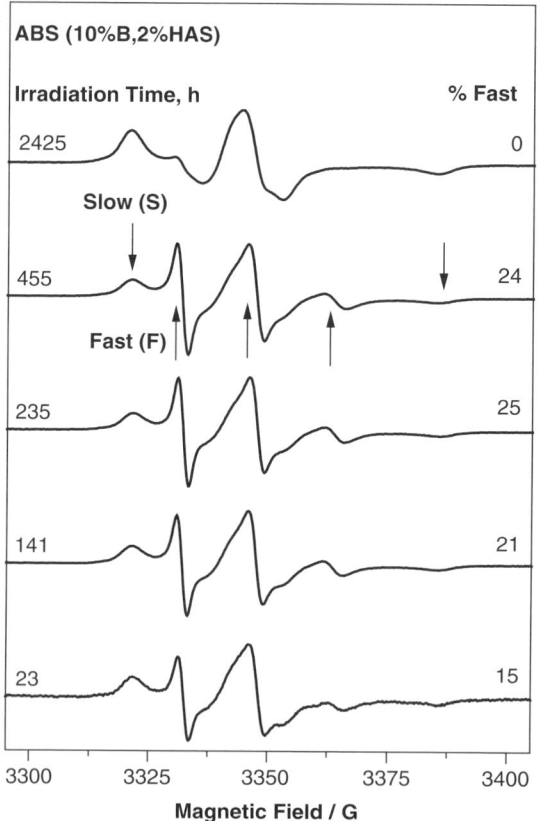

Fig. 4. The ESR spectra measured at 300 K of HAS–NO in ABS containing 10%B and 2% HAS for the indicated irradiation times with a Xe arc in weathering chamber. Upward and downward arrows point, respectively, to the signals from the fast (F) and slow (S) spectral components. The relative intensity of the F component, %F, at 300 K is indicated.

Two components were also detected in the ESR spectra at 300 K of thermally treated ABS (see Fig. 5a). Only one component is seen at 240 K, because both spectral components are close to the "rigid limit" of the nitroxide radical. Two spectral components were also obtained for the thermally treated HPEC1 sample (see Fig. 5b). As for ABS, only one component was detected at 240 K; a similar approach led to the

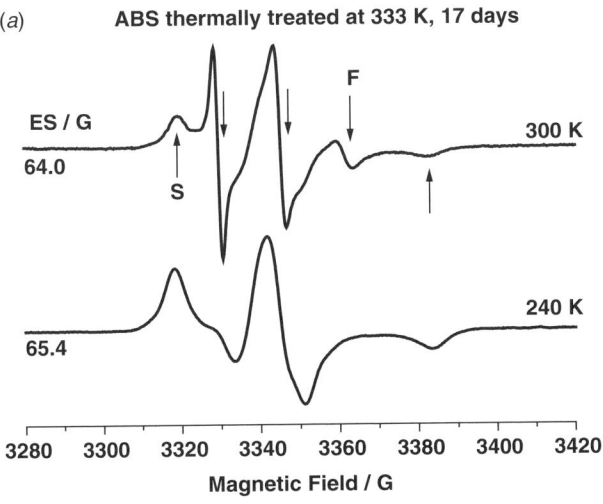

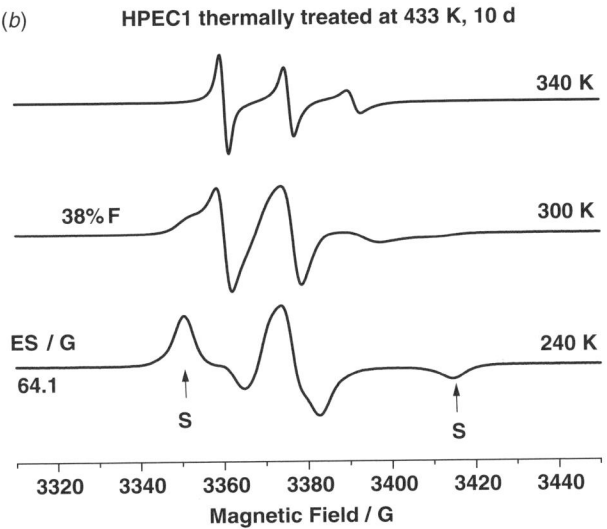

Fig. 5. Temperature variation of the ESR spectra of thermally treated ABS (10%B and 2% HAS) (*a*) and HPEC1 (*b*) containing 2% HAS. The F and S components are indicated in some spectra. The extreme separations (ES) in G of the S component are indicated for spectra of ASB spectra at 240 and 300 K, and for the HPEC spectrum at 240 K.

determination of the %F in the spectrum at 300 K, 39 %. The line shapes at 340 K are typical for rotation of the probe along the long axis of HAS–NO, and can be simulated with a diffusion tilt angle, θ, equal to 90°; a similar simulation was appropriate for the fast component in ABS.[21] The angle θ is between the direction of the N–O bond and the axis of the $2p$ orbital of the unpaired electron. As nitroxide radicals are not expected to intercalate in crystalline domains, the spectra shown above reflect dynamics in the amorphous domains.

The major conclusion that was deduced from the spectra shown in Figs. 4 and 5 is that the two sites detected for the nitroxides in the heterophasic systems studied can serve as a basis not only for describing the morphology of the system, but also to trace the evolution of the nitroxide signal as a function of treatment time, and the time dependence of the degradation process, in the morphologically different domains: in the butadiene-rich and SAN-rich domains in ABS, and in the amorphous PP-rich and EPR-rich domains in HPEC.

Preliminary experiments on cut samples have indicated that as a result of UV treatment, the intensity ratio of the fast and slow components, [S]/[F], varies with sample depth. This information led to the need for 2D spectral–spatial ESRI experiments, which determine the ESR spectrum as a function of sample depth. These experiments allow the determination of the [S]/[F] within sample depth in a nondestructive way. Together with the determination of the total nitroxide concentration and %F as a function of treatment time, the ESRI experiments allow the determination of the nitroxide profile in each morphologically distinct domain.

4. 1D AND 2D SPECTRAL-SPATIAL ESRI IN AGED HETEROPHASIC POLYMERS

4.1. Poly(Acrylonitrile–Butadiene–Styrene) Systems

This section describes ESRI studies on ABS thermally aged at 333, 353, and 393 K, and UV-irradiated ABS containing 10 and 25% B. The HAS content was 1–2%.

4.1.1. Thermal Aging of ABS. Figure 6 presents the 2D spectral–spatial perspective plot of ABS (10% B, 2% HAS) thermally treated at 333 K in a constant temperature bath for 17 days.[22,27,28] The ESR intensity is presented in absorption. "Virtual" slices at the indicated sample depths are presented in the derivative mode at the right side of the figure. The nitroxide distribution is essentially homogeneous along the sample depth, as also seen in 1D profiles (not presented). Furthermore, no spatial variation of the line shapes was detected, and the average %F was 27 ± 4. This F content deduced by 2D ESRI is in agreement with results of sectioning the sample, and weighing and determining the %F in each slice; the average F content in the cut sample was 26 ± 4%.

Completely different results were obtained when the same ABS plaques were treated at higher temperatures, 353[26] and 393 K.[25] Figure 7 presents the concentration profiles of nitroxides along the sample depth for the indicated treatment times at 393 K, deduced from ESR spectra and 1D images measured at 240 K. All profiles are

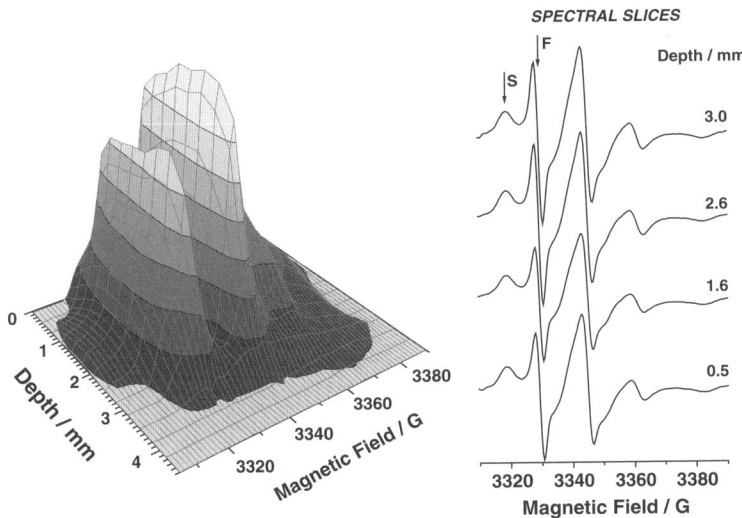

Fig. 6. The 2D spectral–spatial perspective plot of HAS-derived nitroxides in ABS (10% B, 2% HAS) after 17 days of thermal degradation at 333 K, presented in absorption. The "virtual" spectral slices are presented in the derivative mode. The depth in millimeters of the slice from the sample edge is indicated.

symmetrical and are shown with the same maximum height. The profiles represent the evolution from homogeneous (flat) degradation throughout sample depth for treatment up to ≈70 h, to the DLO regime for longer treatment times. The depth of oxygen penetration is reflected in the width of the profiles, which show that nitroxide radicals and consequently the degradation process are limited to narrow layers at the edges of the sample in contact with oxygen. The DLO regime is also clearly seen in the 2D images (not shown), which indicated the negligible amount of nitroxide radicals in the sample interior.

Based on the imaging experiments, it became possible to determine the concentration profile of the F and S components within the sample depth for a given treatment time, as shown in Fig. 8 for ABS aged for 834 h at 393 K. Similar data as a function of treatment time constitute the "elastomer profile":[25,26] the relative intensity of the F component as a function of sample depth and treatment time. For long treatment times, ≈ 240 h and above, the separate profiles for the F and S components and the elastomer profiles indicated the consumption of the HAS on the outer edges of the sample. As clearly seen in Fig. 8, the sample becomes depleted in the F component in outer layers of thickness ≈ 400 μm.

Recently, this approach was applied to the thermal degradation of ABS prepared by emulsion polymerization and containing 25% B.[29] Comparison with results obtained for a parallel study of ABS prepared by mass polymerization and containing 10% wt

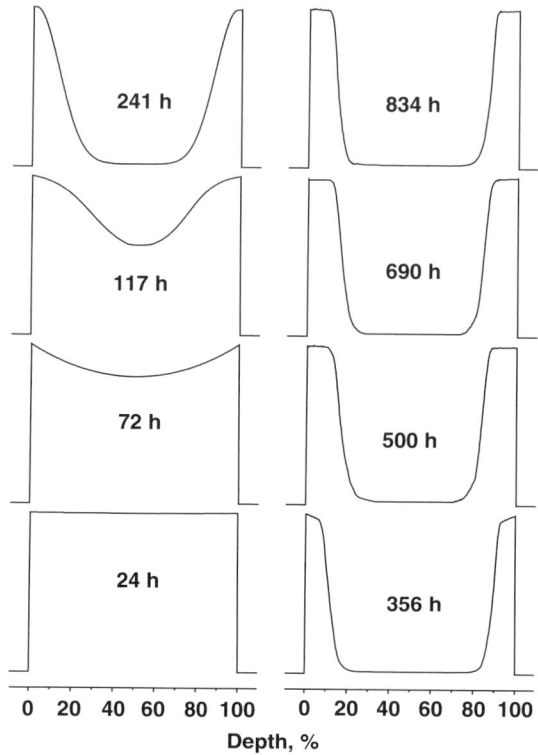

Fig. 7. Evolution of the 1D concentration profiles of HAS-derived nitroxides in ABS (10%B, 2% HAS) for the indicated treatment time at 393 K.

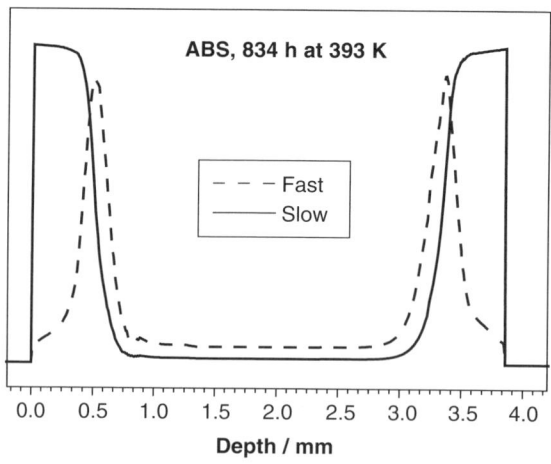

Fig. 8. Concentration profiles for the F and S components after thermal treatment of ABS (10%B, 2% HAS) at 393 K for 834 h.

butadiene (ABS–10B) clearly indicated that the degradation rate of the polymer prepared by emulsion polymerization (ABS–25B) is significantly reduced. This result can be explained by the formation of cross-linked "composite" networks during emulsion polymerization, which lead to greater thermal stability. Several studies have reported that this synthetic method leads to the formation of an extended cross-linked region of butadiene; the B regions are often described as "composites", because they also contain a significant amount of the SAN component. In the polystyrene–polybutadiene (PS–PB) system prepared by emulsion polymerization and containing 20% wt B, for example, the amount of styrene in the composites was large, > 45%, and depended on the molecular weight of PB and reaction conditions.[41] Cross-linking and incorporation of SAN in ABS polymers prepared by emulsion polymerization are expected to reduce both the free volume and the rate of oxygen diffusion, and therefore to lengthen the induction time of oxidation and reduce the oxidation rate.

Additional support for the incorporation of SAN in the B domains in ABS–25B has been obtained in a spin probe study performed in our laboratory, based on 10-doxylnonadecane (10DND) and 5-doxyldecane (5DD).[40] Two spectral components differing in their dynamical properties were detected for both probes in ABS–10B and ABS–25B. The behavior of the fast component in ABS–10B, prepared by mass polymerization, suggested that the low-T_g phase is almost pure butadiene. Data for the corresponding phase in ABS–25B prepared by emulsion grafting, however, also indicated the presence of styrene–acrylonitrile monomers, which act as stabilizers. It is possible that the degradation properties in ABS–25B are affected by the presence of additives, as the polymer was of industrial origin. The spin probe study, however, is not expected to reflect the presence of small amounts of additives, and therefore the results are highly significant for the present study. It appears that emulsion polymerization results in a different morphology and microphase composition, with important effects on the rate of thermal aging.

4.1.2. Photodegradation of ABS. The ESR spectra at 300 K of the HAS-derived nitroxide in UV-irradiated ABS shown in Fig. 4 indicate the absence of the fast component of the nitroxide radicals, which is associated with the butadiene-rich domains, for long irradiation times, 2425 h. The corresponding concentration profiles along the irradiation depth, deduced by deconvolution of images measured at 240 K (not shown) are spatially inhomogeneous, with an initial large nitroxide concentration on the irradiated side, and a gradual increase of the nitroxide concentration at the nonirradiated side.[24] The profiles express the combined effects of oxygen and radiation, and show the regions where the chemistry takes place: If oxygen diffusion is slow compared to the rate of degradation, only surfaces in contact with air are degraded, while the sample interior is little, if at all, affected. The same results were obtained for samples whose back was covered with aluminum foil, indicating that the radicals present on the back side are not due to direct irradiation, for example by scattered light. It is interesting to note that for long irradiation times, 2425 h, the nitroxide concentration on the nonirradiated side is higher than on the directly irradiated side. This effect was assigned to the formation of nitroxide radicals on the back side due to the small, but not negligible, amount of transmitted light (≈1%),[33] and to the fact that nitroxides are not consumed in stabilization processes to the same degree as on the irradiated side.

Figure 9 presents the 2D spatial–spectral perspective plot of nitroxide radicals in ABS (10%B, 2% HAS) UV irradiated by a Xe arc during 6 days in the weathering chamber. The ESR intensity is presented in absorption mode. The spectral slices deduced nondestructively are shown, in the derivative mode, on the right side of Fig. 9 at the indicated distance from the irradiated side. The perspective plot and the spectral slices very clearly show the distribution of the signal intensity and the negligible signal intensity in the sample interior, as also seen in the concentration profiles deduced from 1D ESRI (not shown). The spectral slices indicate not only the line shape variation, but also the relative intensity of each spectral component as a function of depth. For a relatively short irradiation time ($t = 141$ h, ≈ 6 days), the ESR spectrum of the directly irradiated part of the sample exhibits a composite spectrum with %F $\approx$14%. Near the nonirradiated side, %F is significantly larger. After longer irradiation times, > 900 h, the irradiated side does not contain the fast component,[22,23] and %F is significantly lower at and near the nonirradiated side. Even for intermediate irradiation times, 643 h, %F on the irradiated side is only ≈ 3%.[23] The major conclusion deduced from the 2D ESRI experiments on Xe-irradiated ABS is that the nitroxides in the butadiene-rich domains are consumed rapidly on the irradiated side, and their concentration decreases to zero after ≈ 900 h of irradiation. On the nonirradiated side, the degradation is much slower, but even in this area a decrease of %F is detected when the irradiation time increases from 141 to 934 h. As was demonstrated above, the variation with treatment

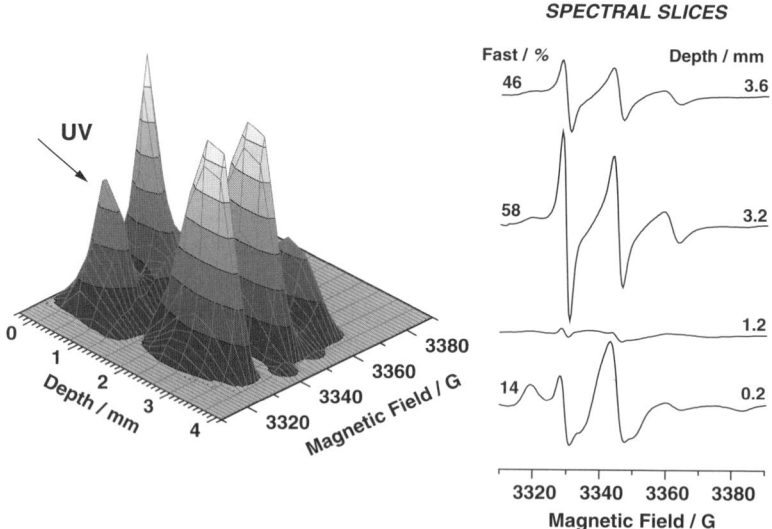

Fig. 9. The 2D spectral–spatial perspective plot of HAS-derived nitroxides after 6 days of UV irradiation in the weathering chamber, presented in absorption. The spectral slices for the indicated depths from the irradiated side are presented in the derivative mode. The relative intensity of the F component, %F, is shown for three of the slices.

time of %F within sample depth, *spectral profiling*, by the ESR imaging method, is an exceptionally sensitive indicator of the extent and location of degradation processes.

4.2. Heterophasic Propylene–Ethylene Copolymers

This section presents the 1D and 2D ESRI study of thermally and UV-treated heterophasic propylene–ethylene copolymers (HPEC) polymers. Two polymers, coded HPEC1 and HPEC2, differing in ethylene content, were investigated. The ethylene content was 25% wt in HPEC1 and 10% wt in HPEC2, within ±2%, as determined by FTIR.[42] Morphologically, HPEC systems are more complex compared to ABS, because of the presence of amorphous *and* crystalline domains.

4.2.1. Thermal Aging of HPEC. The ESR spectra of HAS–NO in HPEC1 (1%HAS) thermally treated at 433 K for 10 days presented in Fig. 5*b* indicated the evolution of the spectra, from the rigid limit in the range 100–240 K, to two components at 300 K, and to a motionally narrowed spectrum at 340 K. As discussed above, the nitroxide radicals reflect the dynamics in the amorphous domains. The relative intensities of the two spectral components at 300 K in the HPEC samples was deduced by deconvolution; at 300 K, %F = 38 in HPEC1, and 40 in HPEC2.

The concentration profiles of HAS–NO in HPEC1 and HPEC2 (both containing 2% HAS) for the indicated treatment times at 393 and 433 K are presented in Fig. 10. Most 1D profiles show the formation of an outside layer that contains a lower amount of nitroxide radicals. The layer is more pronounced for HPEC1 and the effect is seen already after 29 days of treatment at 393 K. For the same conditions, the outside layer in HPEC2 is not as visible, but becomes more pronounced after 168 days. The nitroxide-depleted region is not detected in the 1D profiles of HAS–NO in HPEC1 samples treated at 433 K for 7 days, and in HPEC2 samples treated for 7 and 50 days. All profiles were corrected for the sensitivity profiles of the resonator; for comparison, the resonator profile measured for the 5.65 mm-long cylindrical sample in the region of interest is also shown in Fig. 10, together with the uncorrected profile. The DLO regime is clearly reached for HPEC1 after 10 days of treatment at 433 K (bottom profile in Fig. 10), indicating the advanced stage of oxidation; for HPEC2, the 1D profiles are almost flat even after 50 days of treatment at the same temperature. These results indicate the faster degradation rate for HPEC1, which contains the higher amount of ethylene, 25%.

The 2D spectral–spatial ESRI perspective plots at 300 K and corresponding spectral slices in the derivative mode are shown for HAS–NO in HPEC1 treated at 393 K for 168 days in Fig. 11*a*, and at 433 K for 10 days in Fig. 11*b*. Comparison of the spectral slices in Fig. 11*a* and *b* indicates a lower %F in Fig. 11*b* throughout the sample depth. This result, combined with the DLO regime evidenced in the perspective plot, indicates more advanced degradation for the sample treated at the higher temperature, even though the treatment time was much shorter, 10 versus 168 days. Spectral profiling based on the 2D ESRI experiments for HPEC1 and HPEC2 treated at 393 and 433 K has indicated the formation of a narrow skin, ≈100 μm thick, that contains a lower %F, at both treatment temperatures. It was proposed that the HAS-depleted layer is due to the loss of stabilizer by diffusion (blooming).[32]

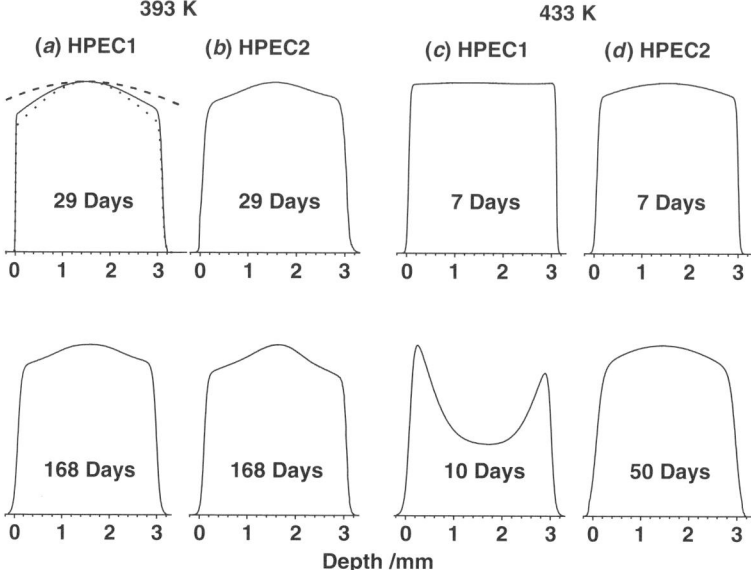

Fig. 10. Concentration profiles of HAS–NO in HPEC1 and HPEC2 (2% HAS) deduced from 1D ESRI experiments, for the indicated treatment times in days at 393 and 433 K. For a treatment time of 29 days, the corrected profile (solid line) is shown together with the sensitivity profile of the resonator (dashed line) and the uncorrected profile (dotted line).

4.2.2. Photodegradation of HPEC. As for the thermally treated HPEC systems, the ESR spectra of the UVA- and UVB-irradiated HPEC (whole samples) containing 1 wt% HAS and measured at 300 K consist of two components, fast (F) and slow (S), which reflect the different dynamics of nitroxide radicals in the amorphous PP and rubber phases of the copolymers. The concentration profiles of HAS–NO in HPEC samples that were irradiated with UVA and UVB sources for the indicated treatment times are shown in Fig. 12 for HPEC1 and in Fig. 13 for HPEC2. The profiles are normalized according to the total radical concentration in whole samples. The profiles for "0 day" in Figs. 12 and 13 represent weak nitroxide signals detected even before the UV treatment. For UVA-treated HPEC1, the concentration of HAS–NO increases with treatment time, and the radicals are almost homogenously distributed within the sample depth (Fig. 12a). In UVB-irradiated HPEC1 samples, radicals are initially formed on the irradiated side, in a layer of thickness <1 mm; longer treatment times led to additional formation of radicals, to the displacement of the maximum intensity on the irradiated side into the sample depth, and also to a small increase of radical concentration on the back side.[33]

The nitroxide distribution profiles shown (Fig. 13a) for UVA-irradiated HPEC2 are dramatically different compared to similarly treated HPEC1 (Fig. 12a): for $t \leq$ 10 days, the radicals are concentrated mostly on the irradiated side of the sample,

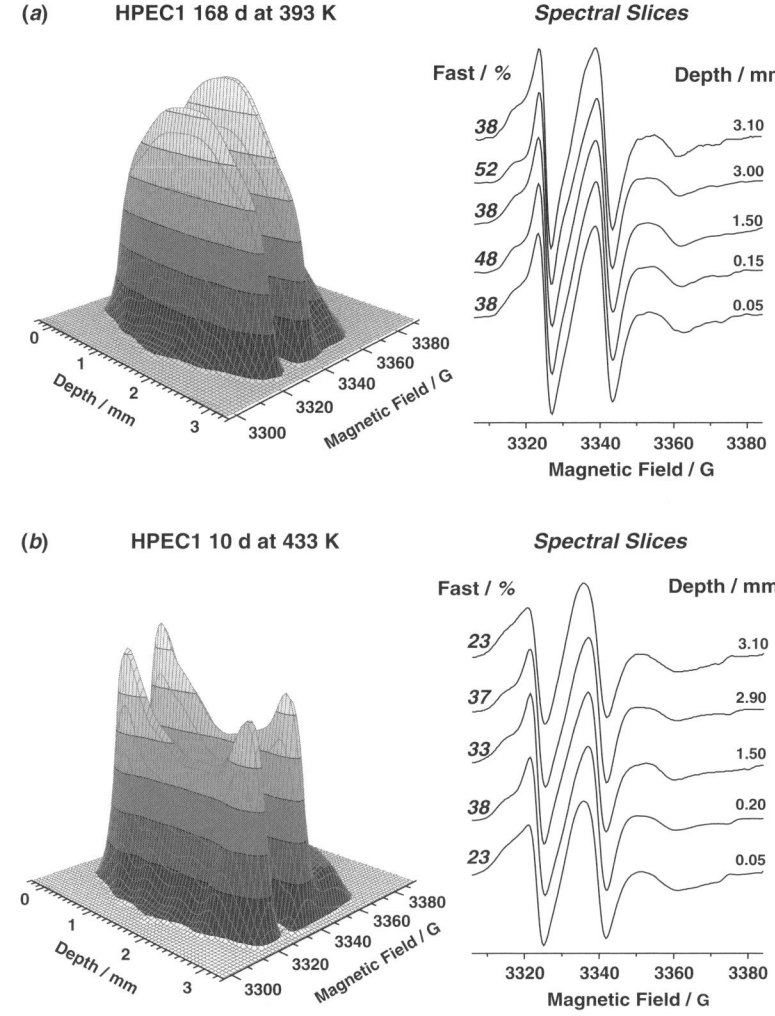

Fig. 11. The 2D spectral–spatial ESRI perspective plots at 300 K and corresponding spectral slices for HAS–NO in HPEC1 (2% HAS). (*a*) Treated at 393 K for 168 days. (*b*) Treated at 433 K for 10 days (Adapted from Ref. 31.)

in a layer of thickness ≈ 1 mm; for $t \geq 20$ days, the radical concentration decreases slightly on the irradiated side and nitroxide are present in increasing concentration on the back side. In the case of UVB-irradiated HPEC2 (Fig. 12*b*), the maximum concentration shifts toward the sample center for $t \geq 15$ days, and radicals are formed on the back side. After 40 days of irradiation, the highest radical concentration is observed on the back side of the sample. No significant differences in

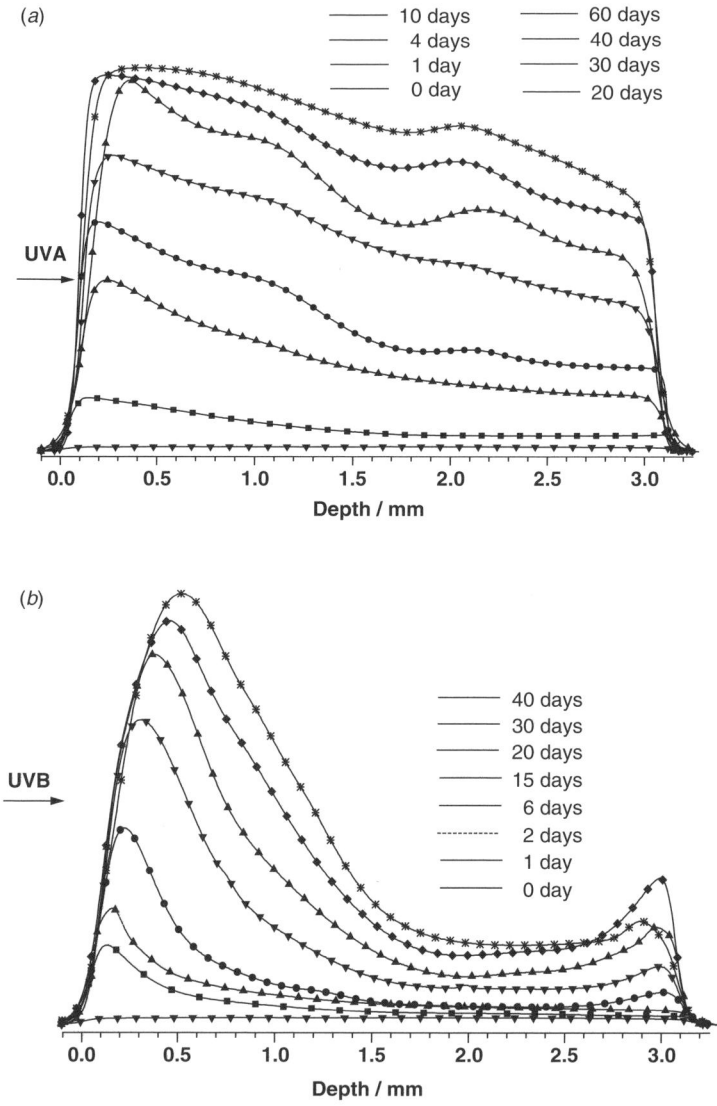

Fig. 12. Normalized concentration profiles of HAS–NO in HPEC1 (2% HAS) deduced from 1D ESRI, for the indicated irradiation times. (*a*) UVA irradiation and (*b*) UVB irradiation. Horizontal arrows indicate the irradiated side of the plaque. (Adapted from Ref. 33.)

ESR spectra and concentration profiles were detected for samples whose back side was covered by aluminum foil, indicating that nitroxides on the back side are not formed due to scattered light, but due to penetration of UV light through the plaque.

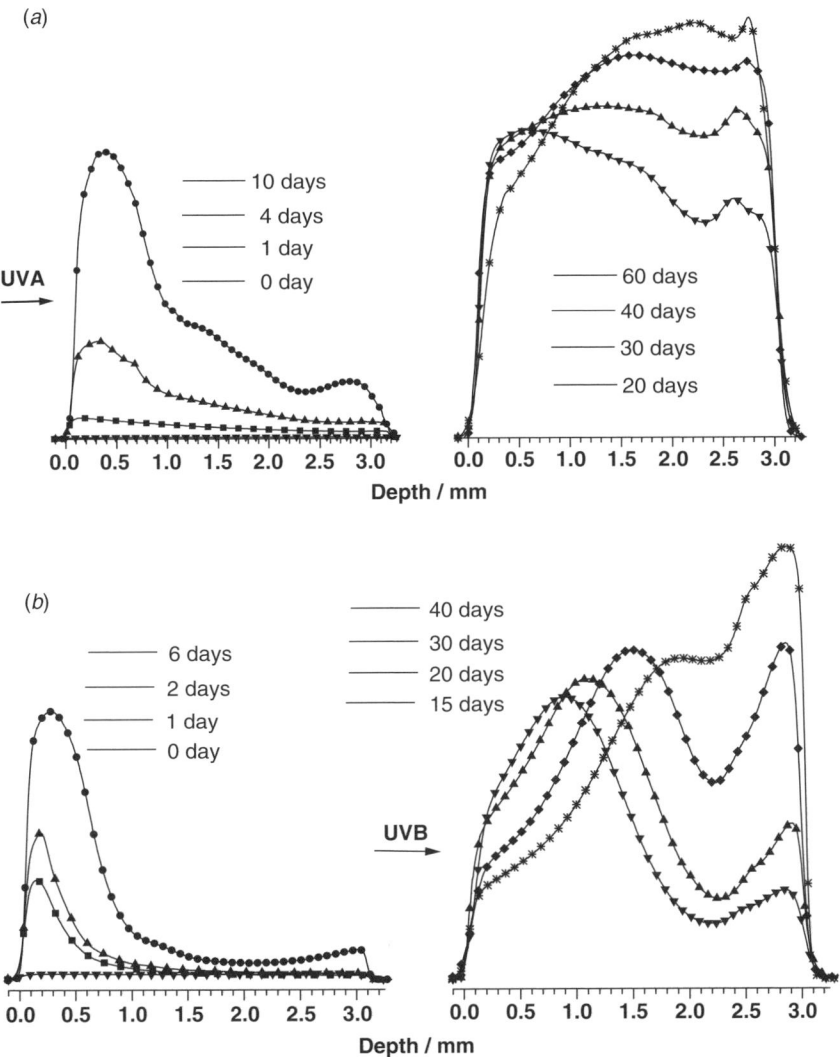

Fig. 13. Normalized concentration profiles of HAS–NO in HPEC2 (2% HAS) deduced from 1D ESRI, for the indicated irradiation times. (*a*) UVA irradiation and (*b*) UVB irradiation. Horizontal arrows indicate the irradiated side of the plaque. (Adapted from Ref. 33.)

The concentration profiles have clearly indicated the faster degradation rates in the UVB-irradiated samples. Moreover, the degradation is faster for HPEC2, which contains the larger amount of the PP component.

The 2D spectral–spatial ESRI contour plot for HAS–NO in UVA-irradiated HPEC1 for $t = 20$ days is shown in Fig. 14, together with spectral "virtual" slices and corresponding %F at several distances from the irradiated side. The [F]/[S] ratios

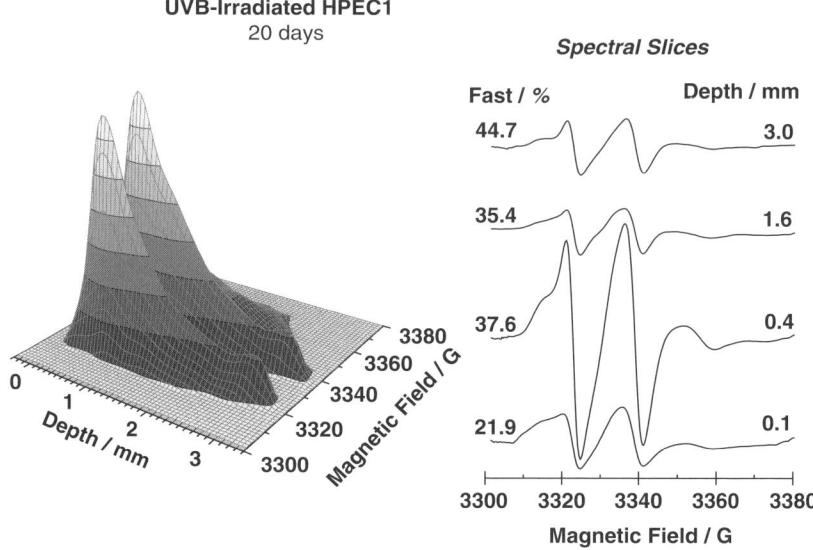

Fig. 14. The 2D spectral–spatial ESRI perspective plot and corresponding virtual slices of HAS–NO in UVB-irradiated HPEC1 (2% HAS), $t = 20$ days. %F as a function of depth from the irradiated side is indicated. (Adapted from Ref. 33.)

vary along sample depth: %F increases from the irradiated side toward the interior, and is about twice higher on the back side compared to the side exposed to the light. The high relative concentration of the F component on the nonirradiated side indicates the low consumption of the nitroxides in stabilization reactions.

The ESRI experiments have made possible the visualization of the profound mechanistic differences between UV and thermal degradation. In thermally treated HPEC systems, the rate of degradation is higher in HPEC1, which contains 25 wt% E. This effect can be explained by the higher diffusion rate of oxygen and reactant mobility at the aging temperatures (393 and 433 K) in copolymers containing more E. The higher degradation rate deduced in UV-irradiated HPEC2, which contained *less* E (10%), showed the dominant effect of PP sensitivity at the point of attack, the tertiary carbon, and suggests a different degradation mechanism in thermally treated and UV-irradiated copolymers, as suggested previously.[33] The effect of HAS content is also different in thermal and UV degradation: increased HAS content leads to a higher rate of degradation for thermally treated samples; but HAS is effective as a light stabilizer, as seen by inspection of UV and thermally treated samples.[33]

4.3. Comparison of ESRI and FTIR Methods

Fourier transform infrared spectroscopy is a well-established method for investigating polymer degradation, because of its ability to detect stable degradation products. In the photooxidation of polystyrene[43] and poly(α-methylstyrene),[44] for example, absorptions in the carbonyl and hydroxyl regions as a function of irradiation time have led to the identification of a number of degradation products and to a proposed reaction scheme. In many cases, the mechanism and extent of degradation, and the corresponding degradation products, depend on the irradiation wavelength. In the case of polystyrene, similar photoproducts, but in different concentrations, have been detected for irradiation with short and long wavelength, $\lambda = 253.7$ and ≥ 300 nm, respectively.[43]

The FTIR spectroscopy has also been used to follow the photodegradation of ABS[45–47] and HPEC[48,49] systems in the presence of oxygen. Jouan and Gardette investigated exposure to radiation $\lambda > 300$ nm of ABS films with thickness in the range 42–211 μm, as well as "packed multilayers", samples composed of several films separated by a layer of cardboard that allowed oxygen penetration.[45,46] The results indicated that the extent of degradation depends on the film thickness, and that directly exposed layers show more damage compared to the interior layers. As expected, the butadiene component was found to be the most susceptible to degradation.[45] The UV-absorbing photoproducts have been assigned to the degradation of polystyrene, initiated by diffusion of radicals formed in the butadiene phase into the SAN phase.[46] Photoacoustic (PA) FTIR spectroscopy has been used to follow chemical changes nondestructively in ABS polymers at depths in the range 5–16 μm from the irradiated side. The spectra were measured after exposure of the samples in the interior of cars, Florida exposure, and Xe-arc irradiation in a weathering chamber.[47] The results suggested that the B and AN components are degraded more extensively than the S component. Infrared microspectroscopy has been used for profiling the thermal oxidative products in nitrile rubber; the spatially heterogeneous degradation has been explained by diffusion-limited oxidation (DLO).[50] The IR intensity of the acrylonitrile component was not affected by thermal treatment up to 413 K.

Published degradation studies on ABS suggested that structural changes depend on the irradiation wavelength. However, the IR work on ABS degradation has been performed on films or "pressed-films",[45,46] and may not reflect DLO processes that occur in ABS polymers used in real applications.

The potential of ESRI studies was evaluated by comparing results obtained by ESRI and FTIR in the same ABS[25,29,51] and HPEC samples.[32] The evolution of signals in the carbonyl and butadiene regions, 1650–1800 and 966 cm^{-1}, respectively, in ABS (0, 1, and 2% HAS) as a function of thermal aging time at 393 K was measured by attenuated total reflectance (ATR)FTIR in an outer layer of thickness 500 μm. Results from the carbonyl region were compared with the total nitroxide concentration in moles per gram of sample; and results from the butadiene region were compared with %F in ABS (1 and 2% HAS). The presence and increase of the carbonyl peak and the decrease of the butadiene peak were detected by ATR–FTIR spectroscopy only in the advanced stages of aging. The ESR results indicated major changes in both nitroxide concentration and %F even in the early stages of degradation, for treatment times of 200 h or even less; under these conditions, the intensity of the carbonyl peak and the decrease of the butadiene peak were negligible. The major advantage of the ESRI approach is the capability to provide details on aging in the early stages of the process. For example, the evolution from the flat profile to the DLO regime was seen by both 1D and 2D ESRI (Figs. 7–9): in the nitroxide profiles deduced by 1D ESRI and in the variation of %F, the spectral component present in the butadiene-rich domains, from 2D ESRI.[25] Additional details were obtained by examination of the total nitroxide concentration and %F in whole samples.

The FTIR spectroscopy can also be used as a direct tool for obtaining morphological information; this information cannot be obtained from ESR or ESRI studies. In our FTIR studies of HPEC, the crystallinity of the PP and EPR domains was determined by inspection of stereoregularity peaks at 1167, 998, 973, and 841 cm^{-1} for isotactic PP (iPP) and peaks at 720 and 730 cm^{-1} for PE. The crystallinity is strongly correlated with the regularity ratio (RR) A_{998}/A_{973}. The peak at 730 cm^{-1} reflects the CH_2 rocking vibration in the crystalline phase, and is an indicator of closely packed (crystalline) methylene chains. The peak at 720 cm^{-1} is associated with both amorphous and crystalline phases. The relative intensity of the two peaks can be used to estimate the relative crystallinity of PE samples.[31,52,53] The corresponding absorbance ratios A_{730}/A_{720} determined by deconvolution of the signal from the CH_2 rocking vibration are 0.59 (HPEC1) and 0.44 (HPEC2), indicating a higher crystallinity in HPEC1.

An increase in the regularity ratio RR due to thermal treatment has been detected,[32] and was assigned to an increase of the degree of crystallinity.[54] The results can be rationalized by assuming that crystallization processes in PP domains take place as the degradation processes advance, due to the formation of shorter chains that more easily rearrange into ordered domains, as has been proposed for pure PP.[55,56]

The ESRI and FTIR data are in agreement on the antiprotective effect of HAS in the advanced stage of thermal aging for both ABS and HPEC systems. The conclusion from the spectroscopic data are in agreement with the visual appearance of the samples: for the same treatment time, considerable more discoloration and shape distortion were seen in samples containing more Tinuvin 770.[25,32]

4.4. Sensitivity of Crystalline Polymers to Microtoming

To study the effect of microtoming on crystalline polymers, we have compared %F in the ESR spectra at 300 K of thermally treated HPEC for 107 days at 393 K in two samples. The first sample was the typical cylinder of diameter and height ≈4 mm, cut from the plaque after treatment. The same sample was subsequently microtomed into 50-μm slices, followed by the transfer of all slices to the ESR sample tube. In the whole sample %F = 41 ± 2, compared to %F = 29 ± 2 in the microtomed sample. The lowering of the relative intensity of the F component in the amorphous phase is related to the increased ordering and further crystallization in the PP domains as a result of microtoming. These domains restrain the mobility of nitroxide radicals located in the vicinal amorphous phases, and to the transformation from F- to S-type nitroxides.[32]

The decrease of %F was not the only result of microtoming: The extreme separation measured immediately after microtoming was unchanged, but increased over a period of several days by ≈1 G at 300 K. The increase was detected in all ESR spectra measured in the temperature range 100–300 K. The lowering of %F and increase of the extreme separation are assigned to increased ordering and further crystallization upon microtoming. The results were interpreted in terms of an additional phase, the "rigid amorphous phase",[57,58] whose extent and dynamics is reflected in the ESR spectra of the nitroxides. This phase is considered to be a part of the amorphous phase that is modified by the proximity to the crystalline phase. In addition to the existence of the interphase, the picture that emerged from the ESR study includes a *gradient* in dynamics of the interphase, which can be visualized directly from the spectra, or by spectral simulations.[32] The ESR spectra have great sensitivity because the effect is observed mainly on the amorphous component with narrow lines (the F component), and leads to large differences in the *height* of the corresponding signals. These conclusions demonstrate that nondestructive methods for polymer degradation are preferable, particularly in the case of crystalline systems.

5. CONCLUSIONS

Polymer degradation and stabilization is a challenging topic of great fundamental and technological importance: Fundamental, because it reflects changes in the properties of polymeric materials due to chemical phenomena that can vary as a function of a complex set of environmental conditions; and technological, not only because use of polymers is increasing steadily, but also because of the increasing sophistication needed in their properties. Historically, materials were used long before their properties were fully understood. In recent years, analytical tools, such as microscopy, imaging, and computational techniques have made possible the determination of exquisite structural and functional details of materials.

The ESRI experiments described in our publications and summarized in this chapter led to spatially resolved information on the effect of treatment conditions, amount of stabilizer, and polymer composition on the degradation rate. In the heterophasic systems studied in our laboratory, ESRI has identified specific morphological domains where chemical processes are accelerated. The combination of 1D and 2D spectral–spatial ESRI experiments led to mapping of the stabilizer consumption on two length scales: within the sample depth on the scale of a few millimeters, and within morphological domains on the scale of a few micrometres.

The ESRI is a nondestructive method for the study of degradation, which is an important advantage, especially for crystalline polymers. The major advantage of ESRI compared with FTIR methods is its sensitivity to early events in the aging process. Further developments of ESRI methods are expected to be of help in the ultimate goal: accurate predictions of lifetimes for polymeric materials and a better understanding of the environmental factors.

The ESRI method requires the presence of a contrast agent, HAS–NO in our work. The implication is that ESRI is an exceptionally sensitive and specific method for observing degradation in HAS-stabilized polymers, but not in polymers in general; this advantage and this limitation is similar to ESR methods, which are specific to and applicable only when radicals are present.

ACKNOWLEDGMENTS

The development of ESRI methods in our laboratory was made possible by the sustained and generous support of the Polymers Program of the National Science Foundation. Early efforts were also supported by a Founders Fellowship of the American Association of University Women (AAUW) to S. Schlick. We acknowledge with gratitude the collaboration with John L. Gerlock, which was supported by the University Research Program of Ford Motor Company. We are grateful to K. Ohno, P. Eagle, and S.-C. Kweon for getting us started in the ESRI field; to J. Pilar and A. Marek from the Institute of Macromolecular Chemistry in Prague, Czech Republic for their major contributions to all aspects of ESRI in the Detroit laboratory; and to Zbigniew Sojka and Tomasz Spalek of the Jagiellonian University, Cracow, Poland for implementation of the Genetic Algorithm in the ESRI software, with support from KBN project 3T0 9A 14726.

REFERENCES

1. O'Donnell, J.H. Radiation Chemistry of Polymers, In *The Effects of Radiation on High-Technology Polymers*; Reichmanis, E., O'Donnell, J.H., Eds.; American Chemical Society: Washington, D.C., 1989; Chapt. 1, pp. 1–13.
2. *Handbook of Polymer Degradation*, Hamid, S.H., Amin, M.B., Maadhah, A.G., Eds.; Marcel Dekker: New York, 1992.
3. *Irradiation of Polymeric Materials: Processes, Mechanisms, and Applications*; Reichmanis, E., Frank, C.W., O'Donnell, J.H., Eds.; American Chemical Society: Washington, DC, 1993.
4. *Polymer Durability: Degradation, Stabilization and Lifetime Prediction*; Clough, R.L.; Billingham, N.C.; Gillen K.T., Eds.; Adv. Chem. Series 249, American Chemical Society: Washington, DC, 1996.
5. Kevan, L.; Schlick, S. *J. Phys. Chem.* **1986**, *90*, 1998, and references cited therein.
6. Alonso-Amigo, M.G.; Schlick, S. *Macromolecules* **1987**, *20*, 795.
7. Rabek, J.F. *Photostabilisation of Polymers, Principles and Applications*; Elsevier: London, 1990.
8. (a) Gerlock, J.L.; Bauer, D.R.; Briggs, L.M. Part I, *Polym. Deg. Stab.* **1986**, *14*, 53; (b) Gerlock, J.L.; Riley, T.; Bauer, D.R. *Polym. Degrad. Stab.* Part II, 73; (c) Gerlock, J.L.; Bauer, D.R. *Polym. Degrad. Stab.* Part III, 97.
9. Faucitano, A.; Buttafava, A.; Martinotti, F.; Bortolus, P. *J. Phys. Chem.* **1984**, *88*, 1190.
10. Faucitano, A.; Buttafava, A.; Martinotti, F.; Greci, L. *Polym. Deg. Stab.* **1992**, *35*, 211 and references cited therein.
11. Pospisil, J. *Adv. Polym. Sci.* **1995**, *124*, 87.
12. Brede, O.; Beckert, D.; Windolph, C.; Gottinger, H.A. *J. Phys. Chem. A*, **1998**, *102*, 1457.
13. (a) Gillen, K.T.; Clough, R.L. *Polymer* **1992**, *33*, 4359. (b) Wise, J.; Gillen, K.T.; Clough, R.L. *Radiat. Phys. Chem.* **1997**, *49*, 565.
14. Gillen, K.T.; Clough, R.L. In *Handbook of Polymer Science and Technology*; N.P. Cheremisinoff, Ed.; M. Dekker: New York, 1989; Chapt. 6, pp.167–202.
15. Gillen, K.T.; Clough, R.L.; Quintana, C.A. *Polym. Degrad. Stab.* **1987**, *17*, 31.
16. Ohno, K. In *EPR Imaging and In Vivo EPR*; Eaton S.S., Eaton, G.R., Ohno, K., Eds.; CRC Press: Boca Raton, FL, 1991; Chapt. 18, p. 181.

REFERENCES

17. Lucarini, M.; Pedulli, G.F.; Borzatta, V.; Lelli, N. *Res. Chem. Intermed.* **1996**, *22*, 581.
18. Lucarini, M.; Pedulli, G.F. *Angew. Makromol. Chem.* **1997**, *252*, 179.
19. Franchi, P.; Lucarini, M.; Pedulli, G.F.; Bonora, M.; Vitali, M. *Macromol. Chem. Phys.* **2001**, *202*, 1246.
20. Marek, A.; Kapralkova, L.; Schmidt, P.; Ptleger, J.; Humlicek, J.; Pospisil, J.; Pilar, J. *Polym. Degrad. Stab.* **2006**, *91,* 444.
21. Motyakin, M.V.; Gerlock, J.L.; Schlick, S. *Macromolecules* **1999**, *32*, 5463.
22. Kruczala, K.; Motyakin, M.V.; Schlick, S. *J. Phys. Chem. B* **2000**, *104*, 3387.
23. Motyakin, M.V.; Schlick, S. *Macromolecules* **2001**, *34*, 2854.
24. Schlick, S.; Kruczala, K.; Motyakin, M.V.; Gerlock, J.L. *Polym. Degrad. Stab.* **2001**, *73/3,* 471.
25. Motyakin, M.V.; Schlick, S. *Polym. Degrad. Stab.* **2002**, *76(1),* 25.
26. Motyakin, M.V.; Schlick, S. *Macromolecules* **2002**, *35*, 3984.
27. Lucarini, M.; Pedulli, G.F.; Motyakin, M.V.; Schlick, S. *Progress Polym. Sci.* 2003, *28*, 331.
28. Schlick, S.; Motyakin, M.V. In *Instrumental Methods in Electron Magnetic Resonance, Biological Magnetic Resonance*, Vol. 21; Bender, C.J., Berliner, L.J., Eds.; Kluwer Academic/Plenum Publishing Corporation: New York, 2004, pp. 349–384.
29. Motyakin, M.V.; Schlick, S. *Polym. Degrad. Stab.* **2006**, *91/7*, 1462.
30. Schlick, S.; Jeschke, G. Electron Spin Resonance, In *Encyclopedia of Polymer Science and Engineering*, Kroschwitz, J.I., Ed.; Wiley-Interscience: New York, 2004; Chapt. 9, pp. 614–651; web edition: 15 July 2004. Hardcopy edition: 15 Aug. 2004.
31. Kruczala, K.; Varghese, B.; Bokria, J.G.; Schlick, S. *Macromolecules* **2003**, *36*, 1899.
32. Kruczala, K.; Bokria, J.G.; Schlick, S. *Macromolecules* **2003**, *36*, 1909.
33. Kruczala, K.; Aris, W.; Schlick, S. *Macromolecules* **2005**, *38*, 6979.
34. Steeman, P.A.M; Meier, R.J.; Simon A., Gast, J. *Polymer* **1997**, *38*, 5455.
35. Aoki, Y.; Hatano, A.; Tanaka, T.; Watanabe, H. *Macromolecules* **2001**, *34*, 3100.
36. Pukansky, B. In *Polymeric Materials Encyclopedia*; Salamone, J.C., Ed.; CRC Press: Boca Raton, FL, **1996**, pp. 6615–6623.
37. Albizzati, E.; Giannini, U.; Collina, G.; Noristi, L.; Resconi, L. In *Polypropylene Handbook*; Moore, E.P., Jr., Ed.; Hanser Publishers: Munich, **1996**. Chapt. 2, p. 92.
38. (a) Mirabella, F.M. *Polymer* **1993**, *34*, 1729. (b) Mirabella, F.M. *J Polym. Sci. B: Polym. Phys.* **1994**, *32*, 1205.
39. Mirabella, F.M.; Jr.; McFaddin, D.C. *Polymer* **1996**, *37*, 931.
40. Varghese, B.; Schlick, S. *J. Polym. Sci. B Polym. Phys.* 2002, *40*, 415 and 424.
41. Li, H.; Ruckenstein, E. *Polymer* **1995**, *36*, 2647.
42. Kruczala, K.; Varghese, B.; Bokria, J.G.; Schlick, S. *Macromolecules* **2003**, *36*, 1899.
43. Mailhot, B.; Gardette, J.L. *Macromolecules* **1992**, *25*, 4127 and references cited therein.
44. Mailhot, B.; Jarroux, N.; Gardette, J.L. *Polym. Degrad. Stab.* **2000**, *68*, 321.
45. Jouan, X.; Gardette, J.L. *J. Polym. Sci. A* **1991**, *29*, 685.
46. Jouan, X.; Gardette, J.L. *Polym. Degrad. Stab.* **1992**, *36*, 91.
47. Carter III, R.O.; McCallum, J.B. *Polym. Degrad. Stab.* **1994**, *45*, 1
48. Delprat, P.; Duteurtre, X.; Gardette, J.-L. *Polym. Degrad. Stab.* **1995**, *50*, 1.

49. Pennington, B.D.; Ryntz, R.A.; Urban, M.W. *Polymer* **1999**, *40*, 4795.
50. Celina, M.; Wise, J.; Ottesen, D.K.; Gillen, K.T.; Clough, R.L. *Polym. Degrad. Stab.* **1998**, *60*, 493.
51. Bokria, J.G.; Schlick, S. *Polymer* **2002**, *43*, 3239.
52. Koenig, J.L. *Spectroscopy of Polymers*; ACS: Washington, DC, 1992; Chapt. 4.
53. Smith, B. *Infrared Spectral Interpretation*; CRC Press: Boca Raton, FL, 1999; pp. 38 and 180.
54. Zhu, X.; Yan, D.; Fang, Y. *J. Phys. Chem. B* **2001**, *105*, 12461.
55. Gensler, R.; Plummer, C.J.G.; Kausch, H.-H; Kramer, E.; Pauquet, J.-R.; Zweifel, H. *Polym. Degrad. Stab.* **2000**, *67*, 195.
56. Rabello, M.S.; White, R.J. *Polymer* **1997**, *38*, 6379 and 6389.
57. Cheng, S.Z.D.; Wunderlich, B. *Macromolecules* **1988**, *21*, 789.
58. Wunderlich, B. *Thermal Analysis*; Academic Press: Boston, 1990.

10

ESR STUDIES OF PHOTOOXIDATION AND STABILIZATION OF POLYMER COATINGS

DAVID R. BAUER AND JOHN L. GERLOCK

Ford Motor Company, Dearborn, Michigan

Contents

1. Introduction — 256
2. Free-Radical Oxidation and Stabilization — 258
3. Experimental — 260
   3.1. Nitroxide Concentration by ESR — 260
   3.2. UV Exposures — 262
   3.3. Sample Preparation for Nitroxide Decay Experiments — 263
   3.4. Sample Preparation for Nitroxide Measurements in HALS Stabilized Coatings — 264
4. Free-Radical Photoinitiation Rates by Nitroxide Decay — 264
   4.1. Derivation of Nitroxide Radical Decay Kinetics and Free-Radical Photoinitiation Rates — 264
   4.2. Photoinitiation Rate Measurements on Model Systems and Nonweathered Coatings — 265
   4.3. Photoinitiation Rate Measurements on Weathered Coatings — 269
5. Nitroxide Kinetics and HALS Stabilization of Coatings — 272
6. Conclusions — 277
Acknowledgments — 277
References — 277

Advanced ESR Methods in Polymer Research, edited by Shulamith Schlick.
Copyright © 2006 John Wiley & Sons, Inc.

1. INTRODUCTION

Modern automotive paint systems comprise many layers, each with specific functions, Fig. 1.[1] The electrocoat layer provides corrosion protection for the metal body while the spray primer improves surface finish, protects the electrocoat from ultraviolet (UV) and visible light, and contributes to chip resistance. The basecoat is the color layer and is responsible for the specific appearance. The clearcoat provides high gloss and resistance to chemicals (e.g., acid rain and fuel) and mechanical stresses (scratch and mar). The clearcoat is also the layer most responsible for maintaining appearance (gloss retention, prevention of peeling and/or cracking) under exposure to UV light, moisture, and other environmental factors. Clearcoats are based on polymer resins that contain reactive functional groups. These functional groups react with cross-linkers to from a cross-linked polymer network during the paint baking process. During the period of time when most of this work was carried out, clearcoats were based on acrylic copolymers. Hydroxy containing monomers, such as hydroxyethylacrylate, are incorporated into the acrylic polymer to provide reactive functionality. In some applications, hydroxyl functional polyesters can also be used. Typical cross-linkers used during this period were either based on melamine formaldehyde resins or on isocyanates.[2,3] Typical structures are shown below.

Fully alkylated melamine   **Partially alkylated melamine**   **Isocyanate**

Two basic types of melamine formaldehyde resins are used: fully and partially alkylated. In fully alkylated melamines, the methoxy group reacts with hydroxyl

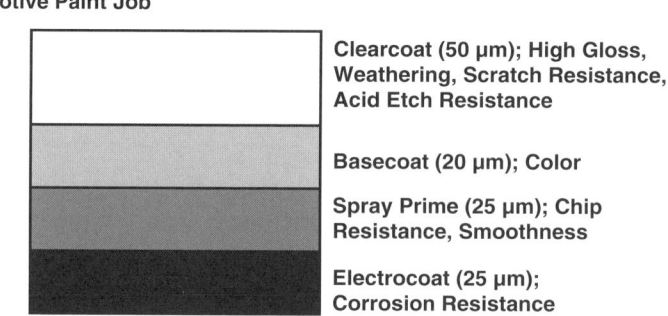

Fig. 1. Modern automotive paint system. The studies in this chapter are primarily concerned with analysis of photooxidation in the very top clearcoat layer.

INTRODUCTION

groups on the acrylic polymer to form ether cross-links. This reaction is catalyzed by strong acid. This same reaction occurs in partially alkylated melamines. In this case, weak acids can be used to catalyze the reaction. In addition, partially alkylated melamines undergo self-condensation to form additional cross-links. The requirements for acid catalysis of cure have implications for the choice of additives that can be used to inhibit oxidation. Isocyanates also react with hydroxyl groups on the acrylic polymer to form urethane cross-links.

Over the past 20 years, automotive paint technologies have changed dramatically in response to various pressures. The need to reduce solvent emissions from painting applications resulted in the replacement of low solids ($>$5.5-lb gal^{-1} solvent) coatings based on higher molecular weight ($>$8000) acrylic polymers with high solids ($<$3.5-lb gal^{-1} solvent) coatings based on lower molecular weight ($<$3000) polymers. The need for coatings with improved acid etch resistance is leading to the total or partial replacement of melamine formaldehyde cross-linked coatings containing hydrolysis prone ether cross-links with more hydrolysis resistant cross-linking technologies including urethanes, carbamates, epoxy-acids, and silanes.[4] Replacing standard monocoat paint systems with clearcoat–basecoat paint systems was one of the most dramatic changes implemented. This substitution substantially improved initial paint appearance and promised to substantially improve the retention of gloss after extended weather exposure. Unfortunately, the weathering chemistry of clearcoat–basecoat paint systems was not sufficiently understood and consequently unacceptable levels of catastrophic failure by cracking and peeling were observed after $<$5 years in service. Clearcoat–basecoat paint systems are intended to retain their excellent appearance for at least 10 years in service. The work described here was initiated to seek relationships between clearcoat–basecoat paint system failure by gloss loss, cracking and/or peeling, and clearcoat photooxidation kinetics.

This chapter describes two applications of ESR to the understanding of photooxidation and stabilization kinetics of clearcoats based on acrylic polymers cross-linked with melamine formaldehyde resins or isocyanates. After summarizing the basic free radical oxidation and stabilization reactions, a technique for measuring the photoinitiation rate of free radicals in cross-linked polymers using a procedure that dopes the polymer with a stable nitroxide radical is described. On exposure to UV light, the free radicals that are formed are scavenged by the nitroxide. The free-radical formation rate can be derived from the rate of loss of nitroxides. Free-radical formation rates have been measured as a function of exposure condition, polymer and cross-linker type, and as a function of exposure time. The free-radical formation rates can be related to other measures of photooxidation to yield detailed understanding of the factors that control photooxidation. Next, we describe how to gain insight into clearcoat photooxidation, stabilizer effectiveness, and longevity by following the concentration behavior of the nitroxides formed when clearcoats stabilized with hindered amine light stabilizer (HALS) are subjected to weather exposure.[5] This measurement has considerably broader application than the former because HALS is widely used to stabilize polymers against oxidation during outdoor exposure. It can be estimated that $> 2 \times 10^7$ kg of HALS is used worldwide every year in polymers intended for outdoor use. Accurate quantification of nitroxide concentrations is critical to both applications and techniques for quantitative ESR measurements are described.

2. FREE-RADICAL OXIDATION AND STABILIZATION

The basic free-radical oxidation process that is used to interpret polymer coating photooxidation kinetics has been described in detail and is presented briefly below:[6–9]

$$\text{Chromophore} + h\nu \xrightleftharpoons[k_{rlx}]{k_{ex}} \text{chromophore}^* \xrightarrow{k_r} 2R\bullet \quad \text{(R1)}$$

$$R\bullet + O_2 \xrightarrow{k_{ox}} ROO\bullet \quad \text{(R2)}$$

$$ROO\bullet + RH \xrightarrow{k_{abs}} ROOH + R\bullet \quad \text{(R3)}$$

$$2ROO\bullet \xrightarrow{k_t} \text{Oxidation} \quad \text{(R4)}$$

$$ROOH + h\nu \xrightarrow{k_{hy,i}} 2R\bullet \quad \text{(R5)}$$

$$ROOH + h\nu \xrightarrow{k_{hy,d}} \text{Oxidation} \quad \text{(R5)}$$

In R1, a chromophore absorbs a photon with a rate constant k_{ex}. The excited (ex) state chromophore can either relax (rlx) back to the ground state (k_{rlx}) or decompose to form free radicals (k_r). In R2, carbon centered free radicals (R•) react with oxygen to form a peroxy radical. This reaction is usually diffusion limited. For surface coatings where the oxygen concentration is high, the concentration of peroxy radicals is much higher than that of carbon centered radicals. The peroxy radicals can either abstract a hydrogen atom (R3) to yield a hydroperoxide and a carbon centered radical, or can terminate with another peroxy radical to form oxidation products, R4. The formation of hydroperoxides is key to the evolution of the photooxidation kinetics since they can become the main source of free radicals in the polymer. Light absorption by a hydroperoxide leads either to the formation of radicals or to the formation of other nonradical products. Some of these products may also be chromophores.

Based on R1–R5, the photoinitiation rate (PIR) of free radicals can be written in terms of the relevant chromophores and rate constants,

$$\text{PIR} = \frac{I*k_r*k_{ex}[A]}{k_r + k_{rlx}} + I*k_{hy,i}[ROOH] \quad (1)$$

where [A] represents the concentration of nonhydroperoxide chromophore, [ROOH] is the hydroperoxide concentration, and I is the flux of UV light. Both the hydroperoxide and nonhydroperoxide chromophore concentration will vary with exposure time. In the case of clearcoats that are based on acrylic polymers, the initial hydroperoxide level is very low and the initial free radical formation results from photolysis of other chromophores (typically ketones).[9-12] Under these conditions and assuming that the radical concentration reach a stationary state, the overall photooxidation rate can be given by,[6,13]

$$\text{PhotooxRate} = k_{abs}[RH]\sqrt{\frac{PIR}{k_t}} \qquad (2)$$

That is, the photooxidation rate increases as the square root of the photoinitiation rate, because the chain length of the oxidation is controlled by the bimolecular termination of peroxy radicals. As the exposure proceeds, the hydroperoxide concentration increases, eventually approaching a relatively constant level.[10] For a stationary hydroperoxide concentration, the photooxidation rate is equal to[8]

$$\text{PhotooxRate} = I^*(k_{hy,i} + k_{hy,d})[ROOH]_{SS} \qquad (3)$$

If hydroperoxides are the dominant chromophore, then the stationary hydroperoxide concentration can be given by the following expression,[8]

$$[ROOH_{SS}] = \frac{(K_{abs}[RH])^2}{k_t} \frac{k_{hy,i}}{(k_{hy,i} + k_{hy,d})^2} \qquad (4)$$

In this case, the photooxidation rate is linearly proportional to the photoinitiation rate, because the kinetic chain length for photooxidation is controlled by hydroperoxide decomposition kinetics rather than bimolecular termination.

Two different types of photostabilizers are typically added to automotive clearcoats, ultraviolet light absorbers (UVA) and HALS. Clearcoat UVA acts to reduce the intensity of light inside the paint system. That is, it reduces clearcoat PIR and screens light sensitive coating layers underlying the clearcoat (e.g., basecoat) from UV light. As a result, clearcoat UVA also acts to extend HALS longevity. Clearcoat UVA performance is not discussed further in this chapter. Typical HALS types that are used in clearcoats are shown below.[14]

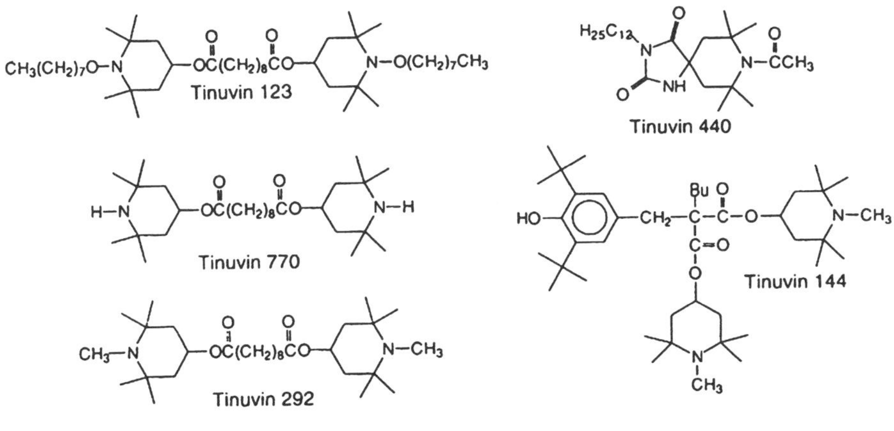

TYPICAL HALS

The main difference in the various HALS is the substituent on the hindered nitrogen of the piperidine ring. Hindered amine light stabilizers reduce the photooxidation rate by scavenging free radicals and thus interfering with the oxidation cycle of R2–R3. The basic reactions that are important are summarized in reactions R6–R10.[15]

$$\rangle NR \xrightarrow{k_{NR}} \rangle NO\bullet \quad (R6)$$

$$\rangle NO\bullet + R\bullet \xrightarrow{k_{NO\bullet}} \rangle NOR \quad (R7)$$

$$\rangle NOR + ROO\bullet \xrightarrow{k_{NOR}} \rangle NO\bullet + ... \quad (R8)$$

$$\rangle NO\bullet + h\nu \xleftrightarrow{k_{NO\bullet,ex}, k_{NO\bullet,rlx}} \rangle NO\bullet^* + RH \xrightarrow{k_{NO\bullet,abs}} \rangle NOH + R\bullet \quad (R9)$$

$$\rangle NOH + ROO\bullet \xrightarrow{k_{NOH}} \rangle NO\bullet + ROOH \quad (R10)$$

The first step, R6, converts the HALS initially added to the clearcoat, parent HALS, into inhibition cycle, R7 and R8, products. These reactions compete with R2 and R3 lowering the stationary radical concentration, which in turn lowers the hydroperoxide concentration and the photooxidation rate. The rate constants and radical concentrations are such that only a small fraction (~5%) of the HALS stabilizer is in the form of nitroxide. Although nitroxides are thermally stable, they are not photolytically stable. Nitroxides absorb light, and excited-state nitroxides can abstract hydrogen atoms to initiate free-radical formation. These reactions have been discussed in detail.[16–18] Reactions R9 and R10 are important both for the nitroxide decay measurement of free radical formation and in limiting the ultimate effectiveness of HALS.[10,19]

3. EXPERIMENTAL

3.1. Nitroxide Concentration by ESR

Both the nitroxide decay measurements of free-radical photoinitiation rates and nitroxide kinetics during HALS stabilization depend on accurate, quantitative measurements of nitroxide concentrations in cross-linked polymers. Quantification of radical concentrations by ESR requires a suitable primary standard, careful sample preparation, a reference standard with which to monitor spectrometer performance, and most important, reproducible positioning of the samples in the resonance cavity of the spectrometer.[20] Most of the experiments described here were carried out with a Bruker-IBM ER 200 D spectrometer equipped with an Aspect 2000 Data System. Because these coatings are cured at temperatures as high as 130°C, the primary nitroxide standard, which was introduced into the coating prior to cure, had to be nonvolatile. The following nitroxide was chosen as the primary standard:[20]

$$CH_3(CH_2)_{16}CH_2NHCNH-\underset{}{\overset{O}{\|}}-\langle N-O^\cdot \rangle$$

Nitroxide I

Known concentrations of nitroxide I were mixed with coating solutions and the solutions smeared on either precision cut quartz plates or disks. The coatings were cured and the coating weight was measured. For various experiments, two different all quartz sample holders were designed for the two sample types, which allow for rapid, accurate positioning of the samples in the resonance cavity, Fig. 2.[20,21] Using either of these holders, samples could be inserted and removed from the cavity with less than $\pm 1\%$ variation in signal intensity. The holders also contained a ruby chip that was used as a reference standard to follow changes in spectrometer performance. Measurements of the first derivative of the nitroxide absorption signal shown in Fig. 3 are double integrated to yield a nitroxide signal area.[22] The nitroxide signal in Fig. 3 is typical of an "immobilized" nitroxide. The relationship between signal area and nitroxide mass was found to depend on the coating thickness for the quartz plate samples. These samples were also affected by nonuniformity of film thickness since only a fraction of the coating is in the detection zone of the cavity. A correction factor for film weight was developed. The corrected signal areas were found to be linear in total nitroxide content in the film and absolute concentrations could be determined to better than $\pm 5\%$ as shown in Fig. 4.[20] The film thickness effect was not observed

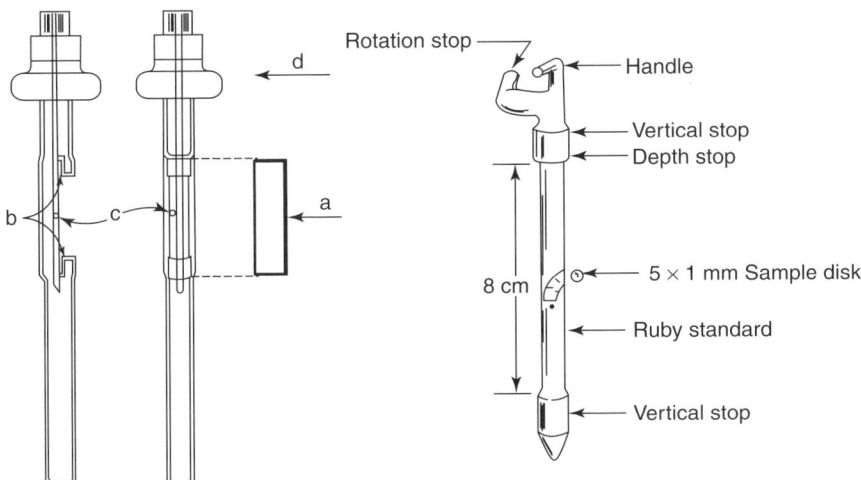

Fig. 2. Flat plate and disk sample holders for rapid accurate positioning of cured films in resonance cavity. Indicated are the plate (*a*), slots to contain plate (*b*), reference sample (*c*), and depth control ring (*d*). (Reprinted with permission from Refs. 20 and 21.)

for the small quartz disks because they fit entirely within the cavity's detection zone. Despite the smaller sample size, accurate nitroxide concentrations could be determined down to 0.2×10^{-6} mol g^{-1}.

3.2. UV Exposures

Two different UV exposure cabinets were employed in these studies. The first used a UV fluorescent bulb that has UV output in the range 270–390 nm with a peak at 313 nm.[23] The second cabinet employed a xenon arc lamp whose light output passed

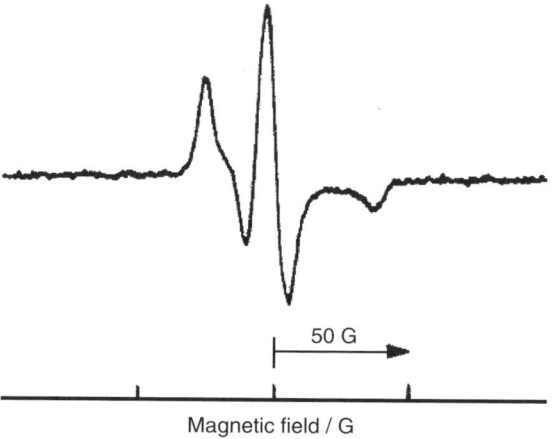

Fig. 3. First derivative X-band ESR spectrum from nitroxide in a cured coating film. Experimental details are given in Ref. 22. (Reprinted with permission from Ref. 22.)

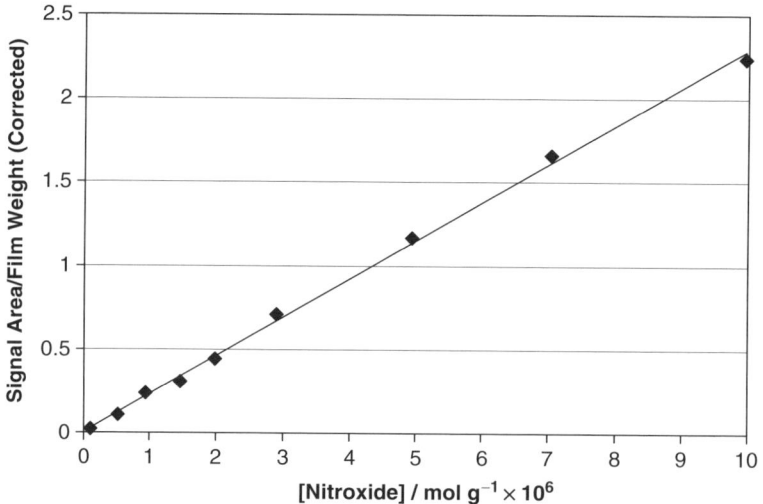

Fig. 4. Corrected signal area versus nitroxide concentration.

EXPERIMENTAL

through two borosilicate glass filters. The wavelength distributions of these sources are compared to that for sunlight in Fig. 5. The fluorescent bulb has substantial intensity at wavelengths shorter than sunlight, which is known to accelerate photochemical reactions. This exposure will be referred to as "harsh". The xenon arc light is a much closer (though not perfect) match to sunlight. This exposure will be referred to as "near ambient".[24] Commercial sample chambers were modified to accommodate accurate sample placement to insure constant, uniform light intensity, temperature, and humidity.

3.3. Sample Preparation for Nitroxide Decay Experiments

For nitroxide decay measurements of the free-radical photoinitiation rates in nonweathered coatings, nitroxide **I** was added to coating solutions to yield concentrations of nitroxide ranging from $1-10 \times 10^{-6}$ mol g^{-1}. For measurements on weathered coatings, the nitroxide had to be introduced to the coating after cure and weathering. This was done by exposing the coating to a vapor of nitroxide, **II** (TEMPOL), by placing the coating in a special container shown in Fig. 6 that also contained a 90:10 mixture of CH_2Cl_2 and CH_3OH.[21]

Nitroxide II

Samples were allowed to equilibrate at 60°C and were found to contain uniform concentrations of nitroxide, **II**. Hydrogen bonding between the nitroxide hydroxyl

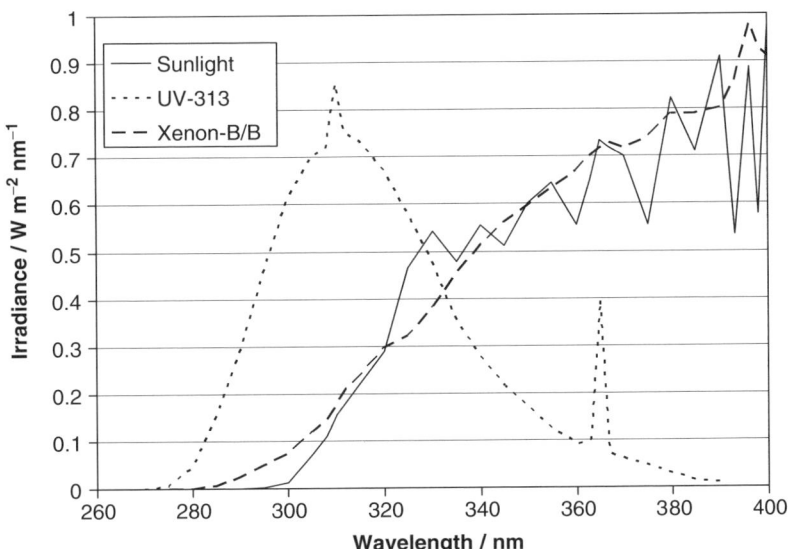

Fig. 5. Wavelength distribution of light sources used in this work compared to sunlight.

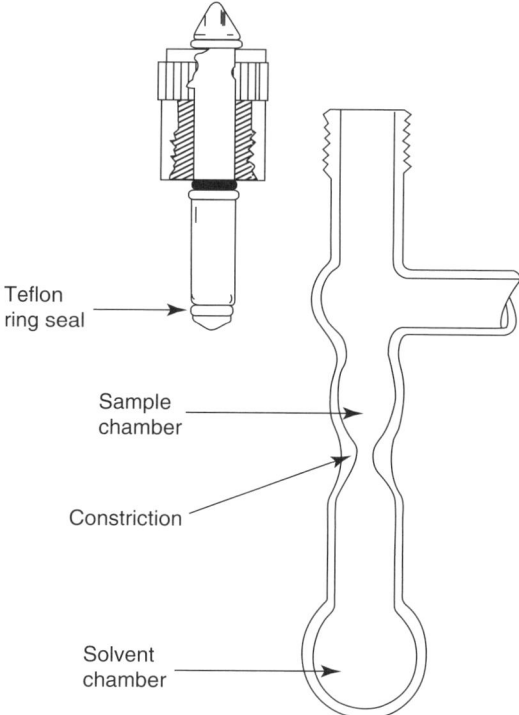

Fig. 6. Nitroxide vapor-phase exposure chamber for introducing nitroxide into weathered coatings. (Reprinted with permission from Ref. 21.)

group of **II** and the amine and other groups in the coating, insured that the radical was not lost from the coating by volatilization.

3.4. Sample Preparation for Nitroxide Measurements in HALS Stabilized Coatings

Different HALS stabilizers were added to coatings and cured on quartz plates or disks described above. Nitroxide concentration was quantified as a function of exposure time via the primary standard, **I**. The total amount of parent HALS and HALS-based inhibition cycle products was determined by oxidizing CH_2Cl_2 extracts with p-nitroperbenzoic acid.[14,25]

4. FREE-RADICAL PHOTOINITIATION RATES BY NITROXIDE DECAY

4.1. Derivation of Nitroxide Radical Decay Kinetics and Free-Radical Photoinitiation Rates

The nitroxide decay measurement of photoinitiation rate involves doping a coating with a known amount of nitroxide **I** or **II** and following its disappearance during photolysis.[19,21] The derivation of the relationship between the decay kinetics and the coating photoinitiation rate is based on the following set of assumptions:

1. Free radicals are produced by photolysis only. This has been confirmed by the fact that the nitroxide concentration does not change if the samples are not exposed to light, even if they are exposed to temperatures as high as 130°C.
2. The formation rate of radicals is essentially constant over the course of the experiment. That is, the chromophore concentration does not change appreciably (either by consumption of chromophore or formation of hydroperoxide) over the course of the experiment.
3. All of the radicals formed are eventually scavenged by nitroxide. Note that this does not require that the rate of Reaction R7 be much larger than that for R2, but it does require that the nitroxide concentration be high enough so that the rate of R7 is much larger than the rate of radical–radical termination, R4.
4. The rate of nitroxide consumption is much larger than the rate of oxidation of inhibition cycle products back to nitroxide (R8 and R10). In practice, this can be assured by stopping the nitroxide decay measurements before there is significant (<25%) build up of inhibition cycle products.

With the above assumptions, the various oxidation and nitroxide reactions can be combined to yield the following expression for nitroxide decay:

$$\frac{-d[\text{NO}\cdot]}{dt} = \text{PIR} + C[\text{NO}\cdot] \qquad (5)$$

where the "C" is related to the rate of abstraction of hydrogen atoms by excited-state nitroxide. For a given doping level of nitroxide, the rate of decrease in nitroxide concentration should be constant until inhibition cycle products build up and compete with nitroxide in scavenging free radicals. The photoinitiation rate is the nitroxide decay rate in the limit of zero nitroxide concentration.

4.2. Photoinitiation Rate Measurements on Model Systems and Nonweathered Coatings [13,19]

A typical plot of nitroxide concentration versus exposure time is shown in Fig. 7. The linear behavior validates assumption 2. Further validation of assumption 2 will be given below in the section of photoinititation rate measurements on weathered coatings. The rate of loss is proportional to the light intensity in agreement with Eq. 1. In addition to measuring nitroxide decay kinetics in coatings, several measurements were performed on model systems including photolysis of ethanol solutions containing 2, 2′-azobisisobutyronitrile (AIBN) free-radical initiator or benzophenone, Fig 8.[26] The results are consistent with the prediction of Eq. 5. In the absence of added AIBN, the intercept (photoinitiation rate) is zero within experimental error. The

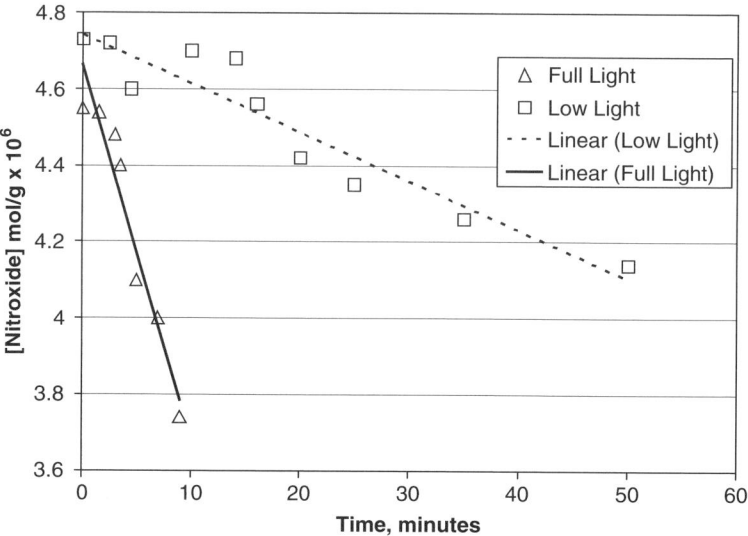

Fig. 7. Nitroxide versus exposure time. Nitroxide (~4.5 × 10⁻⁶ mol g⁻¹) was added to an nonweathered acrylic melamine coating and photolyzed at two different light intensities. The ratio of light intensities was roughly 7:1

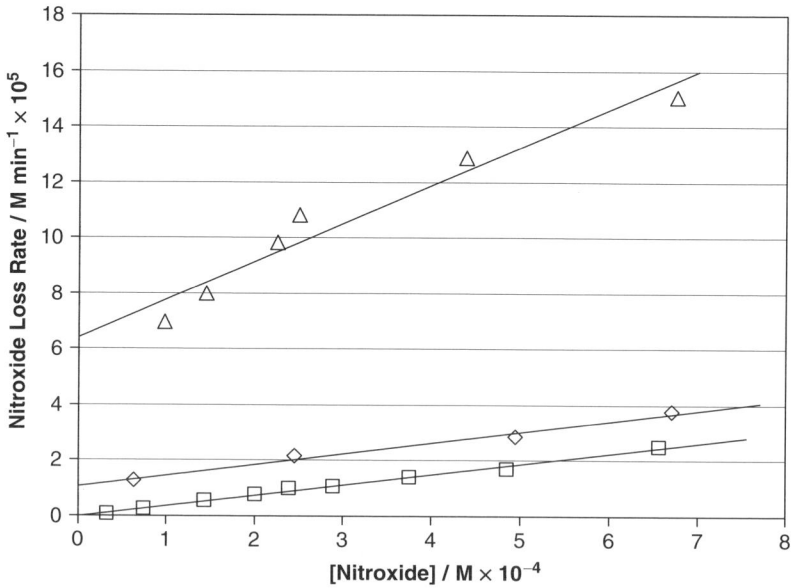

Fig. 8. Nitroxide decay rate versus nitroxide concentration. Pure ethanol (□), 3.3×10^{-3} M AIBN in ethanol (◇), and 2×10^{-5} M benzophenone in ethanol (△). Experimental conditions are described in Ref. 19 and 26.

consumption of nitroxide is a result of excited-state nitroxide hydrogen-atom abstraction. The addition of AIBN yields a photoinitiation rate that is proportional to AIBN concentration, but does not change the rate of excited-state nitroxide hydrogen atom abstraction as measured by the slope of the line. The addition of benzophenone, on the other hand, results in an increase in both the photoinitiation rate and the slope. Nitroxide hydrogen-atom abstraction rates are affected by the formation and decay rates of the nitroxide excited state. This result implies that benzophenone, which has a relatively long excited-state lifetime, can transfer its excitation energy to a nitroxide thus increasing the formation rate of excited-state nitroxide thereby increasing the rate of hydrogen-atom abstraction by nitroxide. At nitroxide concentrations above $10^{-3} M$, deviations from linearity are seen, which suggests that high nitroxide concentrations are sufficient to completely quench the benzophenone excited state. Other factors can influence the relaxation of nitroxide back to the ground state. For example, oxygen has been found to decrease the rate of hydrogen-atom abstraction by nitroxide. The slope of samples exposed in pure oxygen is roughly four times smaller than that for the same sample exposed in pure nitrogen.[13] This is clearly evidence for oxygen quenching of excited-state nitroxide. The UVAs based on benzotriazoles have also been found to quench excited-state nitroxide.[26]

Detailed studies of the dependence of photoinitiation rates of acrylic melamine and urethane cross-linked coatings have been made as a function of composition and exposure conditions. Typical results for a series of acrylic melamine and acrylic urethane coatings are shown in Figs. 9 and 10.[13] Both the photoinitiation rate and the rate of abstraction of hydrogen atoms by excited-state nitroxide vary widely with the choice of polymer. By contrast, the photoinitiation rate of nonweathered coatings does not depend on the choice of cross-linker. This implies that the chromophore(s) responsible for photoinitiation are associated with the acrylic polymer, not the cross-linker. As noted above, photoinitiation rates are proportional to light intensity as expected. The photoinitiation rates for the harsh exposure are roughly a factor 10 times higher than those for the near ambient exposure.[27] This is a result of the shorter wavelength UV light in the harsh exposure. Increasing temperature also increases the photoinitiation rate.[28] This is most likely a result of the "cage effect" and the relative rigidity of the cross-linked polymer matrix. The excited-state chromophore is less likely to form free radicals if the matrix is sufficiently rigid to prevent the diffusion of the radical pair away from one another before they recombine and relax back to the ground state. The activation energy for photoinitiation in these coatings has been found to be ~6.5 kcal mol^{-1}. This is similar to the activation energy for rotational motion of nitroxides as measured by changes in ESR spectral line shapes.[28] The rigidity of the cross-linked polymer matrix depends on the composition of the polymer, cross-linker, and cure state in addition to temperature. For example, lowering the glass transition temperature (T_g) of the acrylic copolymer through changes in monomer selection increases the mobility and the photoinitiation rate.

The biggest factor that accounts for the wide range of photoinitiation rates in Figs. 9 and 10, is the method of polymerization of the acrylic copolymer, namely, the type and amount of initiator and the solvent used in the polymerization.[28] Acrylic polymers are prepared by free-radical polymerization. Typical free-radical initiators

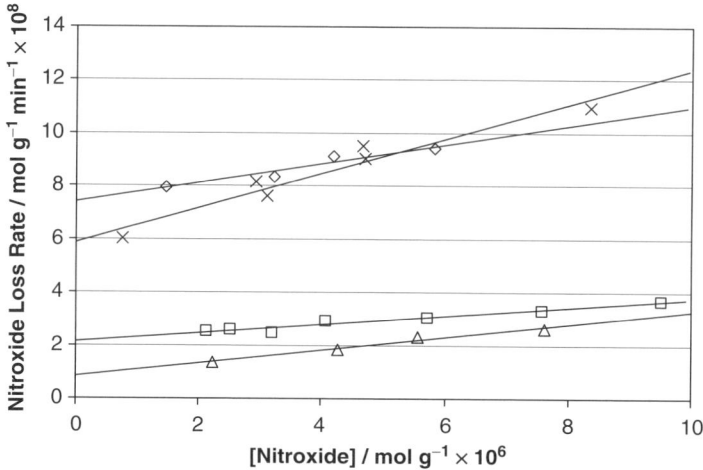

Fig. 9. Nitroxide decay rate versus nitroxide concentration for acrylic melamine coatings exposed to "harsh" weathering conditions. ◇-Copolymer A; □-Copolymer C; △-Copolymer D; X-Copolymer G.

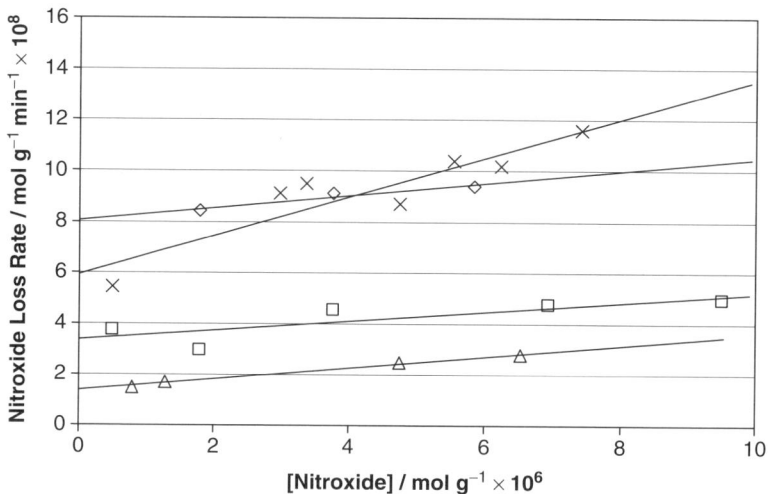

Fig. 10. Nitroxide decay rate versus nitroxide concentration for acrylic urethane coatings exposed to "harsh" weathering conditions. ◇-Copolymer A; □-Copolymer C; △-Copolymer D; X-Copolymer G.

range from AIBN to combinations of peroxides and hydroperoxides. Typical solvents include ketones and aromatics. We have found clear relationships between initiator and solvent used in the polymerization and the incorporation of specific end groups that lead to photoinitiation on exposure. For example, we have found using nuclear magnetic resonance (NMR) spectroscopy that ketone groups can be incorporated onto the acrylic polymer from ketone solvents by a chain-transfer mechanism.[11,12] Ketone groups also result if peroxy initiators are used. Cumene hydroperoxide leads to the incorporation of aromatic ketone groups.[11,12] The level of ketone end groups is proportional to the amount of initiator used in the polymerization step. It is important to note that regulations on solvent emissions from painting operations have resulted in the use of lower molecular weight acrylic polymers. Lower molecular weight polymers require higher levels of initiator and have higher concentrations of end groups. Interestingly, the use of peroxy or hydroperoxide initiators does not lead to the formation of hydroperoxide groups on the polymer or the presence of residual hydroperoxides in the polymer solution. This has been confirmed by measurements of hydroperoxide levels in noncross-linked polymers and cross-linked coatings.[10] It appears that ketone groups are responsible for most of the photoinitiation in nonweathered acrylic copolymer-based coatings. The combination of AIBN as initiator and the use of polymerization solvents, such as p-xylene, essentially eliminates the formation of ketone end groups and results in cross-linked coatings with very low photoinitiation rates. The addition of HALS and UV light absorbers to coatings based on such polymers does not have a significant effect on the photoinitiation rate. The ketone end groups that are responsible for photoinitiation also increase the slope, which suggests that like benzophenone, these ketones can transfer energy to nitroxide. Coatings with styrene-containing acrylic polymers tend to have higher slopes than those without styrene. Styrene does not absorb light at the wavelengths used in these exposures. The most likely explanation is that styrene increases the effectiveness of energy transfer from ketones to nitroxide.

Surprisingly, there is a reasonable correlation between photoinitiation rates measured for nonweathered clearcoats and other more traditional measures of coating degradation for HALS-free acrylic melamine coatings.[13] As shown in Fig. 11, the gloss loss rate measured for a number of HALS-free clearcoats over several hundred hours of "harsh" exposure is roughly proportional to the square root of the photoinitiation rate measured for the same clearcoats during the first few hours of "harsh" exposure. This relationship is consistent with Eq. 2, which, strictly speaking, is only valid in the very early stages of exposure. The correlation shown in Fig. 11 suggests that the relative photoinitiation rates between the different coatings stay constant over the course of the gloss loss measurements. Note that degradation rates in stabilized coatings (carbonyl build up) do not correlate as well with photoinitiation rate measurements on nonweathered clearcoats.[29] These results suggest that measurements of photoinitiation as a function of weathering time would provide significant insight into the overall kinetics of photooxidation in these coatings.

4.3. Photoinitiation Rate Measurements on Weathered Coatings [21,24,27]

As noted above, photoinitiation rate measurements on weathered coatings as a function of exposure time require a different nitroxide doping method than photoinitiation rate measurements on nonweathered coatings. Photoinitiation rate measurements on nonweathered coatings using both nitroxide doping protocols confirmed that the technique of vapor diffusion of nitroxide **II** into the cured coating yields identical

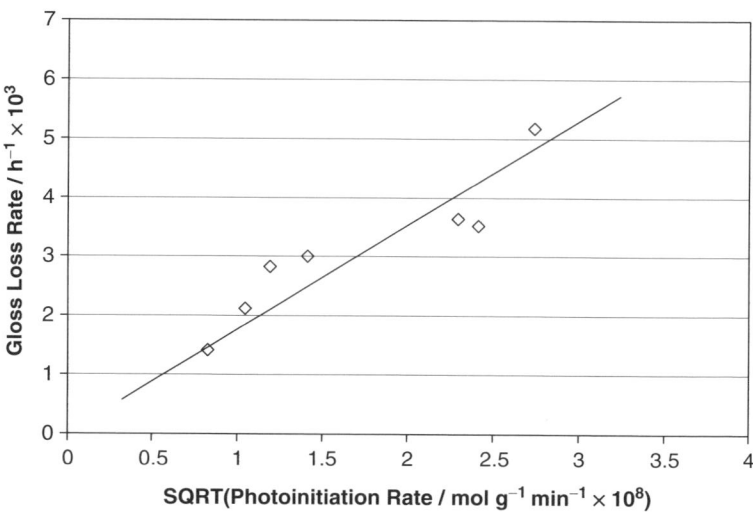

Fig. 11. Gloss loss rate in unstabilized acrylic melamine coatings versus the square root of the nonweathered photoinitiation rate of those coatings. Coatings were exposed to "harsh" exposure conditions.

results as those obtained by mixing nitroxide **I** with polymer resin prior to cure.[21] The major finding of studies of photoinitiation rates in weathered coatings is the role of the cross-linker in the evolution of photoinitiation rate behavior. For melamine cross-linked acrylic coatings, the photoinitiation rate was found to decrease during exposure to a constant value, as shown in Fig. 12.[27] The ratio of photoinitiation rates for weathered coatings to nonweathered coatings vary with exposure condition and coating formulation. For the harsh exposure, photoinitiation rates for weathered acrylic melamine coatings ranged from 20 to 50% of their values obtained for nonweathered samples. Nonweathered coatings with lower photoinitiation rates tend to exhibit smaller decreases on weathering, but the correlation between weathered and nonweathered rates is consistent with the correlation seen between the photoinitiation rates of nonweathered coatings and gloss loss rates on harsh exposure (Fig. 11). The time to reach the plateau value was a function of the exposure conditions. For the harsh exposure, the plateau value was achieved in ~ 10 h while several hundred hours exposure was required for the near-ambient exposure. The observation that acrylic melamine coatings exhibit a near constant photoinitiation rate after a relatively short exposure (note that several hundred hours of near ambient exposure translates to at most a few months in Florida) is consistent with other observations of photooxidation behavior in coatings. For example, measurements of carbonyl growth are also linear in exposure time after a short period of rapid photooxidation.[29]

By contrast to melamine cross-linked coatings, the evolution of photoinitiation rates in urethane cross-linked coatings is very different with photoinitiation rates in weathered coatings tending to increase with exposure time, Fig. 13.[10,24] For coatings

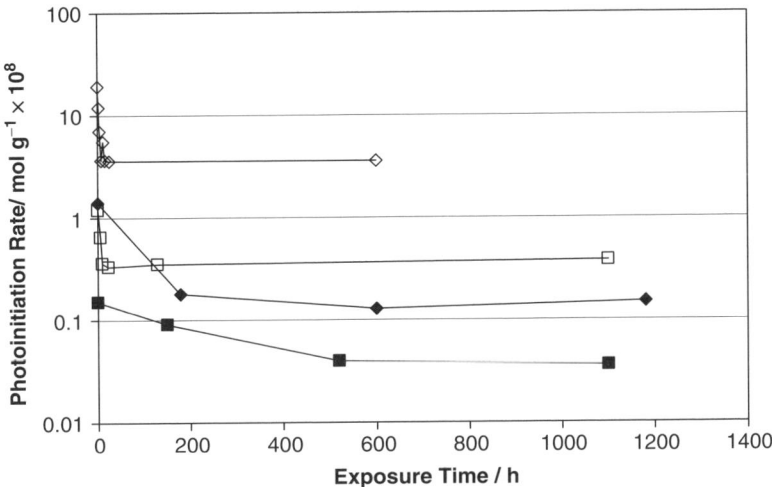

Fig. 12. Photoinitiation rate for acrylic melamine coatings as a function of weathering time. Diamonds-Copolymer A; Squares-Copolymer N. The open symbols represent harsh exposure while the filled symbols represent a near-ambient exposure. (See Ref. 27 for details.)

formulated with identical copolymers and differing only in cross-link chemistry, the weathered photoinitiation rate in the urethane cross-linked coating is > 10 times higher than that for the corresponding melamine cross-linked coating. This observation is consistent with photoinitiation and photooxidation of weathered coatings being dominated by hydroperoxide chemistry. Measurements of hydroperoxide concentrations in these coatings show that the long-time values are 3–10 times higher in urethane cross-linked coatings than in melamine cross-linked coatings. The melamine formaldehyde cross-linker has been found to increase the rate of decomposition of hydroperoxides (i.e., increase in k_d in R5), which leads to a lower steady-state value of hydroperoxide concentration (Eq. 4).[10] After relatively long exposures (1000 h of near ambient exposure) the hydroperoxide levels in the two acrylic urethane coatings with different photoinitiation rates for nonweathered samples become similar. This is also consistent with the long-term photooxidation chemistry being dominated by hydroperoxide chemistry and not initial end-group composition. Infrared (IR) spectroscopic measurements of urethane cross-link scission (one of the key processes in photooxidation of urethanes) are consistent with this observation.[30]

As shown in Fig. 13, the photoinitiation rate of urethane coatings based on polyester resins also increases with weathering time.[24] The rate of increase has been found to depend strongly on the presence of UV light intensity < 300 nm. For example, the addition of a 320–nm cut off filter to the standard near ambient exposure essentially removes the small amount of light < 300 nm that is present in the standard exposure. This has the effect of almost halving the rate of increase of photoinitiation with time. The sensitivity of some polyester coatings to relatively small changes in UV light wavelength distribution ≈ 300 nm appears to result from the fact

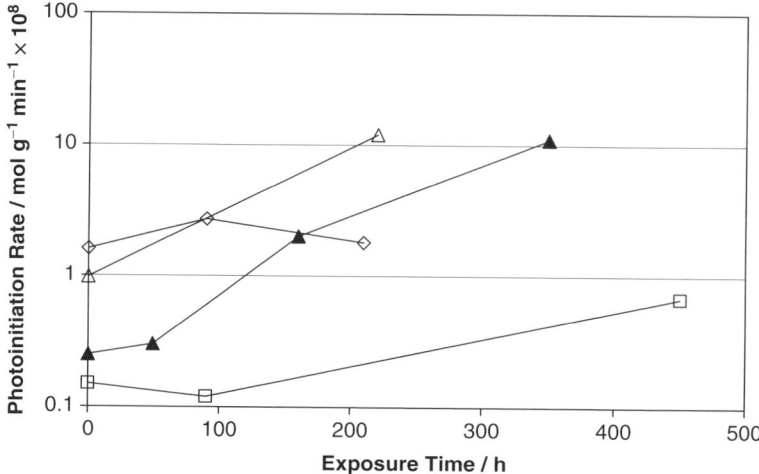

Fig. 13. Photoinitiation rate for acrylic urethane coatings and polyester urethane as a function of near-ambient weathering time. ◇-Copolymer A; □-Copolymer N; △- polyester I. The filled symbols indicate the addition of a 320-nm filter to the near ambient exposure. (See Refs. 10 and 24 for details.)

that isophthalate groups absorb UV light out to close to 300 nm and degrade rapidly on exposure to light < 300 nm.

The higher levels of sustained photoinitiation rates in urethane coatings relative to melamine cross-linked coatings is consistent with the general observation that unstabilized urethane coatings are not intrinsically photooxidation resistant and that their stability is significantly improved with the addition of HALS. Although it is possible to measure nonweathered photoinitiation rates in coatings containing HALS, it is difficult to use nitroxide decay kinetics to measure photoinitiation rates in weathered, HALS stabilized coatings due to the interference of nitroxides arising from the HALS. Although it might be possible to devise means to perform nitroxide decay measurements of photoinitiation rates on HALS-containing coatings, ^{15}N labeled nitroxide dopant, for example, the effort hardly seems worthwhile. In addition, even some unstabilized coatings contain species that appear to interfere with nitroxide doping. For these reasons, the nitroxide decay measurement of photoinitiation is not a practical measurement for photooxidation in all types of coatings. Infrared (IR) spectroscopic measurements of chemical change have become the primary tool for assessing the extent of photooxidation in coatings.[31] As noted in this discussion, however, photoinitiation rate measurements have provided numerous key insights in to the mechanisms of photooxidation in polymer coatings.

5. NITROXIDE KINETICS AND HALS STABILIZATION OF COATINGS

The key issue for HALS stabilization in coatings is choosing the correct HALS type and concentration to achieve effectiveness and longevity. In cross-linked coatings, this is complicated by the necessity that the HALS not interfere with coating cure. Many melamine cross-linked coatings are cured by strong acid catalysts.[2,3] The use of basic HALS, such and TIN 770 or TIN 292 (selected for their effectiveness in many polymer systems), is not possible in these types of coatings. We have used two types of measurements of nitroxide content in HALS-containing coatings. The first is to measure nitroxide concentration as a function of exposure and composition. The second is to convert all HALS-based species to nitroxide using via peracid oxidation to monitor the total amount of parent HALS and HALS-based inhibition cycle products to nitroxide, and thereby determine the total amount of HALS available to inhibit oxidation, "active HALS" as a function of exposure.

Nitroxide concentrations in HALS stabilized coatings depend on exposure conditions, on HALS type and concentration, coating chemistry, and particularly the photoinitiation rate of the coating.[15,32–34] Figure 14 shows the time dependence of nitroxide concentration for an acrylic melamine coating with a high photoinitiation rate that is stabilized with either TIN 770 or TIN 440.[34] The nitroxide concentration starts at a low value (but not zero due to some conversion during cure), rapidly builds to a maximum, and then slowly declines. The time to maximum is much shorter for the harsh exposure although the peak nitroxide concentration is higher for the near-ambient exposure. Even at the maximum, only ~ 5% of the parent HALS added to the system is present as nitroxide. This suggests that under normal circumstances the ratio of R7 and R8 is such that nearly all of the "active HALS" is in the form of aminoether (or other nonradical species). The slow decrease at long exposure times will be shown to be related to the consumption of "active HALS" species. For a given exposure, TIN 770 coatings show more rapid nitroxide

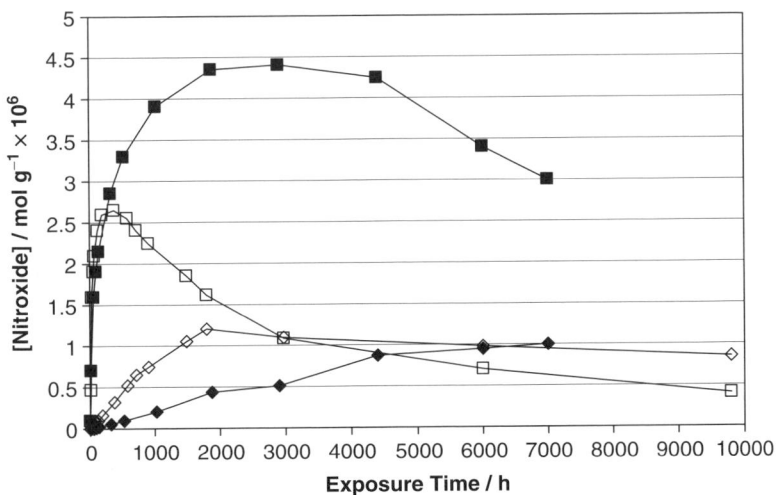

Fig. 14. Nitroxide concentration for acrylic melamine coating "A" versus exposure time. Coatings contained 8.33×10^{-5} mol g^{-1} HALS functionality (equivalent to 2% by weight TIN 770. Squares- TIN 770; Diamonds- TIN 440. Filled symbols are near ambient exposure while unfilled symbols are for harsh exposure.

increases than do coatings containing TIN 440. This is a result of differences in the rate of conversion of HALS to nitroxide, R6. The functionality in TIN 770 is more easily converted to nitroxide than is the functionality of TIN 440 particularly under near-ambient exposure conditions. In fact, after 7000 h of near-ambient exposure, it appears that < 25% of the TIN 440 initially added has been converted into nitroxide and HALS-based inhibition cycle products (R7 and R8). This has profound implications for the relative effectiveness of the two different HALS in this coating as shown in Figs. 15 and 16.[29] In the harsh exposure, Figure 16, TIN 770 and TIN 440 are both reasonably effective at reducing the formation of carbonyl containing oxidation products as measured by IR with TIN 770 being more effective at short exposure times. In the near-ambient exposure, on the other hand, TIN 770 is still very effective at reducing carbonyl growth, but TIN 440 is almost completely ineffective. This clearly illustrates the importance of rapid conversion of HALS into the stabilizing cycle. It also points out the danger of using harsh accelerated tests to optimize formulations. Comparison of nitroxide concentrations with carbonyl growth rate measurements suggests that the ability of HALS to reduce photooxidation increases as the nitroxide concentration increases up to a concentration of ~0.5 $\times 10^{-6}$ mol g^{-1} and then levels off. The exact value will depend on coating composition and exposure conditions. The most important point is that there appears to be a limit to which HALS can reduce photooxidation in both acrylic melamine and acrylic urethane coatings. The addition of higher levels of HALS does not result in

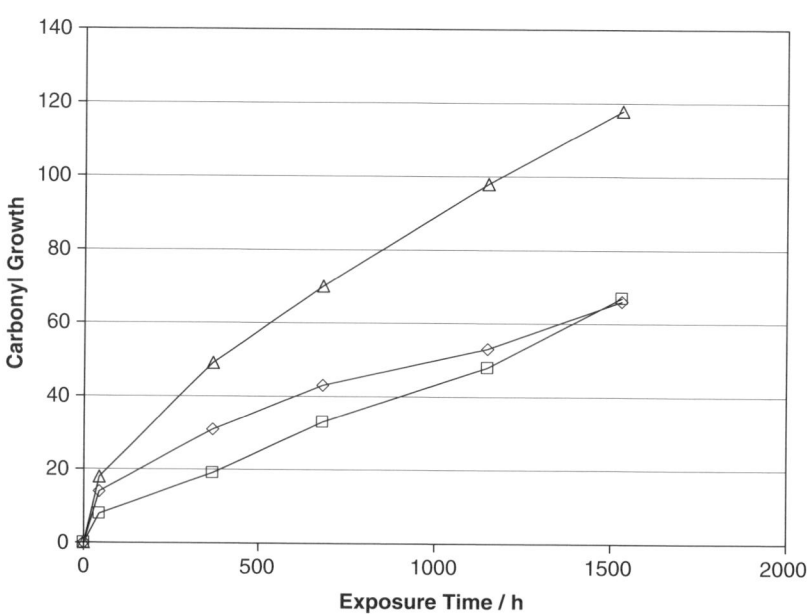

Fig. 15. Carbonyl growth for acrylic melamine coating "A" versus harsh exposure time. △-Unstabilized; □-TIN 770; ◇-TIN 440.

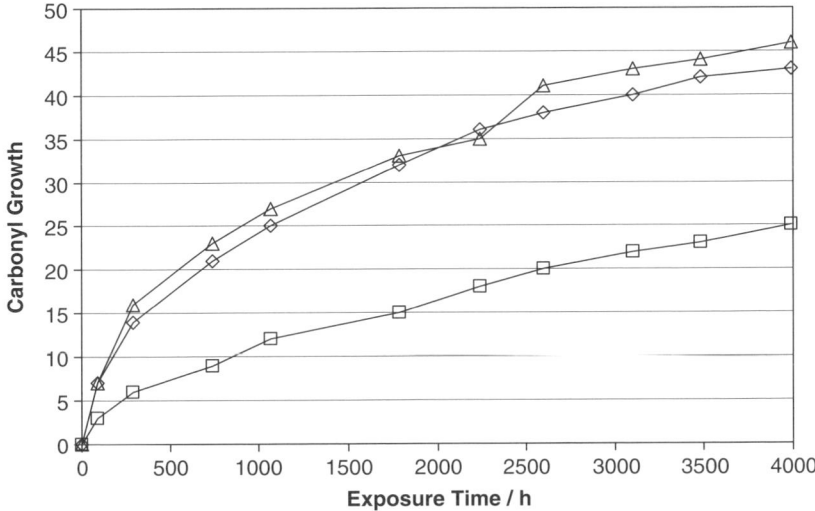

Fig. 16. Carbonyl growth for acrylic melamine coating "A" versus near-ambient exposure time. △- Unstabilized; □-TIN 770; ◇-TIN 440.

further decreases in the photooxidation rate. Kinetic models suggest that this may be a result of free radicals formed by excited-state nitroxides. At some point, the amount of free radical formed by nitroxides will dominate free-radical formation from coating chromophores.[9] When this is the case, nitroxides are simply scavenging radicals that they themselves produce. If the initial photoinitiation rate of free radicals is low enough, then the addition of HALS does not have a measurable effect on coating photooxidation. This is the case for some acrylic melamine coatings.[34] The higher level of weathered photoinitiation rates in urethane coatings means that HALS do improve their photostability.[30]

The stabilizing reaction schemes R6–R8 is deceptively simple. In actuality, there are a large number of coating and exposure factors that influence the balance of the various reactions that ultimately control the nitroxide concentration and stabilizer effectiveness. Some of this complexity is evident from studies of nitroxide formation and hindered amine consumption during exposure of an acrylic melamine coatings containing TIN 770 or TIN 292.[15,32] The initial rate of formation of nitroxide is much lower than the rate of consumption of TIN 770 or TIN 292, as measured by gas chromatography (GC), which in turn is lower than the photoinitiation rate of free radicals. This result is consistent with the premise that the first step in HALS stabilization is the conversion of the parent HALS into nitroxide and inhibition cycle products, such as aminoethers. This conversion is slow compared to the nitroxide scavenging reactions. Therefore nitroxide concentration increases until most of the parent HALS is converted into inhibition cycle products. In acrylic melamine coatings, the time dependence of HALS consumption and nitroxide concentration also has been found to depend on humidity during exposure. For dry exposures, the rate of consumption of TIN 770 is higher, but the net formation of nitroxide and its peak concentration is lower relative to that observed during humid exposure. In addition, specific conversions of HALS N–H groups to HALS N–CH$_3$ groups and regeneration of N–H groups is observed during humid exposures. This behavior has been interpreted in terms of the effect of formaldehyde on HALS chemistry. Formaldehyde is a byproduct of hydrolysis reactions of melamine cross-links. The effects of this chemistry on the degradation and stabilization of melamine cross-linked coatings has been the subject of much investigation and is still not completely understood.[3] Nitroxide kinetics is less sensitive to humidity in urethane cross-linked coatings. One of the most important points of the dependence of nitroxide concentration to exposure variables, such as intensity and humidity, is that in real-world exposures where exposure conditions vary considerably it is difficult to infer loss of HALS "active HALS" from the nitroxide concentration behavior alone. Oxidation of parent HALS and HALS-based inhibition cycle products to nitroxide turns out to be a superior approach to understanding HALS longevity. For example, Fig. 17 shows active HALS measurements on coatings exposed out to 5 years in various locations.[25] There are clear differences in HALS longevity for different coatings in the same exposure and the same coating in different exposures. The rate of loss of "active HALS" is much larger for coatings in harsh environments, such as Florida and N. Australia versus mild environments such as Belgium. This confirms that some photooxidative process is responsible for the slow loss active HALS. When "active HALS" is depleted, the coating photooxidation rate increases.[35]

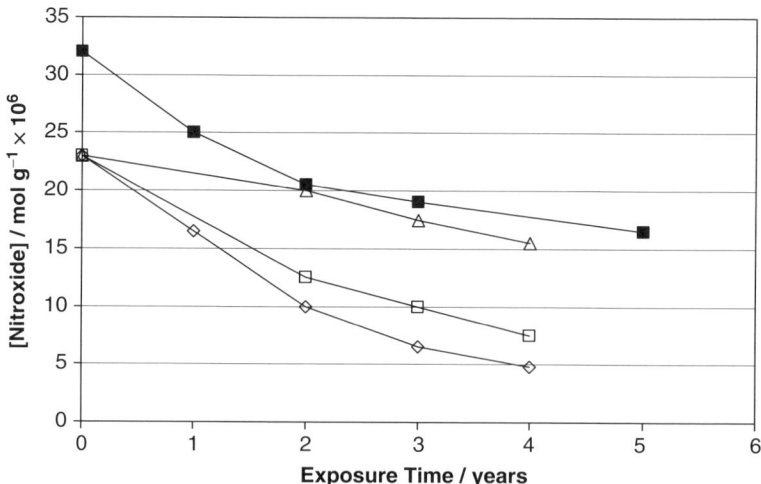

Fig. 17. Active HALS versus outdoor exposure time. The filled and open symbols represent two different commercial coatings. The exposure locations are as follows: □ and ■-S. Florida; ◇-N. Australia; △-Belgium.

6. CONCLUSIONS

This chapter described two basic applications of ESR to understand photooxidation and stabilization of automotive coatings. Both of the applications involve quantification of nitroxides in cured polymer coatings, which is described in detail. In the first application, a stable nitroxide is incorporated into the coating, and the rate of radical formation under controlled exposure conditions is determined by measuring the rate of consumption of nitroxide by ESR. We relate radical formation rates to specific chromophores in the coating and describe how coating composition and exposure affects the evolution of those chromophores, which in turn determines the overall rate of photooxidation. In the second application, nitroxide concentration is monitored in coatings containing hindered amine light stabilizers (HALS). By following the nitroxide formation and decay kinetics, we can assess stabilization mechanisms and understand effectiveness. Converting HALS reaction products into nitroxides by treatment with an organic peracid allows the prediction of HALS longevity.

ACKNOWLEDGMENTS

The authors would like to thank all of the scientists and engineers in Ford Research who contributed to this work. The authors especially acknowledge the many contributions of Linda Briggs, Debbie Mielewski, Roscoe Carter III, Micheline Paputa Peck, and Cindy Peters (née Smith), and Alexei Kucherov.

REFERENCES

1. Bauer, D.R. In *Encyclopedia of Materials: Science and Technology*, Elsevier Press: London, 2001, pp. 421–424.
2. Bauer, D.R. *Prog. Org. Coat.* **1986**, *14*, 193.
3. Bauer, D.R. In *Chemistry, Properties, and Applications of Crosslinking Systems*, Bauer, R., Dickie, R.A., Labana, S.S., Eds.; ACS Symposium Series No. 367: Washington DC, 1988, p. 77.
4. Bauer, D.R. *J. Appl. Polym. Sci.* **1982**, *27*, 3651.
5. Valet, A. *Light Stabilizers for Paints*, Vincentz-Verlag: Hannover, 1997.
6. Emanuel, N.M.; Denisov, E.T.; Maizus, Z.K. *Liquid Phase Oxidation of Hydrocarbons*, Plenum Press: New York, 1967.
7. Bauer, D.R. In *Advances in the Stabilization and Controlled Degradation of Polymers Vol. II*, A. V. Patsis, Ed., Technomic Press: Lancaster PA, 1989, p. 1.
8. Bauer, D.R.; Mielewski, D.F.; Gerlock, J.L. *Polym. Degrad. Stab.* **1992**, *38*, 57.
9. Bauer, D.R. *Polym. Degrad. Stab.* **1995**, *48*, 259.
10. Mielewski, D.F.; Bauer, D.R.; Gerlock, J.L. *Polym. Degrad. Stab.* **1991**, *33*, 93.
11. Carduner, K.R.; Carter III, R.O.; Zinbo, M.; Gerlock, J.L.; Bauer, D.R. *Macromolecules* **1988**, *21*, 1598.

12. Gerlock, J.L.; Mielewski, D.F.; Bauer, D.R.; Carduner, K.R. *Macromolecules* **1988**, *21*, 1604.
13. Gerlock, J.L.; Bauer, D.R.; Briggs, L.M.; Dickie, R.A. *J. Coat. Techno.* **1985**, *57* (722), 37.
14. Kucherov, A.V.; Gerlock, J.L.; Matheson Jr., R.R. *Polym. Degrad. Stab.* **2000**, *69*, 1.
15. Gerlock, J.L.; Bauer, D.R.; Briggs, L.M. *Polym. Degrad. Stab.* **1986**, *14*, 53.
16. Bogatyreva, A.I.; Buchachenko, A.L. *Kinet. Catal.* **1971**, *12*, 1226.
17. Keana, J.F.W.; Dinerstein, R.; Baitis, F. *J. Org. Chem.* **1971**, *36*, 209.
18. Keana, J.F.W.; Dinerstein, R.; Baitis, F. *Tetrahedron Lett.* **1968**, *9*, 365.
19. Gerlock, J.L.; Bauer, D.R. *J. Polym. Sci., Polym. Lett.* **1984**, *22*, 477.
20. Gerlock, J.L. *Anal. Chem.* **1983**, *55*, 1520.
21. Gerlock, J.L.; Mielewski, D.F.; Bauer, D.R. *Polym. Degrad. Stab.* **1988**, *20*, 123.
22. Gerlock, J.L.; Van Oene, H.; Bauer, D.R. *Eur. Polym. J.* **1983**, *19*, 11.
23. Gerlock, J.L.; Bauer, D.R.; and Briggs, L.M. In *Characterization of Highly Crosslinked Polymers,* Labana, S.S., Dickie, R.A., Eds., American Chemical Society: Washington, Symposium Series No. 243, 1984, p. 285.
24. Gerlock, J.L.; Mielewski, D.F.; and Bauer, D.R. *Polym. Degrad. Stab.* **1989**, *26*, 241.
25. Gerlock, J.L.; Kucherov, A.V.; Smith, C.A. *Polym. Degrad. Stab.* **2001**, *73*, 201.
26. Bauer, D.R.; Briggs, L.M.; and Gerlock, J.L.; *J. Polym. Sci., Polym. Phys.* **1986**, *24*, 1651.
27. Bauer, D.R.; Gerlock, J.L.; Mielewski, D.F.; Paputa Peck, M.C.; Carter III, R.O. *Polym. Degrad. Stab.* **1990**, *27*, 271.
28. Gerlock, J.L.; Bauer, D.R.; Briggs, L.M.; Hudgens, J.K. *Prog. Org. Coat.* **1987**, *15*, 197.
29. Bauer, D.R.; Gerlock, J.L.; Mielewski, D.F.; Paputa Peck, M.C.; Carter III, R.O. *Polym. Degrad. Stab.* **1990**, *28*, 39.
30. Bauer, D.R.; Gerlock, J.L.; Mielewski, D.F.; Paputa Peck, M.C.; Carter III, R.O. *Ind. Eng. Chem.* **1991**, *30*, 2482.
31. Gerlock, J.L.; Smith, C.A.; Nunez, E.M.; Cooper, V.A.; Liscombe, P.; Cumming, D.R.; Dusbiber, T.G. In *Polymer Durability — Degradation, Stabilization, and Lifetime Prediction*, Clough, R.L., Billingham, N.C., Gillen, K.T., Eds., ACS Advances in Chemistry Series 249, Washington DC, 1996, p. 335.
32. Gerlock, J.L.; Riley, T.; Bauer, D.R. *Polym. Degrad. Stab.* **1986**, *14*, 73.
33. Bauer, D.R.; Gerlock, J.L. *Polym. Degrad. Stab.* **1986**, *14*, 97.
34. Bauer, D.R.; Gerlock, J.L.; Mielewski, D.F. *Polym. Degrad. Stab.* **1992**, *36*, 9.
35. Gerlock, J.L.; Kucherov, A.V.; Nichols, M.E. *J. Coat. Technol.* **2001**, *73* (918), 45.

11

CHARACTERIZATION OF DENDRIMER STRUCTURES BY ESR TECHNIQUES

M. Francesca Ottaviani

University of Urbino, Urbino, Italy

Nicholas J. Turro

Columbia University, New York

Contents

1. Introduction   279
2. ESR Studies on Dendrimers   282
   2.1. Spin Probes   283
   2.2. Spin Labeling with Nitroxides   289
   2.3. Paramagnetic Metal Ions Added to Dendrimer Solutions   296
   2.4. Metallo-Dendrimers   299
3. Conclusions   302
Acknowledgments   303
References   303

1. INTRODUCTION

The field of highly branched macromolecules emerged during the last several decades of the past millennium due to work done by research groups led by Tomalia,[1] Denkewalter,[2] Vögtle,[3] and Meijer.[4] These scientists devised routes whereby stepwise controlled "polymerizations" could be performed, giving highly branched

---

Advanced ESR Methods in Polymer Research, edited by Shulamith Schlick.
Copyright © 2006 John Wiley & Sons, Inc.

polymers with extremely low polydispersities. Starting from a polyfunctional core, monomeric units (branches) are covalently attached to form subsequent layers (generations). Each monomeric unit must also be polyfunctional. Different functional groups may be attached at the external surface. The first commercially available dendrimers were the polyamidoamine (PAMAM) series synthesized by Tomalia and named "starburst dendrimers" (indicated as SBD), after the Greek word dendra for a tree. The term "dendrimer" is now used almost universally to describe highly branched, often monodisperse compounds. Around the same time, Newkome was developing a series of very highly branched molecules where each generation had a different constitution.[5] These molecules were called "arborols," and were investigated as covalently linked micelle analogues.

Interest in dendrimers mushroomed after this early period to produce the wide range of structures available today. Indeed, a quick scan over the references quoted in this chapter reveals the increased activity in the field of dendrimers in the last few years.

A dendrimer cannot grow indefinitely, generation after generation: The point beyond which the dendrimer cannot grow as a consequence of a lack of space (steric limitations) on the external surface is called the "starburst limit."[6] This limit is a function of a number of factors, such as the core multiplicity, the branching multiplicity, the branch length, the core and branch volumes, and, finally, the type and size of the external functionalities. The final dendrimer structure is characterized by the features depicted in Fig. 1. Figure 2 shows drawings of the PAMAM dendrimers at different generations; the molecular weights and the number of terminal groups are also indicated.

The steric limitation of dendritic wedge length leads to small molecular sizes, but the density of the globular shape leads to fairly high molecular weights and an enormous number of surface groups (Fig. 2). The roughly spherical shape also provides an interesting study in molecular topology. Dendrimers have two major chemical environments, the exterior environment due to the functional groups on the terminal generation, which is the surface of the dendritic sphere, and the sphere's interior environment, which is largely shielded from the exterior environment due to the spherical shape of the dendrimer structure. The existence of two distinct chemical environments in such a molecule implies many possibilities for dendrimer applications.

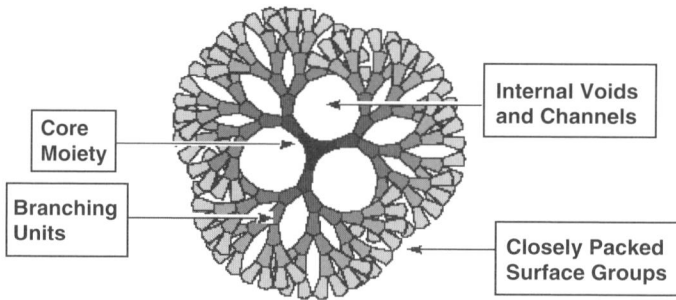

Fig. 1. Sketch of dendrimers: the main features (polyfunctional core, branching units forming internal voids and channels, and closely packed external surface groups) are indicated.

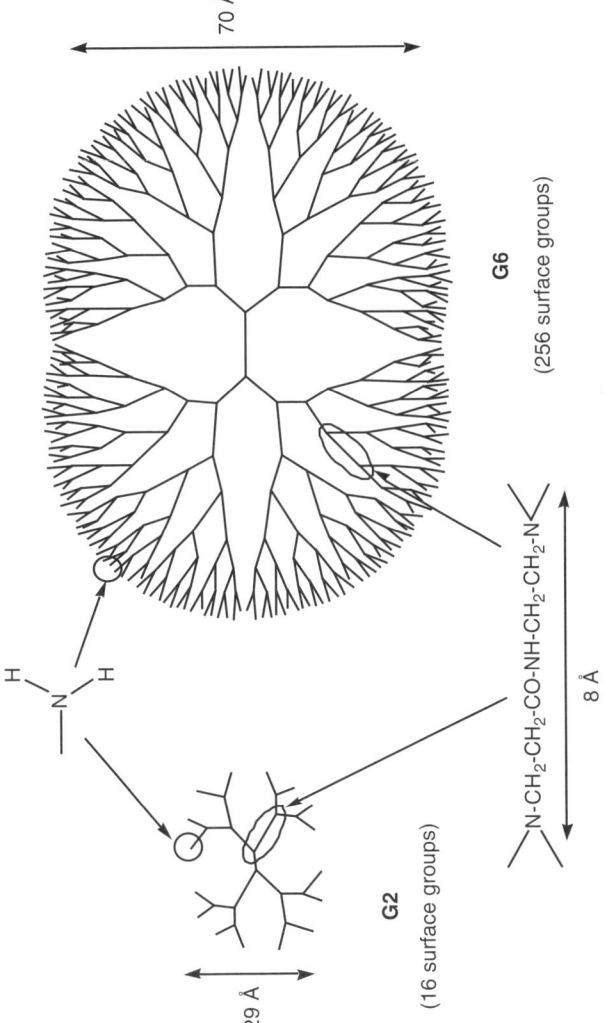

Fig. 2. The PAMAM dendrimers from generation 0 to 7.

Theoretically, hydrophobic–hydrophilic and polar–nonpolar interactions can be varied in the two environments. The existence of voids in the dendrimer interior furthers the possibilities of these two heterogeneous environments to play an important role in dendrimer chemistry. Dendrimer research has confirmed the ability of dendrimers to accept guest molecules in the dendritic voids. Detail on synthesis, properties, and applications of dendrimers can be found in Ref. 7. Two basic methods of dendrimer synthesis are available: divergent and convergent. In the divergent method the molecule is assembled from the core to the periphery; in the convergent method, the dendrimer is synthesized beginning from the outside and terminating at the core. In both methods the synthesis requires a stepwise process, attaching one generation to the last, purifying, and then changing functional groups for the next stage of reaction. Of course the core, the branching units, and the functionalities at the external surface may be chosen according to the desired application. For example, as will be shown in the following, a porphyrin-cored dendrimer was synthesized by Diederich[8] as a model for cytochrome c. The branches can be chosen to model protein structure (for biomedical applications).

Many potential applications for dendrimers have been proposed.[7] Much of the interest in dendrimers involves their use as catalytic agents, based on their high surface functionality and ease of recovery. The behavior of dendrimers as hosts is essential if they are to find success as solubilizing agents,[9] drug delivery,[10] and slow release agents for perfumes, herbicides, and drugs. Research is also active in applications as diverse as polymer additives (for cross-linking and for processing),[11] catalyst supports, thin films, laser-printing toners, and magnetic resonance imaging (MRI) contrast agents.[12] Researchers who synthesize and use the dendrimers in different application fields need to check the dendrimer structure and to characterize their behavior in different media, both in order to investigate the interacting ability of the dendrimer surface, and to understand the physicochemical properties needed for the proposed treatments and uses. The fractal nature of the dendrimers very often impedes the use of X-ray analysis. Destructive analysis, such as mass spectrometry, is of great help to assess the purity and the quality of the products obtained from the synthesis. NMR, IR, and additional spectroscopic methods have also been used to analyze and characterize the dendrimers, as described in the review literature.[7]

As shown in this chapter, ESR provides a very useful technique for the study of dendrimer properties and their local environments, not only when paramagnetic species are directly involved in the dendrimer original structure, but also when paramagnetic units are added as spin probes or spin labels to the dendrimers. The advantage of the ESR technique with respect to other spectroscopic methods of analysis mainly resides in the ability to select the proper spin probe or spin label able to monitor, like a special camera, a selected area internally or externally to the dendrimer. Of course, the basic request is to know if and how the probes or the labels modify the dendrimer properties.

2. ESR STUDIES ON DENDRIMERS

The use of ESR to characterize dendrimers and to investigate their interactions with the surrounding medium is described in this chapter by separately presenting the results from (*1*) radical probes (mainly nitroxides) added to the dendrimers and to their solutions in the absence and presence of other molecules, (2) spin-labeled dendrimers, (3) paramagnetic metal ions as probes or added to the dendrimers, and (4) paramagnetic metal ions as part of the dendrimers system (metallo-dendrimers). Dendrimers will be designated in this report as Gn, where n indicates the generation. For figures from the oldest references about PAMAM dendrimers, the notation nSBD will be retained.

2.1. Spin Probes

The main application discussed for the dendrimers was the possibility to host guest molecules in their interior. Jansen, Meijer, and their co-workers have shown how ESR is helpful in analyzing this host–guest behavior by means of the spin probe technique.[13] A flexible poly(propyleneimine) (PPI) dendrimer with 64 amine end groups and an L-phenylalanine derivative constitutes a "box" (termed 64-L-Phe-Box) able to host different guests. The ESR probe 2,2,3,4,5,5-hexamethyl-3-imidazoliumyloxy methyl sulfate free radical has been employed as a small guest, whereas the dyes Bengal Rose, Rhodamide B, and New Coccine have been used as the large guests. Table 1 shows that, after removal of the protecting groups from the dendrimer surface, the ESR probe is liberated completely, and no ESR signal is detected in the dendrimer after dialysis: The analysis method is based on the spectral integration. The concentration of the probe is detected by integration of the ESR signal of the (perforated) box in $CHCl_3$ in comparison with a sample of the radical with a known concentration in $CHCl_3$. In the case of the completely opened system, the concentration of the probe has not been measured. The concentration of the different dyes is measured using UV–vis spectroscopy of the (perforated) box in $CHCl_3$ and compared with a sample of dye with known concentration. The completely open system has been investigated in water.

In other studies from the group of Jansen and co-workers,[14] the encapsulation of different radicals was studied, that is 0.3–6 pyrrolidinyloxy triplet radical moieties,[14a] and 7,7,8,8-tetracianoquinodimethane (TCNQ),[14b] in a box of polypropylenepolyamine (fifth generation) dendrimer. Intermolecular ferromagnetic alignment of the encapsulated triplet radicals to triplet state radical pairs was detected when more than one radical was trapped into the dendritic box. Similarly,

TABLE 1. Selective Liberation of Dyes from the 64-L-Phe-Box[a]

	Dense Shell		Partly Opened		Completely Opened	
	[ESR Probe]	[Dye]	[ESR Probe]	[Dye]	[ESR Probe]	[Dye]
Bengal Rose	0.8	4.1	0	3.9	0.001	
Rhodamide	0.8	3.9	0	3.8	0.001	
New Coccine	0.4	2.6	0	2.6	0.001	

[a] From Ref. 13. The concentration of the ESR probe is detected by integration of the EPR signal of (perforated) box in comparison with a solution at known concentration. In the case of the completely opened system, the concentration of the probe has not been measured. The concentration of the different dyes is measured using Uv-vis spectroscopy of the (perforated) box and compared with a sample of dye with known concentration. The completely opened system has been investigated in water.

TCNQ radical anions were encapsulated as dimers rather than as monomers. The use of ESR to investigate the interacting ability of dendrimers, on the basis of acid–base interactions, was also presented by Dykes et al.[15] Supramolecular complexes were formed between the carboxylic acid group at the external surface of a dendritic host molecule, based on L-lysine building blocks, and an amine group on the guest molecules. The ESR spectra show the binding between the dendritic branch and 4-amino-TEMPO radicals as inferred by the increase in the rotational correlation times of the radical on binding to the dendritic branch (Table 2). This effect is generation dependent, with larger dendritic branches having a more 1dramatic effect on the tumbling of the radical, and solvent dependent. Effective binding only occurs in nonpolar solvents. In this chapter, attention is focused on acid–base properties; similarly, the electrostatic interactions described by Ottaviani et al.[16] promote the attachment of cationic nitroxide spin probes, bearing hydrophobic chains of different lengths (CATn, where n indicates the chain length), to negatively charged dendrimers (the so-called "half-generation" PAMAM dendrimers, indicated as n.5-SBD), which are characterized by COO^- groups at the surface. Both the increase in chain length of the probe and the increase in generation of the dendrimer promoted interactions between the probes and the dendrimer surface, as inferred by an increase in the correlation time for motion (Fig. 3). It was suggested that the carbon chain of the probe may penetrate into the internal dendrimer structure to interact with sites of low polarity; the larger the dendrimer size, the larger is the anchoring effect of the chain at the internal low polar dendrimer sites. However, analysis of the ESR spectra as a function of pH revealed that the electrostatic interactions between the probe and the external dendrimer surface are the driving forces for closer distance of the probes to the dendrimers. All results are in line with a

TABLE 2. Rotational Correlation Times (τ_c) 4-amino-TEMPO (1.6×10^{-5} M) in the Presence of a Dendritic Additive (3.0×10^{-2} M) in CH_2Cl_2 Solution[a]

Solvent	Dendron	τ_c/s
CH2Cl2	None	0.1×10^{-10}
CH_2Cl_2	G1(COOH)	2.9×10^{-10}
CH_2Cl_2	G2(COOH)	2.9×10^{-10}
CH_2Cl_2	G3(COOH)	4.9×10^{-10}
CH_2Cl_2	G2(COOMe)	0.1×10^{-10}
CH_2Cl_2	G2(COOH)	0.3×10^{-10}

[a] From Ref. 15.

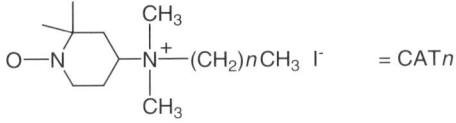

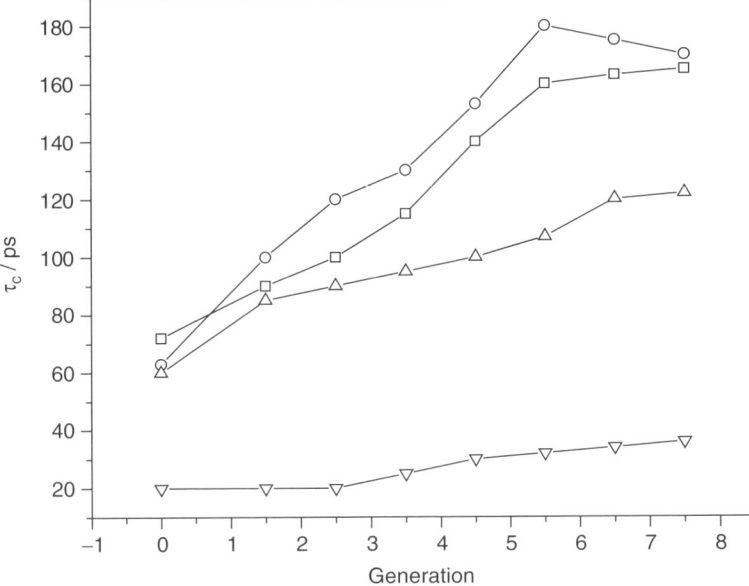

Fig. 3. Variation of τ_c (ps), evaluated from ESR spectra at 298 K, as a function of generation for solutions of n.5-SBD ([SBD–COO–]= 0.32 M) containing (∇) CAT1, (Δ) CAT8, (□) CAT12, and (○) CAT16 ([CATn] =0.16 mM). (From Ref. 16.)

change of the morphology of the SBD from the so-called earlier generations ($G < 4$), characterized by an open structure, to the later generations ($G > 4$), characterized by a densely packed structure.

The long-chain probes CATn behave as surfactants and can aggregate to form micelles. It is of interest to study the aggregation behavior of these surfactants and of their mixtures with other surfactants in the presence of the dendrimers, and the interactions taking place between the dendrimers and the micelles. As the micelles represent the simplest model for biomembranes, the interactions occurring between dendrimers and micelles are also of interest to clarify the delivery mechanism of DNA and drugs eventually played by the dendrimers toward target cells. For CAT16 in the presence of n.5-SBD,[17a] the ESR spectra were analyzed by a subtraction–addition procedure of different components and computation of the experimental line shape (Fig. 4). The computation was performed by the iterative nonlinear least-squares program of Budil et al.[18] By using this method, parameters like the percentage of interacting probes (Fig. 5a), the correlation time for motion (Fig. 5b), and the exchange frequency (Fig. 5c) were plotted as a function of the dendrimer concentration, for the

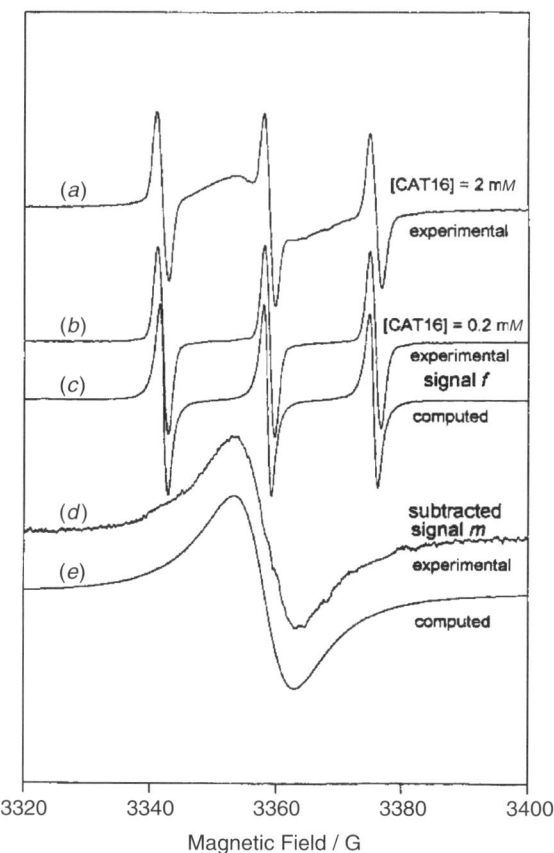

Fig. 4. Experimental ESR signal of CAT16 in water (2 mM) at 333 K (a); Experimental ESR signal of CAT16 in water (0.2 mM) at 333 K (b); simulated b = signal f (c); ESR spectrum obtained by subtracting spectrum c from spectrum a (d); simulation of the subtracted component = signal m. (From Ref. 17a.)

dendrimers at different generations; this analysis led to a model for the interactions between CAT16 and n.5-SBD. A similar analysis was carried out for analyzing the aggregational behavior of cetyl-trimethylammonium bromide and dodecyl-trimethylammonium bromide (CTAB and DTAB, respectively), containing different surfactant spin probes,[17b] and for investigating the interactions between full generation dendrimers (Gn) and SDS micelles.[17c] In this last case, the system was also analyzed by means of pulsed ESR.

A similar analysis, making use of both continuous wave (CW) ESR and pulsed ESR, was performed for amino-terminated-PAMAM dendrimers interacting with a more suitable model of biological membrane if compared to micelles, that is, vesicles, consisting of dimyristoylphosphatidylcoline (DMPC).[19a] In this case, the probes added to the vesicles were doxyl stearic acid spin probes with the doxyl group

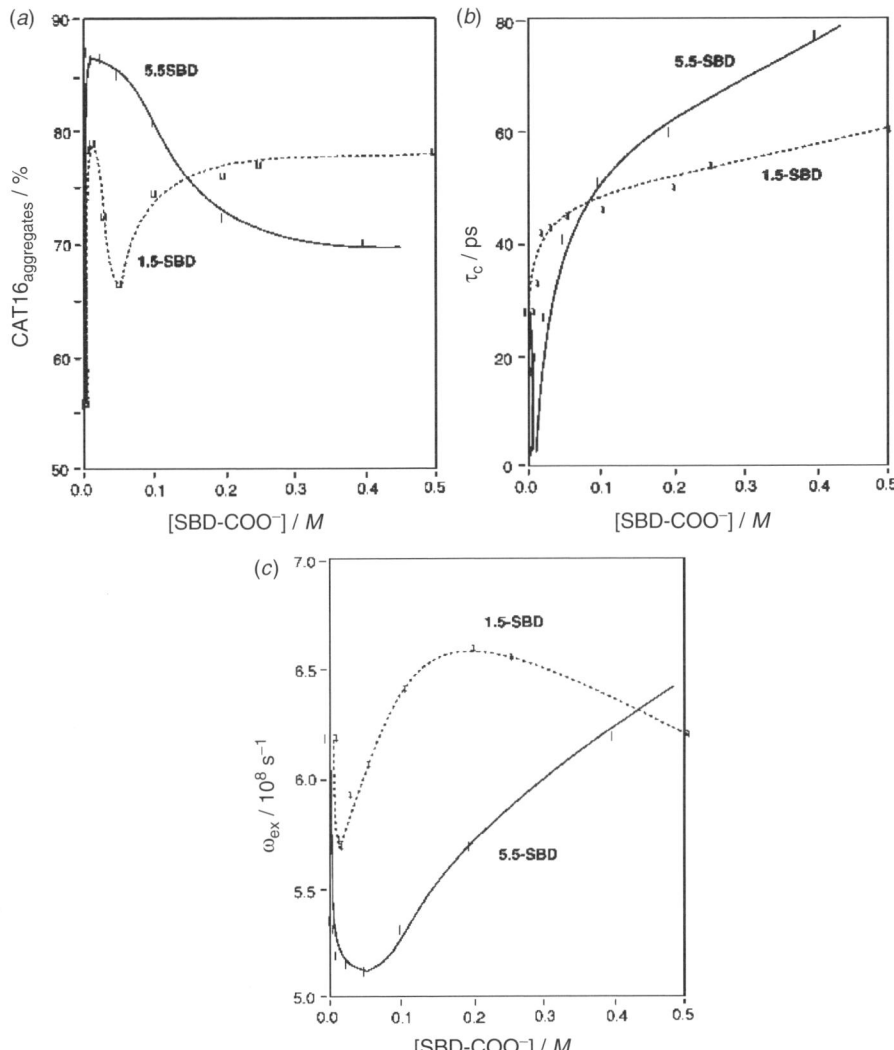

Fig. 5. (*a*) Variation of the percentage of CAT16 aggregates as a function of the concentrations of 1.5- and 5.5-SBD–COO– concentrations. (*b*) Correlation times for the diffusional rotational motion, τ_c (s) for CAT16 solutions (1.0 mM) as a function of 1.5- and 5.5-SBD–COO– concentrations. (*c*) Plot of the Heisenberg spin–spin exchange frequency, $\omega_{ex}(s^{-1})$, versus [SBD–COO–] for 1.5- and 5.5-SBDs at [CAT16]) 1 mM. The parameters were evaluated from simulated ESR spectra at 333 K. (From Ref. 17a.)

attached at different positions (5, 12, and 16) of the stearic chain (termed 5DSA, 12DSA, and 16DSA). Figure 6 shows experimetal (full lines) and simulated (dashed lines) spectra obtained for 5DSA inserted as a probe in DMPC vesicles in the absence and in the presence of G7 at low and high pH (pH 4.5 corresponds to full

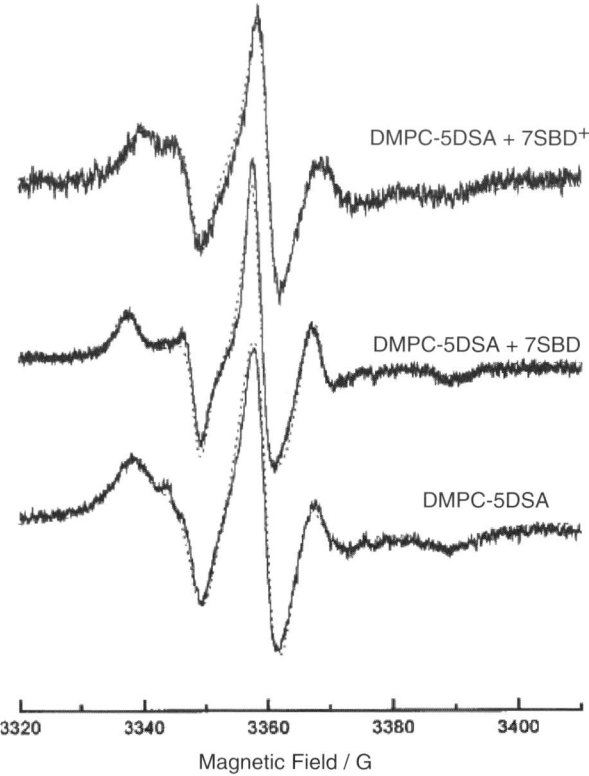

Fig. 6. Experimental (full lines, 303 K) and simulated (dashed lines) ESR spectra of DMPC vesicles containing 1% 5 DSA in the absence and in the presence of 7SBD (G7-low protonated) and 7SBD$^+$ (G7-highly protonated). (From Ref. 19a.)

protonation of the external amino groups, and is indicated in the figure as 7SBD$^+$; pH 8.5 corresponds to ≈ 30 % protonation of these groups, indicated in the figure as 7SBD). The computation indicated an increased ordering of the phospholipids when interacting with the dendrimers (the order parameter was $S = 0$ in the absence of the dendrimers, $S = 0.53$ for low protonated dendrimers, and $S = 0.38$ for highly protonated ones).

If mixed vesicles of DMPC and its salt DMPA-Na are used,[19b] the order disappears and the DSA probes are no longer able to monitor the dendrimer–vesicle interactions. The decay and modulation of the electron spin–echo (2-pulse- and 3-pulse-ESE) signal was also computed and provided information on the structural environment of the paramagnetic center, such as the number and the distances of protons and deuteria (from deuteriated water). The results indicated that the water interface was compressed due to dendrimer–vesicle interactions.

The Ru(II) polypyridyl complexes are good photochemical probes. The opportunity to combine the photophysical and ESR techniques became possible with nitroxide-labeled Ru(II) polypyridyl complexes, where one of the ligands is a TEMPO-labeled phenantroline;[20] the radical (T) is tethered to the ligand either through a single –CH$_2$– group (phen-T) or a –(CH$_2$)$_8$– chain (phen-C8-T). The analysis of the ESR spectra was performed as a function of generation, temperature, and pH, and evaluated the structural and mobility parameters. Finally, the translational diffusion coefficients were extracted by photophysical measurements (quenching experiments) and compared with the rotational diffusion coefficients obtained by ESR. Results showed that the rotational and the translational mobilities of the probe are influenced in a similar manner by the binding of the probe to the dendrimer surface.

2.2. Spin Labeling with Nitroxides

Spin labeling of dendrimers allows the analysis of the interactions occurring between the dendrimers and their environment, and to follow the fate of the dendrimers themselves. In addition, labeling of differently functionalized dendrimers may help in understanding the location and distribution of the different chemical groups and of the labels. The crucial point is to quantify the amount of labeling and the distribution of the labels by the ESR analysis. Bosman et al. performed a fundamental ESR study of lower generation poly(propylene imine) (DAB) dendrimers, fully functionalized with 2,2,5,5-tetramethylpyrrolidin-1-oxyl (PROXYL).[21] In this case, the Hamiltonian that describes the ESR line shape also includes a term accounting for the exchange interactions. The increase in the number of spins with increasing size of the dendrimers (DAB-*dendr*-(NH$_2$)$_n$; $n = 2, 4, 8, 16, 32, 64$) produced variations in the ESR line shape. The nitroxide radicals at the periphery of the dendrimers exhibit a strong exchange interaction resulting in thermally populated high-spin states. For the lower generations ($n = 2, 4$, and to some extent 8), the number of transitions in the ESR spectrum and their spacing can be directly related to the number of end groups. For the higher generations, such an analysis is hampered by the decreasing value of the splitting (a_N/n) and the increasing number of lines ($2n + 1$). In these cases, however, the ESR spectrum gives direct spectral evidence for interaction between the end groups from the exchange narrowing. This result is in contrast with previous studies on dendrimers with paramagnetic end groups in which no interaction[22–24] or formation of a diamagnetic state was observed.[25] An interesting aspect of the exchange interaction is the fact that it is modulated by the dynamic behavior of the dendritic branches, since they probe conformations with strongly different values of J in time. Since this modulation of the exchange interaction can be influenced by the temperature and the nature of the solvent, the ESR spectra give evidence of the dynamic behavior of the dendritic branches. Interestingly, the authors found an increased flexibility of the dendritic polyradicals in polar solvents compared to nonpolar solvents due to the intramolecular hydrogen-bonded network for poly(propylene imine) dendrimers. In this study, the lowest generation dendrimers provide an example of a dinitroxide, which exhibits an $S = 1$ state. With increasing generation, the spin multiplicity increases. Figure 7 shows the spectra obtained for the proxyl-labeled-DAB dendrimers **4n** (*n* is the number of surface groups, and *m* is

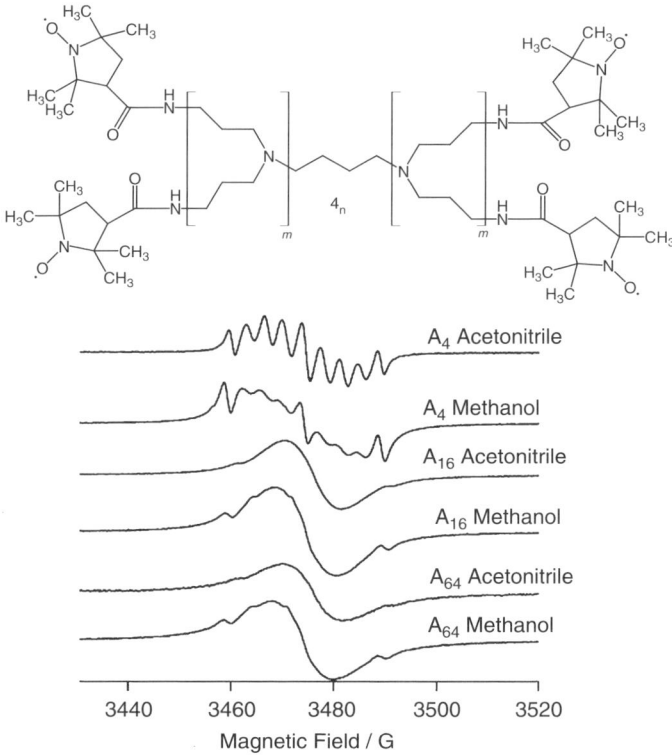

Fig. 7. The ESR spectra of the PROXYL-labeled-DAB dendrimers **4n**, where $n = 4$ corresponded to $m = 1$, $n = 16$ to $m = 3$ and $n = 64$ to $m = 5$. (From Ref. 26.)

the generation). Therefore $n = 4$ corresponded to $m = 1$, $n = 16$ to $m = 3$ and $n = 64$ to $m = 5$.[26] The ESR spectrum for **4₄** in acetonitrile exhibits a nine-line pattern ($2I + 1$ lines, with $I = 4$), indicating strong exchange and suggesting the possibility of a $S = 2$ state. This result indicates fast exchange on the ESR time scale for this compound. The larger dendrimers show broadened signals, and splittings are no longer resolvable, but if the trend continues, the possibility exists that spin states higher than $S = 2$ could result.

It is of interest to describe spin systems that are completely different from nitroxide-labeled dendrimers, but that also bear in their organic structure a high-spin multiplicity (with $S > 1$).[27] High-spin polycarbenes with π-conjugation, which are the first and second generations of the spin-mediated dendrimers based on phenylacetylenes, were studied by a two-dimensional electron–spin transient nutation (2D-ESTN) method based on pulsed-ESR to determine the spin multiplicity. The contour plots of the 2D-ESTN spectra of the dendrimer at 3.5 K allowed to identify the nutation frequencies, which were attributed to $|4,\pm 4\rangle \leftrightarrow |4,\pm 3\rangle$, $|4,\pm 3\rangle \leftrightarrow |4,\pm 2\rangle$, $|4,\pm 2\rangle \leftrightarrow |4,\pm 1\rangle$, and $|4,\pm 1\rangle \leftrightarrow |4,\pm 0\rangle$. The ESR allowed the detection of transitions due to the nonet state ($S = 4$). By analyzing the transition moment, a septet ground state of the triscarbene system, which is the first generation of the phenylacetylene dendrimers, was unequivocally identified.[27] The authors indicated that the stable diphenylcarbene serves as a useful building block and the phenylacetylene dendrimer as a ferromagnetic coupling unit with robust spin polarization for super high-spin polymeric systems and super paramagnetic materials. The fine-structure parameters of the triscarbene and the molecular structure were obtained by means of a semiempirical treatment of a fine-structure tensor.[27]

Louie and Hartwig described relatively stable (half-life 1 h) radical cations formed for the tryarilamine dendrimers, giving an ESR broad signal with a spectral width of 34 G.[28] Proton hyperfine coupling broadened the spectrum so that clear resolution of the ^{14}N line pattern was not observed, but a multiline pattern indicates multiple couplings to nitrogen. Similarly, radical cations were detected by ESR also for dendralene-type vinylogs of tetrathiafulvalene (TTF) bearing a 1,3-diselenole moiety.[29]

Returning to nitroxide-labeled dendrimers at a high degree of labeling (when spin–spin interactions prevail), ESR line broadening effects were found for sugar (mannose) functionalized PAMAM dendrimers with nitroxide attached to the surface.[30] The ESR spectra of the differently functionalized dendrimers were analyzed through empirical parameters directly calculated from the spectra, that is, the peak heights of the first and second peaks. The ratio between these peak heights increases as a function of the radical percentage monitoring the distribution of the labels. The results of this analysis indicate that both the acetylated mannose groups and the (deprotected) mannose residues are randomly distributed on the dendrimer surface. The authors also found that the binding of the dendrimers to Concavaline A was not perturbed by the labeling. Recently, a similar ESR study from the same group, was performed on G4-PAMAM dendrimers functionalized with TEMPO and another functional group.[31] The relative locations of the spin labels were determined. At all loadings studied (5–95% TEMPO), the spin label was found to be randomly distributed on the surface of the dendrimer, and therefore the other group (2,2,6,6-tetramethylpiperidine, propanol, *tert*-butane, and phenol) must also present a random surface display.

The spin labeling of G6-PAMAM dendirmers with TEMPO radicals was used to follow the redox reaction of TEMPOL and TEMPOL-H.[32] The intensity of the ESR signal of TEMPOL, formed by reoxidation of TEMPOL-H with spin-labeled PAMAM dendrimers, increased as a function of time, after addition of the dendrimer at different labeling degrees, demonstrating that spin-labeled G6 dendrimers are potential candidates as contrast agents in MRI.

The ESR studies of labeled dendrimers intended to clarify their role as MRI contrast agents have been described in several papers.[33–35] Spin labeling of several different dendrimers (DAB and PAMAM, at different generations) let the authors calculate the rates of nitroxide reduction due to superoxide, by observing the change of the low-field spectral peaks in the ESR spectra as a function of time (Table 3). Dendrimers labeled with nitronylnitroxides work well as contrast agents in MRI, and therefore their characterization by means of ESR is of interest.[36,37]

An interesting medical application of the nitronylnitroxide-labeled dendrimers is spin trapping of nitric oxide.[37] Nitronyl nitroxides can spin trap NO, but are unstable

Table 3. Bioreduction Rate Calculated for Labeled Dendrimers[a]

Number of Nitroxides	Bioreduction Rate µM/min^{-1}
No dendrimer	0.01 ± 0.01
DAB-3	7.93 ± 0.21
DAB-4	7.54 ± 0.15
DAB-8	0.00 ± 0.03
DAB-16	0.00 ± 0.02
DAB-32	0.00 ± 0.01
PAMAM-16	0.00 ± 0.01
PAMAM-32	0.00 ± 0.01

[a] (Polypropyleneimine = DAB; polyamidoamine = PAMAM). From Ref. 35.

in a biological medium. Poly(propyleneimine n-amine) dendrimers (DAB-Am)$_n$, where n = 2,3,4, and 8, functionalized with nitronyl nitroxide, also behaved as spin traps for NO, leading to the corresponding imidazolidinoxyls. The ESR spectra indicated a slow rate of trapping and a high stability of the adducts in biological media, thus making possible the use of these spin labeled dendrimers to effectively detect NO *in vivo* and in real time.

As mentioned at the beginning of this section, spin labeling of the dendrimers may serve to follow the fate of the dendrimers themselves and to monitor their interactions with molecules and structures belonging to the environment, in solution, or at solid or liquid surfaces. Following this rationale, TEMPO-labeled PAMAM dendrimers can monitor the interactions of the dendrimer surface with different biomolecules, or structures mimicking the behavior of biostructures, such as vesicles,[38–19b] polynucleotides, and DNA,[39, 40] amino acids and proteins.[41] The interactions of TEMPO-labeled PAMAM dendrimers (indicated as nSBD-T) with DMPC vesicles were studied by a computer-aided analysis of CW ESR and pulsed ESR spectra.[38] CW-ESR spectra as a function of temperature, pH, and generation were computed using the procedure of Freed and co-workers.[18] These CW-ESR spectra often consisted of two components with different mobility, due to interacting and noninteracting dendrimers. Table 4 presents the main parameters used for computation of the ESR spectra recorded at 277 K. It was found that interactions (mainly due to charged surface groups) are favored when the dendrimers are protonated at the surface, and for high generation dendrimers. Interestingly, the computation needed a tilt of the main rotation axis, which indicated that the label undergoes constraints when dendrimers and vesicles interact. The ESE spectra were analyzed by the procedure described by Romanelli and Kevan.[42] The calculation provided the number and the distances of the protons and deuteria (from deuteriated water) in the nitroxide environment. These

TABLE 4. Main Parameters for the Computation of ESR Spectra at 277 K[a,b]

System	A_{zz} (G)	τ_c ($\times 10^9$/s)	Main Rotation Axis/ Tilt Angle	% Signal
6SBD-T	39	2.3	y / 50°	50
	36	6.0	z / 50°	50
6SBD-T-DMPC	39	2.3	y / 50°	30
	36	15.0	z / 50°	70
6SBD-T$^+$	39	1.5	y / 50°	100
6SBD-T$^+$-DMPC	39	2.3	y / 50°	30
	36	17.0	z / 50°	70
2SBD-T	39	2.0	y / 40°	70
	36	4.0	z / 50°	30
2SBD-T-DMPC	39	2.0	y / 40°	85
	36	6.0	z / 50°	15
2SBD-T$^+$	39	0.9	y / 50°	100
2SBD-T$^+$-DMPC	39	2.3	y / 50°	35
	36	6.0	z / 50°	65

[a] From Ref. 38.
[b] Accuracy: A_{ii} and $<\tau>$: 2%.

data showed that a "complex" is formed between a single solvent molecule and the nitroxide radical, and that the interactions between the dendrimer surface and the vesicle partially compressed the hydration layer of the N–O group.

A similar analysis of the ESR spectra was performed for TEMPO-labeled PAMAM (nSBD-T, or Gn-T) dendrimers interacting with mixed DMPC–DMPA-Na vesicles.[19b] In this case too, the main direction of the label rotation was modified by the dendrimer–vesicle interactions, but the interactions were weaker compared to pure DMPC vesicles. Figure 8a shows examples of ESR spectra (258 K) obtained for G6-T interacting with different polynucleotides: Calf Thymus DNA (C.T.DNA); poly(deoxyadenylic-deoxythymidylic acid) [Poly(AT)] and poly(deoxyguanylic-deoxycytidylic acid) [Poly(GC)], which contain only adenine-thymine base pairs or cytosine-guanine base pairs, respectively; DNA with 12 base pairs (DNA-12mer).[39] Here also the ESR analysis (computation of the spectra are the dashed lines in Fig. 8a) showed the presence of two components (interacting and noninteracting, or free) constituting the spectra; the mobility conditions of the interacting component and its relative intensity indicated the following sequence of interactions with the dendrimers: DNA-12mer > CT-DNA > Poly(GC) > Poly(CT). A stronger interaction of the dendrimer with DNA-12mer is due to the easier adaptation of this short DNA

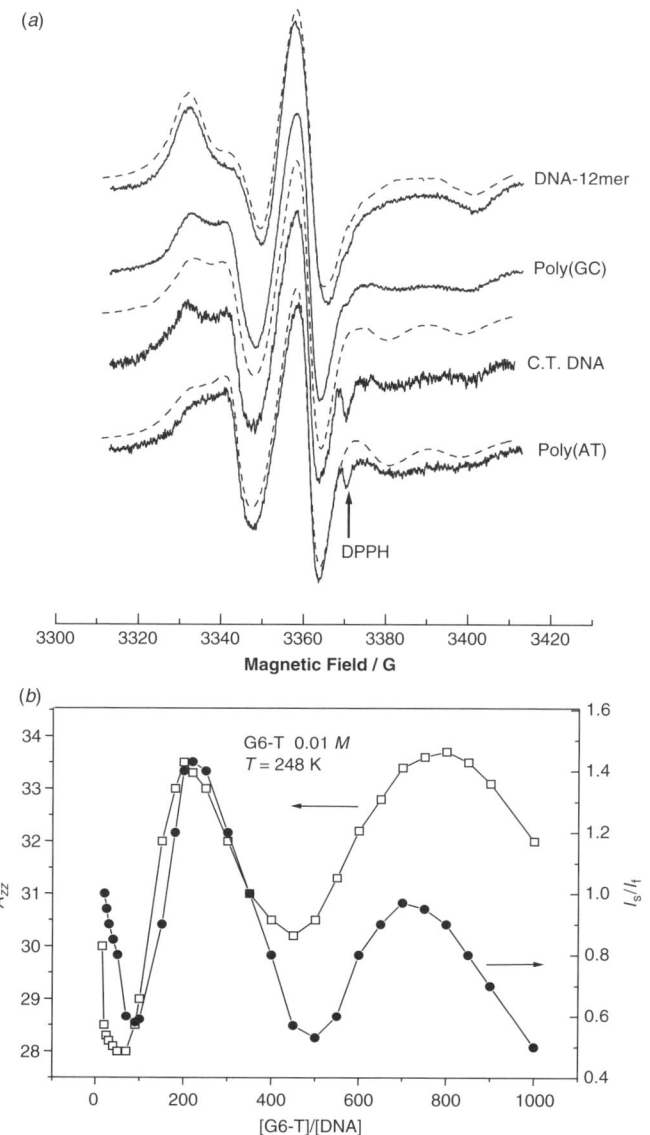

Fig. 8. (*a*) The ESR spectra (258 K; solid lines, experimental spectra; dashed lines, computed spectra) for G6-T interacting with different polynucleotides: Calf Thymus DNA (C.T.DNA); poly(deoxyadenylic-deoxythymidylic acid) [Poly-(AT)] and poly (deoxyguanylic-deoxycytidylic acid) [Poly(GC)], which contain only adenine-thymine base pairs or cytosine-guanine base pairs, respectively; DNA with 12 base pairs (termed DNA-12mer). (From Ref. 39.) (*b*) Plot of A_{zz}', extracted from the interacting component in the ESR spectra, and I_s/I_f (intensity ratio between the interacting and the free components) as a function of r = [G6-T](surface groups)/[DNA](base pairs); ESR spectra recorded at T = 248 K and at a fixed dendrimer concentration of 0.01 M in phosphate buffer solution. (From Ref. 40.)

sequence to the dendrimer surface.[39] Fig. 8b shows the variation of the parameters indicative of the dendrimer-DNA interactions as a function of the increase in the ratio r = [G6-surface groups]/[DNA-base pairs].[40] The maxima of these parameters correspond to the formation of supramolecular structures due to the dendrimer–DNA interactions.

The interactions between labeled PAMAM dendrimers (Gn-T) and amino acids and proteins,[41] were studied with the biomolecules selected on the basis of acid–base properties. The protonated dendrimers, which are acidic, interact well with the basic Arginine. The relatively strong interactions with the hydrophobic Leucine indicated a synergistic effect between hydrophilic and hydrophobic interactions. The ESE spectra confirmed the results obtained from CW ESR, since the protons of Leucine reside in vicinity of the nitroxide group.

The same TEMPO-labeled PAMAM dendrimers (Gn-T) were used to investigate the interactions of these dendrimers with different solid surfaces.[43] Table 5 reports the main parameters (A_{zz}', the z component of the hyperfine coupling; the correlation time for motion, τ; the percentage of interacting labels, %); these parameters increase with the increase in the strength of interactions evaluated from the analysis of the ESR spectra of G2-T and G6-T interacting with homoporous silica gel (S40, 60, 100, 200, 500 nm of pore size), and acid, neutral, and basic aluminas at different dendrimer concentrations. It was found that the dendrimer–solid surface interactions increase with decrease of the silica pore size, and with the basicity of the aluminas.

TABLE 5a. Main Parameters Extracted from ESR Spectral Analysis of Dendrimers (0.005 M) Interacting with Silica Gels[a,b]

Dendrimer	Silica	τ_c^a/s Free Labels	%[c] Interacting Labels	A_{zz}'/G[c] Interacting Labels
G2-T	S40	1.9×10^{-9}	65	33.3
G2-T	S60	1.5×10^{-9}	55	33.3
G2-T	S100	9.6×10^{-10}	40	
G2-T	S200	1.0×10^{-10}	37	
G2-T	S500	2.0×10^{-10}	20	
G6-T	S40	2.1×10^{-9}	73	34.5
G6-T	S60	1.1×10^{-9}	32	
G6-T	S100	9.0×10^{-10}		
G6-T	S200	6.0×10^{-10}		
G6-T	S500	5.0×10^{-10}		

[a] From Ref. 43.
[b] Accuracy 5 %.
[c] Unreported values are not measurable

TABLE 5b. Main Parameters Extracted from ESR Spectral Analysis of Dendrimers Interacting with Alumina[a,b]

Dendrimer	[Gn-T]	Alumina	$\tau_c \times 10^{10}$/s	%	A_{zz}'/G
G2-T	0.04 M	Acid	8.2	38	29.15 G
G2-T	0.04 M	Neutral	12.0	60	30.9 G
G2-T	0.04 M	Basic	12.2	62	30.4 G
G2-T	0.005 M	Acid	5.5		
G2-T	0.005 M	Neutral	8.5	46	
G2-T	0.005 M	Basic	8.7	42	
G6-T	0.04 M	Acid	16.4	84	34.75
G6-T	0.04 M	Neutral	17.2	87	33.6
G6-T	0.04 M	Basic	17.2	88	33.65
G6-T	0.005 M	Acid	10.0	(95)	
G6-T	0.005 M	Neutral	14.3	73	34.0
G6-T	0.005 M	Basic	19.5	85	33.7

[a] Ref 43.
[b] Accuracy 5%.

2.3. Paramagnetic Metal Ions Added to Dendrimer Solutions

Analysis of the ESR spectra of organic dendrimers containing paramagnetic metal ions is a useful tool to obtain information about structure, conformation, and interacting ability of dendrimers, and to assess different applications, mainly in the biomedical and electronic fields. This section provides examples on the ESR studies carried out by using paramagnetic metal ions to probe the dendrimer properties.

An ESR study of Cu(II) complexes formed with the *n*.5 PAMAM dendrimers (from G = 0.5 to G = 7.5), henceforth called *n*.5 G, was performed at different pH conditions, corresponding to different degrees of protonation of the COO^- groups that form the external dendrimer layer.[44] The availability of both surface COO^- and internal –N– and –NH–CO– binding sites makes possible the formation of different complexes, due to the distribution of Cu(II) at different sites. Analysis of spectra recorded at ambient and low temperature allowed extraction of the magnetic and mobility parameters of the different complexes formed by Cu(II) coordinating internal and external dendrimer sites. Four different complexes were formed as a function of generation and pH, corresponding to different sites of Cu(II). An interesting feature was found in the spectra at room temperature: the unpaired electron partly delocalizes from copper to the nitrogen ligands, giving rise to a resolved structure due to coupling with nitrogen and protons. A similar ESR analysis was performed for the amino terminated PAMAM dendrimers as a function of generation, pH, and temperature.[45] In this case too, computer-aided analysis of the ESR spectra led to identify different complexes for Cu(II). This study served as a reference for following the formation of silver nanocomposites with PAMAM dendrimers of different generations and differently functionalized at the surface ($-NH_2$, $-NH-C(CH_2OH)_3$, $-NH-COC(CH_3)_3$, that is amino, tris, and pivalate, respectively, and termed En-A; En-T, and En-P, respectively, where n indicates the generations):[46] Ag(I), adsorbed in the internal structure of the dendrimers, was reduced by X-rays to Ag, and Cu(II) was added to monitor, by means of ESR, the site and distribution of the silver nanocomposites. In this case too, different components identify different coordinations of Cu(II). The ESR analysis revealed that the internal structure of the dendrimer–silver ion complexes changed during their transformation into nanocomposites. The ESR analysis of Cu(II)–Gn system also demonstrated that, in spite of the stability of the Cu(II)–Gn complexes, uranyl ions are able to displace copper ions from the dendrimer interior.[47] This conclusion opened the interesting possibility to use these dendrimers as uranium sponges. However, the strong interacting ability of Cu(II) toward the nitrogen sites diminishes the reliability of these ions to probe the real structure of the dendrimer. Mn(II), which is known to preferentially coordinate oxygen sites, proved to be a good probe of the dendrimer structure and interacting ability, thus providing complementary information to Cu(II).[48] The computation of the ESR spectra of Mn(II) needed inclusion of the zero-field splitting (ZFS) contributions to both transitions and relaxation. Table 6 lists the main parameters used for computation: τ_c, the ZFS parameter; the isotropic hyperfine coupling constant, A_{iso}; and the relative percentages of the components A and B, where B corresponds to Mn(II) directly interacting with the dendrimer surface at the COO– groups.

Copper(II) complexes of Schiff base derivatives of DAB Am-8 dendrimer and different aldehydes (3-hydroxybenaldehyde, 3,5-dichlorosalicylaldehyde, and 5-nitrosalicylaldehyde) have been investigated using the ESR technique in the 3.5–300 K temperature range.[49] The spectra were computer analyzed to obtain

TABLE 6. Parameters Used for the Computation of the ESR Spectra of Mn(II) Interacting with Half-Generation PAMAM Dendrimers[a]

Generation	Signal	τ_c/ps	ZFS/G^2	A_{iso}/G	%
0	A	7	24,000	96.4	100
2.5	A	10	30,000	92.8	20
	B	7	250,000	92.8	80
5.5	A	10	30,000	92.5	15
	B	7	250,000	92.5	85
7.5	A	10	30,000	94.2	15
	B	7	250,000	94.2	85

[a] Correlation time for motion, τ_c (±5%), ZFS parameter, (±10%); isotropic hyperfine coupling constant, A_{iso} (±5%); and relative percentages of the components A and B (±5%).

anisotropic g-values and ΔB_i line widths, which revealed a Curie–Weiss type behavior evidencing ferromagnetic and antiferromagnetic interactions between copper ions. The authors described the formation of different copper clusters and different local symmetries of the crystal field around the copper ions

Poly(propylene imine) (DAB) dendrimers functionalized with bis[2-(2-pyridyl) ethyl]-amine (PY2) ligands also formed different complexes with Cu(II).[50] The analysis of the ESR spectra showed the formation of tetranuclear complexes in which the metal centers are complexed by the PY2 parts of the dendrimer and are further weakly coordinated by solvent molecules. There was no evidence for participation of the tertiary amine N atoms of the dendrimer core in the complexation. The ESR titration with an increasing amount of copper showed that the pyridine dendrimers at generations 2, 3, and 4 were saturated with 8, 16, and 32 copper ions per dendrimer, respectively: larger amounts of Cu(II) resulted in a free Cu(II) component superimposed on the "complexed-Cu" component.

The molecular dynamics of vanadyl complexes with the chelate 2-(4-isothiocyanatobenzyl)-6-methyldiethylenetriaminepentaacetic acid, covalently attached to ammonia core poly(amidoamine) (PAMAM) dendrimers (PAMAM-TU-DTPA), was also studied by ESR.[51] The spectra were computed with modification of the slow-motional line-shape theory. From this analysis the rotational correlation time for motion was evaluated. Table 7 lists both the isotropic τ values, at different temperatures (7a); and the anisotropic τ values (7b). In the table, $\tau_\perp$ is the perpendicular component, $\tau_{//}$ is the parallel component, and $\tau_{av} = (\tau_\perp \tau_{//})^{1/2}$. The anisotropic parameters were compared to τ_{Riso}, the rotational correlation time determined with the isotropic tumbling model. At near freezing, the isotropic model and the anisotropic model give nearly identical simulated spectra. The results indicated that the rotational correlation times of the surface chelate increase with molecular weight and resemble those of "internal" segmental motions found in PAMAMs. The comparison of isotropic and anisotropic tumbling models in Table 7 indicates anisotropic tumbling of the ion–chelate complex at physiological temperatures, which is consistent with a model that incorporates segmental motions of the dendrimer side chains.[51]

An interesting family of dendrimers are those forming liquid–crystal structures. Liquid crystals of copper(II)-containing PPI dendrimers were studied using ESR.[52,53] 4-(4-N-Alkoxybenzoyloxy)salicylaldehydes attached at the peripheral amino groups

TABLE 7a. Rotational Correlation Times(τ)[a] of PAMAM-TU-DTPA-VO(II) at Different Temperatures[b]

	20°C	25°C	35°C
Free chelate	0.15	0.14	0.12
G = 2 PAMAM-TU-DTPA	0.93	0.64	0.49
G = 6 PAMAM-TU-DTPA	2.5	2.1	0.64

[a] τ is in nanoseconds (ns)
[b] From Ref. 51.

TABLE 7b. Anisotropic Rotational Correlation Times ($\tau_\perp$, Perpendicular Component; $\tau_{//}$, Parallel Component; $\tau_{av}=(\tau_\perp\tau_{//})^{1/2}$; τ_{Riso}, Determined with the Isotropic Tumbling Model) for G=6 PAMAM-TU-DTPA.

T(°C)	$\tau_\perp$	$\tau_{//}$	τ_{av}	τ_{Riso}
1.9				5.6
8.0				3.9
13.9	62	0.26	4.0	3.0
19.7	39	0.26	3.2	2.5
25.4	31	0.24	2.8	2.1
32.5	20	0.22	2.1	1.7
37.5	10	0.21	1.5	0.64
52.0	4.7	0.16	0.86	0.40

[a] τ is in nanoseconds (ns)
[b] From Ref. 51.

of the dendrimers, as mesogenic moieties, give rise to dendromesogens with a different number of mesogenic branches.[52] Different copper complexes were characterized by ESR and the orienting effect of a high magnetic field provided information on the dendrimeric metallomesogens.[53] The ESR spectra of Cu(II)–PPI complexes at different orientations with respect to the magnetic field were calculated with a distribution function. This approach allowed the resolution of the anisotropic superhyperfine structure due to amido nitrogen ligands in dendrimers, and led to geometric information on the complexation of the copper metal ions. At the lowest Cu(II) content, the copper complexes have an approximately square-planar geometry, in which each Cu(II) ion is bonded to two nitrogen and two carbonyl oxygen atoms from the amido groups in the outer core of the dendrimers (N_2O_2 coordination). For intermediate concentrations, two different kinds of copper complex geometry exist: an N_2O_2 pseudotetrahedral geometry with coordination of two amido nitrogen and two carbonyl oxygen atoms, and a five-coordinate, square-pyramidal geometry, presumably N_3O_2, by additional bridging to a tertiary amino nitrogen atom from the inner core of the dendrimer. Higher Cu(II) loadings lead to increased exchange coupling between the complex sites.[53]

2.4. Metallo-Dendrimers

Two main classes of metallo-dendrimers can be distinguished: the classic organic dendrimers (e.g., DAB and PAMAM dendrimers) that include in their structure some metallic elements in different oxidation states, and the emergent inorganic dendrimers.[54,55]

Incorporation of transition metal complexes in both organic and inorganic dendrimers leads to new and interesting material with specific properties, such as the capability to absorb visible light, to exhibit luminescence, and to undergo reversible multielectron redox processes.[55] These materials may find applications as components in molecular electronics and as photochemical molecular devices for solar energy conversion and information storage.[55] Metal-containing dendrimers can be classified into three categories (Fig. 9): Dendrimers with coordination centers at the core (a), dendrimers with coordination centers through all layers (b), and dendrimers with coordination centers on the periphery (surface groups) (c). In the a category, dendrimers bearing a metalloporphyrin as a core have a special role, since they mimic the biological function of heme proteins: the ability to reversibly binding O_2. The combination of metalloporphyrin with dendrimer chemistry led to an interesting new class of model systems, with dendritic superstructures that mimic the encapsulation of the heme in a natural protein environment. When a paramagnetic metal ion is involved in these structures, the ESR analysis is helpful in determining the local structure of the metal environment. Interesting examples of this analysis are reported for dendrimers with a $Fe^{(III)}$–porphyrin core, which possess a vacant coordination site available for ligand binding and catalysis.[56] All dendritic Fe(III) complexes display a predominant axial signal in the ESR spectrum, typical for high-spin Fe(III) porphyrins, and a weak rhombic signal indicative of a low-spin Fe(III) species. These paramagnetic properties indicate a possible weak coordination of the chloride counterion at the vacant coordination site of the Fe(III) porphyrin, as, in these cases, the formation of head-to-tail dimers is improbable, considering the steric bulkiness of the dendritic shells.[56]

The use of different ESR techniques permitted a deeper understanding of the binding properties of the dendrimers with a metallo-porphyrin core.[57] Dendritically functionalized Co(II) porphyrins in the presence of 1,2-dimethylimidazole (DiMeIm), pyridine, and 1-methylimidazole, were studied by means of CW and pulsed ESR, and electron nuclear double resonance (ENDOR) techniques, which revealed specific information on the oxygenated forms of these porphyrins. Both ENDOR and hyperfine sublevel correlation (HYSCORE) spectra demonstrated that no hydrogen-bond forms between the bound O_2 and a dendritic NH moiety. This hydrogen bond had earlier been proposed on the basis of the large dioxygen affinity of the corresponding Fe(II) complex. The ENDOR experiments indicated that the dendritic branches are closely packed in toluene and that the Co–O_2 bond has a highly ionic character. This observation is linked to the packing and the polarity of the dendritic branches and can be related to the O_2 and CO affinity of the corresponding Fe(II) complexes.[57]

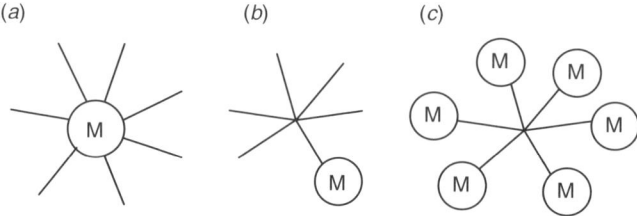

Fig. 9. Schematic representation of the three classes of metallo-dendrimers.

Another interesting ESR study of inorganic dendrimers, of the *c* category (Fig. 9), reported the synthesis and ESR characterization of new inorganic dendrimers, based on cyclophosphazene units, containing single metal complexes on the surface.[58] The complexes containing Fe ions exhibit intense signals both in the solid state and in dichloromethane solutions. The spectra exhibit the typical ESR signal of high-spin Fe(III) complexes (g near 2); the Fe(III) ion is a ground state $6S$ ($6A_{1g}$ in the complexes) and is not split in an octahedral or even a lower symmetry ligand field. In addition, there can be no splitting by spin–orbital interaction and a single isotropic resonance line at $g = 2$ is observed. Hyperfine interaction of the unpaired electron with the ^{14}N nucleus ($I = 1$) of the cyanide group was not observed, although the broad signal in the solid state could contain some unresolved splitting. The ESR spectrum of the copper complex with a PF_6 counterion exhibits the splitting typical of a tetragonal distortion of the copper complex. The cobalt complexes are mainly ESR silent. The Mn complexes show an intense ESR spectra in the case of a PF_6 contraion.[58]

As the first example of a "metallodendrimer catalyst", the catalytic behavior of nickel-containing carbosilane dendrimers (belonging to category *c*) was studied by ESR.[59] These dendrimers are catalytically active in the atom-transfer radical addition (ATRA) reaction by using methyl methacrylate (MMA) as substrate and carbon tetrachloride (CCl_4) as the polyhalogenated alkane. The catalytic deactivation of [G1]-Ni_{12} and [G2]-Ni_{36} is caused by irreversible formation of catalytically inactive Ni(III) sites on the periphery of these dendrimers. This hypothesis is supported by the ESR analysis. Theoretical modeling showed that the ESR spectrum of the G1–Ni complex most probably consists of two phases. In one phase, the nickel(III) sites are more closely packed leading to a faster relaxation mode, and give rise to a signal with a relative high dipolar broadening. In the other phase, the Ni(III) sites are more separated and this "spectrum" is comparable to the ESR spectrum of a model Ni(III) complex measured at low temperature.[59]

As possible biochemical applications of metallo-dendrimers, we present two original examples for metallo-dendrimers belonging to the category *a* and *c*, respectively.[60,61] Polyether amine dendrimers with a Mo core were analyzed by ESR to get information about the dimethyl sulfoxide (DMSO) reductase family of enzymes.[60] The authors did not find any significant variation in the g values of the axial resolved spectrum of Mo(V) for the dendrimer with respect to the free Mo complex, indicating that the integrity of the square-pyramidal $MoOS_4^-$ core (C_{4V} local symmetry) is retained. The addition of bulky and symmetric dendritic architecture does not impose any further distortion at the metal center; therefore, the electronic structure remains unaltered. Lysine-based dendrimers containing gadolinium complexes at the external surface, termed Gadomer 17 and used as contrast agents in MRI, were also investigated by ESR.[61] The electronic transverse relaxation rate from the spectral line widths at different frequencies (X-band, 150 and 225 GHz) were plotted as a function of temperature. The experimental $1/T_{2e}$ values were treated as the sum of ZFS and intramolecular dipole–dipole contributions. The fit provided an effective Gd–Gd distance of $r_{Gd-Gd} = 12.6$ Å.

3. CONCLUSIONS

The dendrimers are a special class of polymer with a "ball" shape, due to the radial symmetry of their structure. The possibility to modify the synthesis with various cores, various polymerization units, and various external surface functions offer an infinite range of different dendrimer structures. Their use as catalytic agents, as selective boxes for host molecules and ions, as gene carriers and MRI contrast agents, and as light antennas has been demonstrated, and increased the interest in this kind of polymers.

The ESR studies performed on the dendrimers have demonstrated the utility of this technique to characterize their structure and properties and to clarify their behavior in specific applications. This chapter presented ESR studies describing the use of radical spin probes, spin labels, paramagnetic ions as spin probes, and metallo-dendrimers with paramagnetic ions in their structure.

The intensity of the ESR signal provides a measure of the radical concentration and is evaluated by double integration of the ESR signal, based on an intensity reference. Variation of the ESR intensity indicated the entrance–exit of a paramagnetic probe from the dendritic box, or the oxidation–reduction of a paramagnetic label or ion belonging to the dendrimer. Variation of relative intensity of a spectral component provided a measure of the distribution of paramagnetic species in different locations and environments at the external and/or internal surface of the dendrimers.

The magnetic parameters, such as the components of the g- and A-tensors or their isotropic values, were calculated for copper complexes at dendrimer surface sites in order to identify the type of complexes.

When the relaxation mechanism is the modulation of the magnetic g and A components due to the rotational diffusion of the paramagnetic group (mainly for the nitroxide radicals, and for $S = \frac{1}{2}$ paramagnetic ions), the analysis of the spectra in the fast–slow motion regime provides the correlation time for the rotational motion. An increase in the correlation time corresponds to a decrease in mobility of the paramagnetic probe or label. The evaluation of the correlation time for the rotational motion was performed by simple methods or by computation of the spectra. Different diffusion models can be considered, such as Brownian or jump diffusion models, and the rotational mobility may be considered isotropic or anisotropic. In this latter case, for nitroxide radicals, the main information was obtained from the perpendicular component of the correlation time. Furthermore, a shift of the main rotational axis accounts for the compression of the labels due to other molecules approaching the label at the dendrimer surface.

The analysis of the correlation time for motion provided information on the interactions of probes with the dendrimer surface, or the interactions of the dendrimers with other species, such as biomolecules, surfactant aggregates, and different types of liquid and solid surfaces. In the case of interactions of the dendrimer with surfactant aggregates, the use of radical surfactants inserted in the aggregates also provided information about the type and the strength of the interactions. The order parameter is extracted from the analysis of the ESR spectra.

Parameters related to spin–spin coupling and/or exchange interactions (such as the J coupling, ZFS parameters, and the exchange frequency) become the main parameters when paramagnetic species approaches each other, or when the spin multiplicity increases. These cases are accomplished in the case of increased loading of paramagnetic probes or labels inside the dendrimer or at the dendrimer surface, or when species with $S > \frac{1}{2}$ are doped in the dendrimer system. However, in these cases very often the analysis could be performed by simply evaluating the line width of the single line, which originated from high local concentration of spin or high-spin density. A more correct line width is calculated by spectral simulation, but the empirical peak-to-peak line width also provided a good indication of spin–spin interactions. Similarly, empirical parameters, such as ratios of the heights of the hyperfine lines, were used to calculate the partition of labeled dendrimers in different environments and the effects of weak spin–spin interactions (dipole broadening).

In addition to CW ESR, ENDOR, and pulsed techniques (HYSCORE spectra, 2D-ESTN, and ESEEM) were also used for dendrimer analysis. The ESR techniques have provided useful information on the dendrimers structure, properties, and applications, and their interactions with selected species.

ACKNOWLEDGMENTS

We are grateful to the National Science Foundation and the Italian Ministry of University and Research for financial support. We thank D. A. Tomalia for the gift of the PAMAM dendrimers and for his contagious enthusiasm. We appreciate fruitful collaborations with S. Jockush and A. Moscatelli at Columbia University, with M. Brustolon, A. Barbon, and A. Zoleo at University of Padua, Italy, and with M. Cangiotti at University of Urbino. MFO thanks E. Cossu, P. Favuzza, R. Daddi, F. Furini, P. Matteini, F. Montalti, G. Amantini, S. Calici, and all the students who contributed to the dendrimer studies. The authors are grateful to S. Schlick for her great help in revising this chapter.

REFERENCES

1. (a) Tomalia, D.A.; Banker, H.; Dewald, J.; Hall, M.; Kallos G. Martin, S.; Roeck, J.; Ryder, J.; Smith, *Polym. J.* **1985**, *17*, 117. (b) Tomalia, D.A.; Baker, H.; Dewald, J.; Hall, M,; Kallos, G.; Martin, S.; Roeck, J.; Ryder, J.; Smith, P. *Macromolecules* **1986**, *19*, 2466. (c) Tomalia, D.A.; Hall, M.; Hedstrand, D. *J. Am. Chem. Soc.* **1987**, *109*, 1601. (d) Tomalia, D.A.; Berry, V.; Hall, M.; Hedstrand, M. *Macromolecules* **1987**, *20*, 1164. (e) Tomalia, D.A. *US Pat.*, **1986**, 4, 587, 392; 4, 568, 737; 4, 599, 400; 4, 631, 337 **1985**, 4, 558, 120; 4, 507, 466. (f) Tomalia, D.A.; Naylor, A.M.; Goddard III, W.A. *Angew. Chem. Int. Ed. Engl.*, **1990**, *29*, 138.
2. Denkewalter, R.G.; Kolc, J.; Lukasavage, W.J. *US Pat.*, **1981**, 4, 289, 872.
3. (a) Moors, R.; Vögtle, F. *Chem. Ber.*, **1993**, *126*, 2133. (b) Wörner, C.; Mülhaupt, R. *Angew. Chem. Int. Ed. Engl.* **1993**, *32*, 1306.
4. de Brabander-van den Berg, E.M.M.; Meijer, E.W. *Angew. Chem. Int. Ed. Engl.* **1993**, *32*, 1308.

5. (a) Newkome, G.R.; Yao, Z.; Baker, G.R.; Gupta, V.K..; Russo, P.S.; Saunders, M.J. *J. Org. Chem.* **1985**, *50*, 2004. (b) Newkome, G.R.; Yao, Z.; Baker, G.R.; Gupta, V.K.; Russo, P.S.; Saunders, M.J. *J. Am. Chem. Soc.*, **1986**, *108*, 849. (c) Newkome, G.R.; Baker, G.R.; Saunders, M.J.; Russo, P.S.; Gupta, V.K.; Yao, V.K.; Miller, J.E.; Bouillion, K. *J. Chem. Soc., Chem. Commun.*, **1986**, 752.

6. de Gennes, P.G.; Hervet, H. *J. Phys. Lett.*, **1983**, 351.

7. (a) *Dendrimers and Other Dendritic Polymers*; Frechet, J.M.J.; Tomalia, D.A.; Eds.; John Wiley & Sons, Inc.: West Sussex, UK, 2001; (b) Bosman, A.W.; Janssen, H.M.; Meijer, E.W. *Chem. Rev.* **1999**, *99*, 1665. (c) Newkome, G.R., Ed. *Advances in Dendritic Macromolecules*; JAI Press: Greenwich CT, 1993; (d) Zeng, F.; Zimmerman, S.C. *Chem. Rev.* **1997**, *97*, 1681; (e) Issberner, J.; Moors, R.; Voegtle, F. *Angew. Chem. Int. Ed. Eng.* **1994**, *33*, 2413; (f) Alper, J. *Science* **1991**, *251*, 1562; (g) Amato, L. *Sci. News* **1990**, *138*, 298; (h) Frechet, J.M.J. *Science* **1994**, *263*, 1710. (k) Ardoin, N.; Astruc, D. *Bull. Soc. Chim. Fr.* **1995**, *132*, 875.

8. Dandliker, P.J.; Diedrich, F.; Gross, M.; Knobler, C.B.; Louati, A.; Sanford, E.M. *Angew. Chem. Int. Ed. Engl.* **1994**, *33*, 1739. For another 'Dendritic model of a redox protein', see Chow, H.-F.; Chan, I.Y.-K.; Chan, D.T.W.; Kwok, R.W.M. *Chem. Eur. J.* **1996**, *2*, 1085.

9. Hawker, C.J.; Wooley, K.L.; Fréchet, J.M.J. *J. Chem. Soc. Perkin Trans. 1* **1993**, 1287.

10. (a) Liu, M.; Frechet, J.M.J. *Pharm. Sci. Technol. Today* **1999**, *2*, 393. (b) Sideratou, Z.; Tsiourvas, D.; Paleos, C.M. *Langmuir* **2000**, *16*, 1766. (c) Denning, J.; Duncan, E. *Rev. Mol. Biotechnol.* **2002**, *90*, 339.

11. de Brabander, E.M.M.; Put, J.A. *Polym. Mat. Sci. Eng. (Proc. ACS Div. PMSE)* **1995**, *73*, 79.

12. Wiener, E.C.; Brechbiel, M.W.; Brothers, H.; Magin, R.L.; Gansow, O.A.; Tomalia, D.A.; Lauterbur, P.C. *Magn. Res. Med.* **1994**, *31*, 1.

13. Jansen, J.F.G.A.; Meijer, E. W.; de Brabander-van den Berg, E.M.M. *J. Am. Chem. Soc.* **1995**, *117*, 4417.

14. (a) Jansen, J.F.G.A.; Janssen, R.A.J.; De-Brabander-Van Den Berg, E.M.M.; Meijer, E.W *Adv. Mat.* **1995**, *7*, 561. (b) Bosman, A.W.; Jansen, J.F.G.A.; Janssen, R.A.J.; Meijer, E.W. *Polym. Mater. Sci. Eng.* **1995**, *73*, 340.

15. Dykes, G.M.; Smith, D.K.; Caragheorgheopol, A. *Org. Biomol. Chem.*, **2004**, *2*, 922.

16. Ottaviani, M.F.; Cossu, E.; Turro, N.J.; Tomalia, D.A. *J. Am. Chem. Soc.* **1995**, *117*, 4387.

17. (a) Ottaviani, M.F.; Turro, N.J.; Jockusch, S.; Tomalia, D.A. *J. Phys. Chem.* **1996**, *100*, 13675. (b) Ottaviani, M.F.; Andechaga, P.; Turro, N.J.;Tomalia, D.A. *J. Phys. Chem.* **1997**, *101*, 6057. (c) Ottaviani, M.F.; Daddi, R.; Brustolon, M.; Turro, N.J.;Tomalia, D.A. *Appl. Magn. Reson.* **1997**, *13*, 347.

18. (a) Schneider, D.J.; Freed, J.H. In *Biological Magnetic Resonance. Spin Labeling. Theory and Applications*, Berliner, L.J., Reuben, J., Eds.; Plenum Press: New York, 1989, Vol.8, p. 1. (b) Budil, D.E.; Lee, S.; Saxena, S.; Freed, J.H. *J. Magn. Res. A* **1996**, *120*, 155.

19. (a) Ottaviani, M.F.; Daddi, R.; Brustolon, M.; Turro, N.J.;Tomalia, D.A. *Langmuir* **1999**, *15*, 1973. (b) Ottaviani, F.M.; Favuzza, P.; Sacchi, B.; Turro, N.J.; Jockusch, S.; Tomalia, D.A. *Langmuir* **2002**, *18*, 2347.

20. Ottaviani, M.F.; Turro, C.; Turro, N.J.; Bossmann, S.H.; Tomalia, D.A. *J. Phys. Chem.* **1996**, *100*, 13667.

21. Bosman, A.W.; Janssen, R.A.J.; Meijer, E.W. *Macromolecules* **1997**, *30*, 3606.

22. Bryce, M.R.; Devonport, W.; Moore, A.J. *Angew. Chem., Int. Ed. Engl.* **1994**, *33*, 1761.

23. (a) Fillaut, J.-L.; Linares, J.; Astruc, D. *Angew. Chem., Int. Ed. Engl.* **1994**, *33*, 2460. (b) Moulines, F.; Djakovitch, L.; Boese, R.; Gloaguen, B.; Thiel, W.; Fillaut, J.-L.; Deville, M.-H.; Astruc, D. *Angew. Chem. Int. Ed. Engl.* **1993**, *32*, 1075.
24. Alonso, B.; Moran, N.; Casado, C.M.; Lobete, P.; Losada, J.; Cuadrado, I. *Chem. Mater.* **1995**, *7*, 1440.
25. (a) Miller, L.L.; Hashimoto, T.; Tabakovic, I.; Swanson, D.R.; Tomalia, D.A. *Chem. Mater.* **1995**, *7*, 9. (b) Duan, R.G.; Miller, L.L.; Tomalia, D.A. *J. Am. Chem. Soc.* **1995**, *117*, 10783.
26. Maliakal, A.J.; Turro, N.J.; Bosman, A.W.; Cornel, J.; Meijer, E.W. *J. Phys. Chem. A* **2003**, *107*, 8467.
27. (a) Sato, K.; Shiomi, D.; Takui, T.; Hattori, M.; Hirai, K.; Tomioka, H *Synthetic Metals* **2001**, *121*, 1816. (b) Sato, K.; Shiomi, D.; Takui, T.; Hattori, M.; Hirai, K.; Tomioka, H *Mol. Cryst. Liq. Cryst. Sci. Technol., Sect. A: Mol. Cryst. Liq. Cryst.* **2002**, *376*, 549.
28. Louie, J.; Hartwig, J.F. *J. Am. Chem. Soc.* **1997**, *119*, 11695.
29. Amaresh, R.R.; Liu, D.; Konovalova, T.; Lakshmikantham, M.V.; Cava, M.P.; Kispert, L.D. *J. Org. Chem.* **2001**, *66*, 7757.
30. Samuelson, L.E.; Sebby, K.B.; Walter, E.D.; Singel, D.J.; Cloninger, M. *J. Org. Biomol. Chem.* **2004**, *2*, 3075.
31. Walter, E.D. *Diss. Abstr. Int., B* **2004**, *65*, 240.
32. Yordanov, A.T.; Yamada, K.-i.; Krishna, M.C.; Mitchell, J.B.; Woller, E.; Cloninger, M.; Brechbiel, M.W. *Angew. Chem., Int. Ed.* **2001**, *40*, 2690.
33. Rosen, G.M., Rauckman E.J. *Biochem. Pharmacol.* **1977**, *26*, 675.
34. Pou, S; Davis, P.L.; Wolf, G.L.; Rosen, G.M. *Free Radic. Res.* **1995**, *23*, 353.
35. Winalski, C.S; Shortkroff, S.; Mulkern, R.V.; Schneider, E.; Rosen, G.M. *Magn. Res. Med.* **2002**, *48*, 965.
36. Francese, G.; Dunand, F.A.; Loosli, C.; Merbach, A.E.; Decurtins, S. *Magn. Res. Chem.* **2003**, *41*, 81.
37. Rosen, G.M.; Porasuphatana, S.; Tsai, P.; Ambulos, N.P.; Galtsev, V.E.; Ichikawa, K.; Halpern, H.J. *Macromolecules* **2003**, *36*, 1021.
38. Ottaviani, M.F.; Matteini, P.; Brustolon, M.; Turro, N.J.; Jockusch, S.; Tomalia, D.A. *J. Phys. Chem. B* **1998**, *102*, 6029.
39. Ottaviani, M.F.; Sacchi, B.; Turro, N.J.; Chen, W.; Jockusch, S.; Tomalia, D.A. *Macromolecules* **1999**, *32*, 2275.
40. Ottaviani, M.F.; Furini, F.; Casini, A.; Turro, N.J.; Jockusch, S.; Tomalia, D.A.; Messori, L. *Macromolecules* **2000**, *33*, 7842.
41. Ottaviani, M.F.; Jockusch, S.; Turro, N.J.; Tomalia, D.A.; Barbon, A. *Langmuir* **2004**, *20*, 10238.
42. Romanelli, M.; Kevan, L. *J. Magn. Reson.* **1991**, *91*, 549.
43. Ottaviani, M.F.; Turro, N.J.; Jockusch, S.; Tomalia, D.A. *J. Phys. Chem. B* **2003**, *107*, 2046.
44. Ottaviani, M.F.; Bossmann, S.; Turro, N.J.; Tomalia, D.A. *J. Am. Chem. Soc.* **1994**, *116*, 661.
45. Ottaviani, M.F.; Montalti, F.; Turro, N.J.; Tomalia, D.A. *J. Phys. Chem. B* 1997, *101*, 2, 158.
46. Ottaviani, M.F.; Valluzzi, R; Balogh, L. *Macromolecules* **2002**, *35*, 5105.
47. Ottaviani, M.F.; Favuzza, P.; Bigazzi, N; Turro, N.J.; Jockusch, S.; Tomalia, D.A. *Langmuir* **2000**, *16*, 7368.

48. Ottaviani, M.F.; Montalti, M; Romanelli, M.; Turro, N.J.; Tomalia, D.A. *J. Phys. Chem.* **1996**, *100*, 11033.
49. (a) Leniec, G.; Typek, J.; Wabia, L.; Kolodziej, B.; Grech, E.; Guskos, N. *Mol. Phys. Rep.* **2004**, *39*, 154. (b) Leniec, G.; Typek, J.; Wabia, L.; Kolodziej, B.; Grech, E.; Guskos, N. *Mol. Phys. Rep.* **2004**, *39*, 159.
50. Gebbink, R.J.M.K.; Bosman, A.W.; Feiters, M.C.; Meijer, E.W.; Nolte, R.J.M. *Chem.—A Eur. J.* **1999**, *5*, 65.
51. Wiener, E.C.; Auteri, F.P.; Chen, J.W.; Brechbiel, M.W.; Gansow, O.A.; Schneider, D.S.; Belford, ƒnR. L.; Clarkson, R.B.; Lauterbur P.C. *J. Am. Chem. Soc.* **1996**, *118*, 7774.
52. Barbera, J.; Marcos, M.; Omenat, A.; Serrano, J.-L.; Martinez, J.I.; Alonso, P.J. *Liq. Cryst.* **2000**, *27*, 255.
53. Domracheva, N.; Mirea, A.; Schwoerer, M.; Torre-Lorente, L.; Lattermann, G. *ChemPhysChem* **2005**, *6*, 110.
54. Majoral, I.P.; Caminade, A.M. *Chem. Rev.* **1999**, *99*, 845.
55. Balzani, V; Campagna, S.; Denti, G.; Juris, A.; Serroni, S.; Venturi, M. *Acc. Chem. Res.* **1998**, *31*, 26.
56. (a) Weyermann, P.; Diederich, F.; Gisselbrecht, J.-P.; Boudon, C.; Gross, M. *Helv. Chim. Acta* **2002**, 85, 571. (b) Weyermann, P.; Diederich, F. *Helv. Chim. Acta* **2002**, *85*, 599. (c) Felber, B.; Diederich, F. *Helv. Chim. Acta* **2005**, *88*, 120.
57. Van Doorslaer, S.; Zingg, A.; Schweiger, A.; Diederich, F. *ChemPhysChem* **2002**, *3*, 659.
58. Diaz, C.; Valenzuela, M.L. *Polyhedron* **2002**, *21*, 909.
59. Kleij, A.W.; Gossage, R.A.; Gebbink, R.J.M.K.; Brinkmann, N.; Reijerse, E.J.; Kragl, U.; Lutz, M.; Spek, A.L.; van Koten, G. *J. Am. Chem. Soc* **2000**, *122*, 12112.
60. Mondal, S.; Basu, P. *Inorg. Chem.* **2001**, *40*, 192.
61. Nicolle, G.M.; Toth, E.; Schmitt-Willich, H.; Raduchel, B.; Merbach, A.E. *Chem. — A Eur. J.* **2002**, *8*, 1040.

12

HIGH-FIELD ESR SPECTROSCOPY OF CONDUCTIVE POLYMERS

Victor I. Krinichnyi

Institute of Problems of Chemical Physics, Chernogolovka, Russia

Contents
1. Introduction 307
2. Magnetic Parameters of Charge Carriers in Conductive Polymers 310
3. Relaxation and Dynamics of Charge Carriers in Slightly Doped Conductive Polymers 317
4. Charge Transfer in Highly Doped Conductive Polymers 326
5. High-Field Saturation Transfer ESR Method in the Study of Conductive Polymers 328
6. High-Field Spin Probe Method in the Study of Conductive Polymers 330
7. Concluding Remarks 333
Acknowledgments 334
References 334

1. INTRODUCTION

Various low-dimensional compounds can be considered as organic conductors; organic conductive polymers are of special interest,[1,2] due to their potential applications in molecular electronics. A major scientific goal is to reinforce the human brain with the ability of a computer. However, a convenient computer is based on three-dimensional (3D) silicon technology, whereas human organisms consist of low-dimensional biological

* This chapter is dedicated to my wife, Tatiana Krinichnaya.

Advanced ESR Methods in Polymer Research, edited by Shulamith Schlick.
Copyright © 2006 John Wiley & Sons, Inc.

systems. So, the future computer based on organic conductive polymers, combined with biosystems, is expected to increase considerably the power of human comprehension.

Initially, π-conjugated polymers are insulators. Their conductivity can be controlled by the chemical–electrochemical introduction of dopants, for example, acids or metal ions, which accept from a chain or inject to a chain an elemental charge. Consequently, the conductivity of a polymer increases up to $\approx 10^{-5}$–1 S cm^{-1} (semiconductor) or even up to $\approx 10^2$–10^6 S cm^{-1} (metal).[1]

Polyacetylene (PA) is the simplest conjugated polymer. It can exist in cis- and trans-forms (cis- and trans-PA isomers). The latter is thermodynamically more stable. The transition between C–C and C = C bonds in trans-PA does not require energy change, so the Peierls distortion[3] opens up a substantial gap in the Fermi level. This twofold degeneration leads to the formation of mobile solitons with a length of ≈ 15 C–C units and spin $S = \frac{1}{2}$ on trans-PA chains. These correspond to a break in the pattern of bond alternation,[4] and thus determine the fundamental properties of the polymer.

Poly(p-phenylene) (PPP), polythiophene (PT), polypyrrole (PP), polyaniline (PANI), and poly(tetrathiafulvalene) (PTTF) are other conductive polymers with pentameric and hexameric unit rings in which two neighboring units are tilted with respect to one another by a torsion angle, for example, equal in PPP to $\approx 23°$.[5] This angle is a compromise between the effect of conjugation and crystal-packing energy, which would lead to a planar conformation, and the steric repulsing between ortho-hydrogen atoms, which would lead to a non-planar conformation. For these polymers a resonance form can also be derived, which corresponds to a quinoid structure; however, in contrast to trans-PA, benzenoid and quinoid forms are not energetically equivalent, with the quinoid structure being substantially higher in energy. As a result, the solitons are trapped in the slightly doped PPP by the charges in polymer structure and the other mobile nonlinear excitations, namely polarons, are formed.[6] These quasiparticles, with the size of about five polymer units,[7] possess an elemental charge e and spin $S = \frac{1}{2}$. With increase of the doping level y (the number of the dopant molecules per polymer unit), the polaron pairs can combine to form spinless bipolarons with charge 2e.

Some of the most studied conductive polymers, as well as nonlinear quasiparticles formed in their chains, are schematically shown in Fig. 1.

A highly anisotropic quasi-one-dimensional (Q1D) π-conjugated structure with the above topological distortions as charge carriers makes such systems fundamentally different from traditional inorganic semiconductors, for example, silicon and selenium, and insulating polymers, for example, polyethylene. Their direct current (dc) conductivity σ_{dc} can be controlled by chemical or electrochemical oxidation or reduction from the insulator to the semiconductor and then to the metal.[1]

The electronic properties of conductive polymers correlate with their structure, morphology, and quality, as well as with the nature and number of the dopant molecules introduced into the polymer matrix.

Different methods were widely used for the study of fundamental structure and dynamics properties of conductive polymers: optical and X-ray photoelectron spectroscopy, scanning electron microscopy (SEM), chromatography, dc and alternating current (ac) conductometry, microwave dielectrometry, Faraday balance and alternating force magnetometry, and thermoelectric power.[1,2] As the electronic properties of

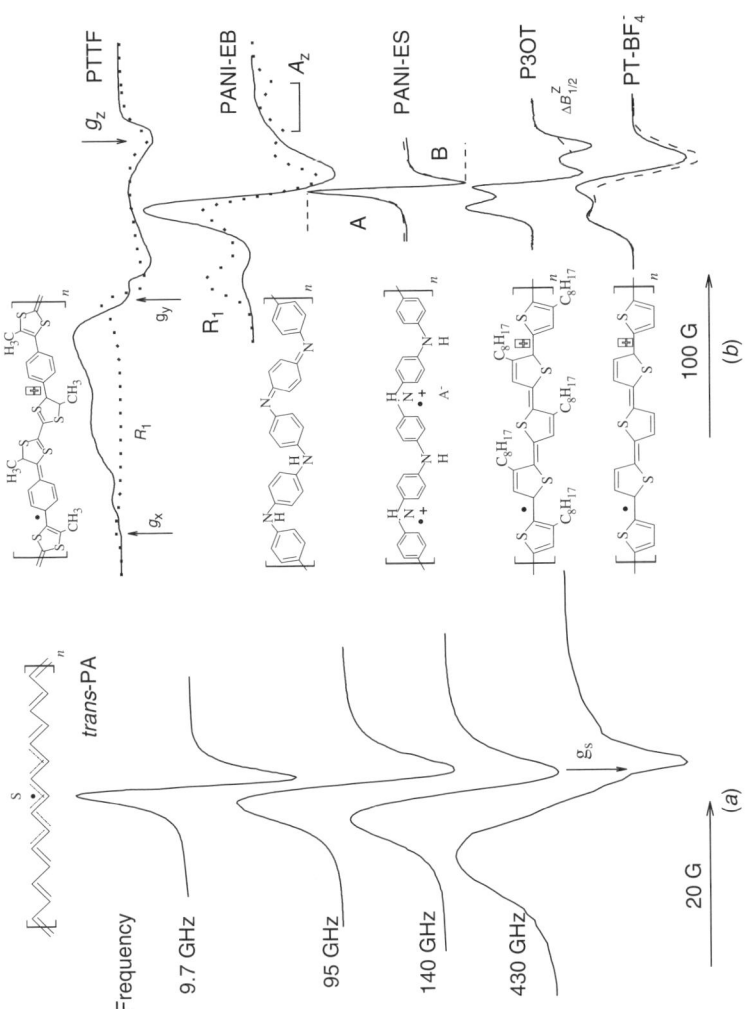

Fig. 1. (a) The ESR spectra of *trans*-PA registered at different frequencies at room temperature. (b) The D-band ESR spectra of PTTF with $y = 0$ (dotted line) and 0.08 (solid line), PANI-EB whose R_1 PC is shown by dotted line, PANI-ES (in the formula A^- is an anion, and dashed lines show the spectrum calculated from Eq. 5 with D/A = 0.39 and ΔB_{pp} = 4.89 G), P3OT, and PT-BF$_4^-$ registered at $T = 300$ (solid line) and 200 K (dashed line). The components of anisotropic g-factors of PC stabilized in the samples are summarized in Table 1. The soliton S in *trans*-PA and the polaron in other conductive polymers spread out over a larger number of units than shown schematically here. [From Refs. 12(b), 19, 24, 30(b), 44(b), and 45(a) with permission.]

309

conductive polymers are related to the existence of nonlinear excitations possessing electron and nuclear spins, various magnetic resonance methods, such as nuclear magnetic resonance (NMR), electron–nuclear double resonance (ENDOR), electron spin–echo envelope modulation (ESEEM), and ESR were also used for the study of these systems.[8,9] Electron spin resonance spectroscopy has the ability to analyze directly the nature and dynamics of nonlinear charge carriers stabilized in conductive polymers.

The ESR investigations are usually carried out at a frequency of $\omega_e/2\pi = \nu_e \leq 10$ GHz and a field $B_0 \leq 3.4$ kG. However, these conditions result in single-line spectra with information limited to the concentration and line width of the paramagnetic centers (PC). In addition, strong cross-relaxation of PC exists at low magnetic fields.[10] Thus, the ESR study of organic conductive systems at $\nu_e \leq 10$ GHz faces considerable limitations.

It was shown[11,12] that the measurement of organic radicals in different solids, especially in conductive polymers at D-band ESR ($\omega_e/2\pi \approx 140$ GHz and $B_0 \approx 50$ kG) enables the absolute sensitivity, precision, and informativity of the method to increase considerably. At high frequencies, the main advantage of the method is the higher spectral resolution of the g-factor due to the basic relation $g_i - g_e \propto \omega_e$ (here, $g_e = 2.00232$ is the g-factor for a free electron). The transition to D-band widens the range of molecular motions in condensed systems by more than an order of magnitude. This sensitivity of the ESR method, that is, the minimum number of spins detected, increases with the increase of ω_e as $N_{min} \propto \omega_e^{-\alpha}$, where $\alpha \approx 0.5$–4.5.[13] Finally, the probability P_{cr} of cross-relaxation (cr) of PC decreases strongly with the increase of a polarizing microwave quantum energy $\hbar\omega_e \propto B_0$ as $P_{cr} \propto \exp(-B_0^2)$,[14] so the spin packets become noninteracting at D-band ESR, and therefore can be saturated at lower values of the magnetic term B_1 of polarizing microwave radiation. The electron relaxation time of PC stabilized in some solids can increase with an increase in ω_e, which is another reason for the appearance of fast passage effects that make an efficient study of their relaxation and dynamics properties possible.

In our laboratory, ESR studies at D-band have investigated the structure, dynamics, and other specific characteristics of PC and their local environment in various biological[11] and conductive[12,15] polymers. The main results for conductive polymers are summarized in this chapter.

2. MAGNETIC PARAMETERS OF CHARGE CARRIERS IN CONDUCTIVE POLYMERS

Trans-PA can be obtained by thermal, chemical, or electrochemical treatment of *cis*-PA.[16] Such a treatment leads to the increase of the polymer dc conductivity σ_{dc} from $\approx 10^{-12}$ up to 10^{-5} S cm^{-1} and is accompanied by the increase in spin concentration from $\approx 10^{18}$ up to 10^{19} spin g^{-1} and by strong line narrowing.[8]

Numerous ESR studies of the nature, composition, relaxation, and dynamics of PC in *cis*- and *trans*-PA were carried out at convenient $\omega_e/2\pi \leq 10$ GHz.[8] Figure 1 exhibits the X-band ($\omega_e/2\pi = 9.7$ GHz) ESR spectra of *trans*-PA as an example: a single symmetric line with effective $g = 2.0026$ and line width $\Delta B_{pp} = 2.2$ G. A small

ESR line width broadening (0.5 G) was observed on cooling to 77 K, probably due to a smaller libration motion of different parts of the polymer chains. The data obtained have no simple interpretation because at this frequency it is impossible to discern the difference between localized and mobile π-radicals with close magnetic parameters.

The increase of the frequency ω_e leads to a change in the line shape, as shown in Fig. 1a.[17–19] Analysis of the spectra shows[19] that two PCs are stabilized in *trans*-PA, namely, neutral solitons pinned on short polymer chains with $g_\perp = 2.00283$ and $g_\parallel = 2.00236$, and those moving along the long polymer axis with $g = 2.00269$ and relative contributions of 1.1×10^{-3}:6×10^{-5} spins per carbon atom, respectively. The latter value is two orders of magnitude smaller than that reported by Goldberg[20]. By replacing hydrogen atoms with SR-groups (R is alkyl substitute), that is, the transition from *trans*-PA to poly(bis-alkylthioacetylene) (PATAC), causes a drastic increase in the anisotropy of magnetic parameters of the polymer.[21]

The $g_\perp$ value of *trans*-PA differs from g_e by $g_\perp - g_e = 5 \times 10^{-4}$. According to a perturbation theory,[22] the shift of the g-factor is

$$g_{x,y} - g_e = \frac{g_e \rho(0) \lambda}{\Delta E n - \pi^*, \sigma - \pi^*} \quad (1)$$

where $\rho(0)$ is the spin density at the nucleus position, λ is the spin–orbit coupling constant, $\Delta E_{n-\pi^*}$ and $\Delta E_{\sigma-\pi^*}$ are the energies of the unpaired electron $n \to \pi^*$ and $\sigma \to \pi^*$ transitions, respectively. Therefore, the difference obtained corresponds to an unpaired electron excitation from the σ_{c-c} orbital to an antibonding π^* orbital with $\Delta E_{\sigma-\pi^*} = 14$ eV, which is near $\Delta E_{\sigma-\pi^*} = 14.5$ eV calculated for a normal C–C bond in π-conducting systems.[23] The isotropic g-factors of delocalized PC lies near to that of the pinned one. This is evident from the averaging of the **g** tensor components of delocalized solitons due to their mobility with the rate[13]

$$D_{ID}^0 \geq \frac{(g_\perp - g_e)\mu_B B_0}{\hbar} \quad (2)$$

that exceeds 10^9 rad s^{-1} in *trans*-PA.

Figure 1b also exhibits a typical D-band ESR spectra of both neat and iodine-doped PTTF sample with more pronounced anisotropic magnetic parameters.[24,25] This polymer with methyl (Me) and ethyl (Et) substituents in which TTF units with R = H, Me, or Et are linked via phenyl (PTTF-R-Ph) or tetrahydroanthracene (PTTF-THA) bridges is a *p*-semiconductor with dc conductivity $\sigma_{dc} \approx 10^{-4}$–$10^{-3}$ S cm^{-1}.[26] The anisotropic ESR spectrum of localized PC R_1 in, for example, PTTF-Me-Ph, is characterized by g-factors presented in Table 1. In addition, more mobile polarons R_2 with $g_x = 2.00928$, $g_y = 2.00632$, $g_z = 2.00210$ also appear upon polymer doping. The spectrum of PTTF-Et-Ph was analyzed in terms of two contributions, of localized PC R_1 with magnetic parameters presented in Table 1, and PC R_2 with a nearly symmetric spectrum and $g_{eff} = 2.00706$ diffusing along the polymer long axis with the rate (see Eq. 2) of $D_{ID}^0 \geq 3 \times 10^{10}$ rad s^{-1}. The main components of the g-tensor of PC R_1 localized in PTTF–THA are summarized in Table 1 as well. The spectrum of delocalized PC R_2 have $g_\parallel^p = 2.00961$ and

$g_\perp^P = 2.00585$ (R_2). The concentration ratio $[R_1]/[R_2]$ is 20:1 in PTTF–Me–Ph, 3:1 in PTTF–THA, and 1:1.8 in iodine-doped PTTF–Et–Ph.

Polaron motion along the polymer chain induces interactions between electron spins and between electron and proton spins, which depend on the frequency.[27] The line width of the ESR spectral components of polarons immobilized and delocalized in, for example, PTTF–Et–Ph, increases from 2.8 to 3.8 G and then to 39 G, and from 10.2 to 11.5 and then to 175 G, when the microwave frequency increases from 9.5 to 37, and then to 140 GHz, respectively, at room temperature.[24] It is then possible to determine a real line width of these PCs at the $\omega_e \to 0$ limit, 2.4 and 8.3 G, respectively.

Polythiophene and its derivatives, poly(3-alkylthiophenes) (P3AT), with sulfur in the backbone are a suitable system for understanding the electronic and optical properties of Q1D systems with nondegenerate ground states.[28] Normally, these compounds are semiconductors, whose mid-gap is determined by the presence of the π-orbital conjugation along the main polymer axis. The presence of polarons in polythiophene and its derivatives was revealed by optical absorption measurements and ESR spectroscopy.[8] At X-band ESR pristine PT and P3AT have a single symmetric line with $g \approx 2.0026$ and $\Delta B_{pp} \approx 4$–8 G, evidencing that the spins do not belong to a sulfur-containing moiety and are localized on the polymer chains.[8]

The D-band ESR spectra of PT and P3OT are more informative.[29,30] The D-band ESR spectrum of P3OT is a superposition of Gaussian and Lorentzian lines with the anisotropic g-factor typical for PC in some other conductive polymers with heteroatoms.[12] The main components of the g-tensor of PC in P3OT are presented in Table 1. The structure of a polymer should affect the distribution of an unpaired electron in polaron, changing the principal values of the PC g-tensor and hyperfine structure. The lower limit of the polaron motion in P3OT can be determined from the shift of the g_x and g_y values compared to g_e, by using Eq. 1 with $\lambda_s = 0.047$ eV to be $D_{ID}^0 \geq 3.4 \times 10^9$ rad s^{-1}.

In contrast to X-band ESR, the high spectral resolution achieved at D-band ESR allows us to register the structure and/or dynamic changes in all spectral components separately. Fig. 2 shows that the g_x and g_y values clearly reflect the properties of the radical microenvironment. These values of the initial P3OT decrease with the

Table 1. Principal g-Tensor Components g_i and Averaged g-Factors of PC in Conductive Polymers Determined from D-Band ESR at Room Temperature

Sample	g_x	g_y	g_z	$<g>$
PTTF-Me-Ph	2.01189	2.00544	2.00185	2.00639
PTTF-Et-Ph	2.01424	2.00651	2.00235	2.00770
PTTF-THA	2.01292	2.00620	2.00251	2.00721
P3OT	2.00409	2.00332	2.00235	2.00325
PANI-SA	2.00603	2.00382	2.00239	2.00408
PANI-HSA	2.00522	2.00401	2.00228	2.00384

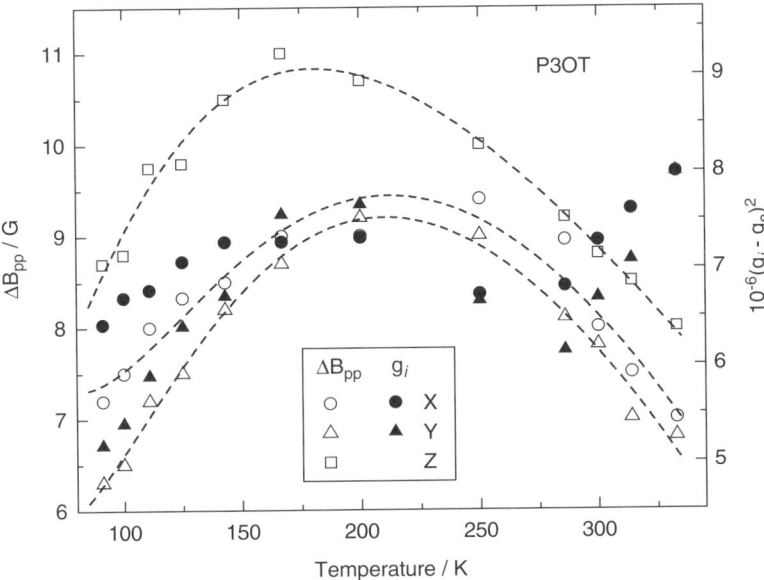

Fig. 2. Temperature dependence of the line width of the main spectral components, and of the squared shifts of g_x and g_y from $g_e = 2.00232$ for P3OT at D-band. From the top down, the dashed lines show the dependences calculated from Eq. 3 with $\Delta B_{pp}^0 = 7.3$ G, $\omega_{hop}^0 = 8.7 \times 10^{14}$ s^{-1}, $E_a = 0.014$ eV, $J_d = 0.30$ eV, $\Delta B_{pp}^0 = 12.3$ G, $\omega_{hop}^0 = 1.8 \times 10^{14}$ s^{-1}, $E_a = 0.022$ eV, $J_d = 0.29$ eV, and $\Delta B_{pp}^0 = 8.7$ G, $\omega_{hop}^0 = 8.5 \times 10^{14}$ s^{-1}, $E_a = 0.020$ eV, $J_d = 0.27$ eV. [From Ref. 30(b) with permission.]

temperature decrease from 333 down to 280 K, possibly due to the transition to a more planar conformation of the polymer chains. Below 280 K, these values increase at the sample freezing down to 160–220 K, and then decrease with a further temperature decrease. The decrease of g_x and g_y values at low temperatures can be explained by a harmonic vibration of macromolecules that reflects the crystal-field modulation and is characterized by the $g(T) \propto T$ dependence.[31] At the same time, the line width of the P3OT spectral components increases with a temperature increase from 90 K up to the phase transition characteristic temperature $T_c \approx 200$ K, and then decreases as the temperature increases. This effect of the line width decrease below T_c can be due to molecular motion of alkyl groups and/or acceleration of relaxation processes. If one assumes that molecular dynamics and/or electron relaxation lead to broadening of the X, Y, and Z spectral components, the respective activation energies 3.6, 4.9, and 4.3 meV are determined from the slopes of these dependences at $T \geq T_c$.

The extremal variation of the line width can be interpreted in the framework of the Houzé–Nechtschein model[32] of dipole–dipole interaction between different PC. According to this model, the collision of mobile PC with other spins should broaden ESR lines as

$$\Delta B_{pp} = \Delta B_{pp}^0 + \delta(\Delta B_{pp}) = \frac{16\omega_{hop} C}{27\gamma_e}\left(1 + \frac{\hbar^2 \omega_{hop}^2}{144 J_d^2}\right) \quad (3)$$

where ΔB_{pp}^0 is the line width of noninteracting polarons, ω_{hop} is the frequency of the polaron hopping along a polymer chain, C is the number of guest PC per aniline ring, and J_d is the constant of the spin dipole–dipole interaction in the system. If the ratio $J_d/\hbar$ exceeds the frequency of collision of both PC types, the condition of strong interaction is realized in the system that leads to a direct relation of the spin-spin interaction and polaron diffusion frequencies, so that $\lim[\delta(\Delta B_{pp})] = \frac{16}{27} C\omega_{hop}$. In an opposite case, if the condition of weak interaction prevails, the result is an inverse dependence of these frequencies, $\lim[\delta(\Delta B_{pp})] = \frac{4}{3} (C/\omega_{hop})(J_d/\hbar)^2$. According to the spin exchange fundamental concepts,[33] the extreme $\delta(\Delta B_{pp})(T)$ dependence should reflect both types of spin–spin interaction, at $T \leq T_c$ and $T \geq T_c$, respectively. An additional plausible reason for line broadening can be higher spin localization with a temperature decrease at $T \geq T_c$.

Assuming an activation character of the spin–spin interaction with activation energy E_a, when $\omega_{hop} = \exp(-E_a/k_BT)$, the temperature dependence of the line widths for the X, Y, and Z spectral components can be calculated from Eq. 3 with $E_a =$ 0.022, 0.020, and 0.014 eV, respectively (Fig. 2). The $J_d \approx 0.3$ eV value obtained exceeds the corresponding spin-exchange constant for nitroxide radicals with paramagnetic ions, $J_d \leq 0.01$ eV.[34]

The data presented in Fig. 2.3 evidence that the correlation of line width with the spin–orbit coupling according to the Elliot mechanism is [35]

$$\sigma \propto \Delta B_{pp}^{-1} = \frac{\tau_s}{\alpha(g-g_e)^2} \qquad (4)$$

where σ is the conductivity, α is the constant, $g-g_e \propto \lambda/\Delta E_{i-\pi^*}$ is the shift of the g-factor, and τ_s is the electron scattering time. Such a mechanism plays an important role in the charge transfer in organic ion–radical salts[36] and conductive polymers with pentameric rings.[8] Indeed, $\Delta B_{pp}^x \propto \Delta g_x^2$ and $\Delta B_{pp}^y \propto \Delta g_y^2$ dependences are valid for P3OT at least at $T \leq T_c$. This means that different mechanisms affect the individual components of the P3OT spectrum, and that the scattering of charge carriers (see below) should be governed by the potential of the polymer backbone.

PANI differs from other conducting polymers in several important aspects. In contrast to PA- and PPP-like polymers, it has no charge conjugation symmetry. Besides, both phenyl rings and nitrogen atoms are involved in polymer conjugation. The rings of PANI can rotate or flip, significantly altering the nature of electron–phonon interaction in this polymer. During doping of PANI or transition from its emeraldine base form (PANI-EB) to emeraldine salt (PANI-ES), its conductivity increases by > 10 orders of magnitude, whereas the number of electrons on the polymer chains remains constant in the PANI-ES.[37–41] Such doping is accompanied by appearance of the Pauli susceptibility,[38,39] characteristic for classic metals, due to formation of high-conductive completely protonated or oxidated clusters with a characteristic size of ≈ 5 nm in amorphous polymer.[40] In some cases, diamagnetic bipolarons[41] and/or antiferromagnetic interacting polaron pairs,[42] each possessing two elemental charges, can also be formed in heavily doped polymer.

PANI-EB and PANI-ES doped with surfuric (PANI-SA), hydrochloric (PANI-HCA), camphorsulfonic (PANI-CSA), 2-acrylamido-2-methyl-1-propanesulfonic

(PANI-AMPSA), and p-toluenesulfonic (PANI-PTSA) acids up to $y \leq 0.60$ were also studied in detail by multifrequency ESR.[25,43–48]

Figure 1 shows typical D-band ESR spectra of PANI-EB and PANI-ES samples. The analysis of the data obtained showed that two types of PC are stabilized in PANI-SA: polarons R_1 localized on short polymer chains, $-(Ph-HN^{+*}-Ph)-$, with $A_x = A_y = 4.5$ G, $A_z = 30.2$ G, and g-factors presented in Table 1; and polarons R_2 with $g_\perp = 2.00439$ and $g_\parallel = 2.00376$ moving along the polymer chains with a rate (see Eq. 2) $D_{ID}^0 = 6.5 \times 10^8$ rad s^{-1}. Assuming a McConnell proportionality constant for the hyperfine interaction of the spin with nitrogen nucleus $Q = 23.7$ G,[49] a spin density is estimated on the heteroatom nucleus of $\rho_N(0) = (A_x + A_y + A_z)/(3Q) = 0.55$. The lowest excited states of the localized PC were determined from Eq. 1 to be $\Delta E_{n\pi^*} = 2.9$ eV and $\Delta E_{\sigma\pi^*} = 7.1$ eV at $\rho_N^\pi = 0.56$.[50] In PANI-HCA, R_1 also demonstrates the strongly anisotropic spectrum with the principal components of the g-tensor summarized in Table 1, and hyperfine coupling constant $A_z = 22.7$ G. For radicals R_2 $g_\perp = 2.00463$ and $g_\parallel = 2.00223$.

Highly doped PANI,[44,45,47,48] PT,[29] PPP,[51] and other conductive polymers demonstrate a Dyson-like[52] spectra with the line asymmetry factor A/B (Fig. 1) due to the formation of a skin layer on the polymer surface at D-band ESR. This effect was also detected at $\omega_e/2\pi \leq 10$ GHz.[8] The above line shape distortion is accompanied by its shift to higher magnetic fields; therefore, in order to determine correctly all the magnetic parameters, both the A and D terms should be considered.

Generally, the first derivative of the Lorentzian line can be calculated as

$$\frac{d\chi}{dB} = A\frac{2x}{(1+x^2)^2} + D\frac{1-x^2}{(1+x^2)^2} \quad (5)$$

where $x = 2(B-B_0) / \sqrt{3}\Delta B_{pp}^L$. The line asymmetry A/B (see Fig. 1) correlates with A and D coefficients of Eq. 5 as $A/B = 1 + 1.45$ D/A.

The PC stabilized in PT-BF$_4^-$ has an axially symmetric D-band ESR spectrum with $g_\parallel = 2.00412$, $g_\perp = 2.00266$, $\Delta B_{pp} = 15.2$ G, $\Delta E_{\sigma-\pi^*} = 4.0$ eV, typical for a radical localized on a polymer backbone (Fig. 1). The ClO$_4^-$ doped PT is characterized by $g_\parallel = 2.00230$, $g_\perp = 2.00239$, $\Delta B_{pp} = 26.3$ G, and $\Delta E_{\sigma-\pi^*} = 7.1$ eV.[29]

As the doping level y of PANI-SA exceeds 0.21, the g-factor of PC R_2 becomes isotropic and decreases from $g_{R2} = 2.00418$ down to $g_{iso} = 2.00314$. This effect is accompanied by a narrowing of the R_2 ESR spectrum, and was explained by a depinning of Q1D spin diffusion along the polymer chain and formation of the areas with high spin density in which a strong exchange of spins on neighboring chains occurs. This is in agreement with the assumption[40] of formation in amorphous PANI-EB of metal-like domains with Q3D delocalized electrons.

Doping of conducting polymers leads to the increase of the paramagnetic susceptibility, χ. This parameter consists of the temperature-independent Pauli susceptibility of the Fermi gas χ_P, a temperature-dependent contribution of localized Curie PC χ_C, and the term χ_{ST} "becoming" due to a possible singlet–triplet spin equilibrium in the system,[53]

$$\chi = \chi_P + \chi_C + \chi_{ST} = N_A\mu_{eff}^2 n(\varepsilon_F) + \frac{N_s\mu_{eff}^2}{3k_BT} + \frac{k_1}{T}\left[\frac{\exp(-J/k_BT)}{1 + 3\exp(-J/k_BT)}\right]^2 \quad (6)$$

where N_A is the Avogadro's number, N_s is a number of spins with $S=\frac{1}{2}$, $\mu_{eff}=\mu_B g \sqrt{S(S+1)}$ is the effective magneton, k_B is the Boltzmann constant, $n(\varepsilon_F)$ is the density of states per unit energy (eV) for both spin directions per monomer unit at the Fermi level ε_F, $N_s/3k_B = C$ is the Curie constant per mol-C(mol-monomer)$^{-1}$, k_1 is a constant, and J is the antiferromagnetic exchange coupling constant.

For example, PC in PANI-SA demonstrates the extreme temperature dependence of the effective paramagnetic susceptibility (Fig. 3). As in the case of polyaniline perchlorate,[54] this indicates a strong antiferromagnetic spin interaction due to a singlet–triplet equilibrium included in the total paramagnetic susceptibility. Indeed, Fig. 3 shows that the paramagnetic susceptibility experimentally determined for the PANI-SA samples is well reproduced by Eq. 6 with $J \geq 0.051$ eV. The J value is close to that obtained for the ammonia treated PANI.[54] Note that $n(\varepsilon_F) = 0.9$–1.4 states per electroviolet per one ring determined for PANI-SA, is consistent with the value determined earlier for PANI heavily treated with other dopants.[55,38] With assumption of a metallic behavior, one can estimate that the energy of N_P Pauli spins,[56] $\varepsilon_F = 3N_p/2n(\varepsilon_F)$, for example in PANI-ES with $0.21 \leq y \leq 0.53$, will be 0.1–0.51 eV.[44] This value is close to that (0.4 eV) obtained, for example, for PANI-CSA.[57] From this value, the number of charge carriers with mass $m_c = m_e$ in heavily doped PANI-SA, $N_c = (2m_c\varepsilon_F/\hbar^2)^{3/2}/3\pi^{2,56}$ is evaluated to be 1.7×10^{21} cm^{-3}. The

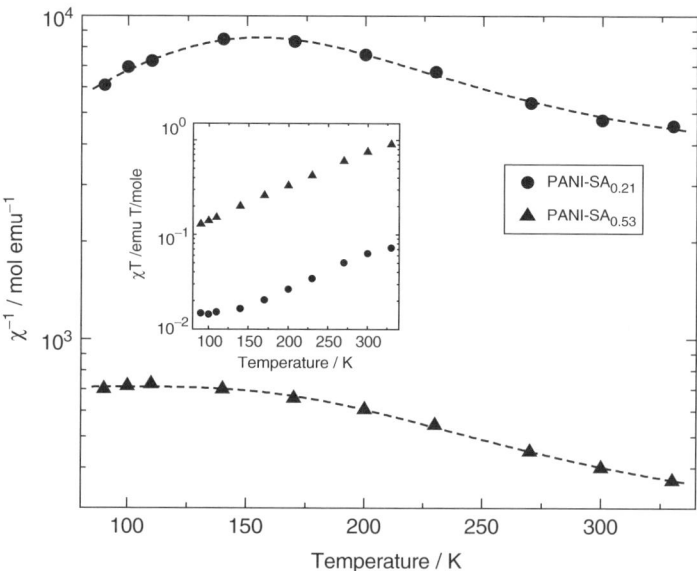

Fig. 3. Temperature dependence of inverse paramagnetic susceptibility and χT product (inset) of PANI-SA samples with different doping levels. Upper and lower dashed lines show the dependences calculated from Eq. 6 with, respectively, $\chi_P = 3.1 \times 10^{-5}$ emu mol^{-1}, $C = 1.2 \times 10^{-2}$ emu K mol^{-1}, $k_1 = 4.2$ emu K mol^{-1}, $J = 0.051$ eV, and $\chi_P = 1.4 \times 10^{-3}$ emu mol^{-1}, $C = 1.6 \times 10^{-2}$ emu K mol^{-1}, $k_1 = 48.6$ emu K mol^{-1}, $J = 0.057$ eV. [From Refs. 12(b) and 45(a) with permission.]

value obtained is close to a total volume spin concentration in PANI-SA. This fact leads to the conclusion that all PC take part in polymer conductivity. For heavily doped PANI-SA samples, the concentration of spin charge carriers is less than that of spinless ones, due to the possible collapse of pairs of polarons into diamagnetic bipolarons. The velocity of the charge carrier near the Fermi level, $v_F = 2c_{1D}/\pi n(\varepsilon_F)$,[56] can then be calculated, $v_F = (3.3–7.2) \times 10^7$ cm s^{-1}. Note that the paramagnetic susceptibility depends weekly on the measuring frequency; therefore this parameter for other conductive polymers is discussed only in general terms.

3. RELAXATION AND DYNAMICS OF CHARGE CARRIERS IN SLIGHTLY DOPED CONDUCTIVE POLYMERS

The line width of mobile PC in PANI-AMPSA$_{0.6}$ depends on ω_e: $\Delta B_{pp}(\omega_e) = 1.5 + 6.5 \times 10^{10}\omega_e^{-0.84}$ G at room temperature. As in the case of other polymers, it is possible to estimate a correct line width at the $\omega_e \to 0$ limit and to reveal the dependence of spin–spin relaxation time on the microwave frequency. As emphasized above, the cross-relaxation between spin packets decreases drastically at high magnetic fields and causes saturation at lower values of B_1. This effect can be used for the study of relaxation and dynamics parameters of PC in different solids,[11] conducting polymers among them.[12]

Generally, the first derivative of a dispersion signal U is[58]

$$U(\omega_e t) = u_1 g^I(\omega_e)\sin(\omega_e t) + u_2 g(\omega_e)\sin(\omega_e t - \pi) + u_2 g(\omega_e)\sin(\omega_e t \pm \pi/2) \quad (7)$$

where u_1, u_2, and u_3 are, respectively, the in-phase and quadrature dispersion terms with $g(\omega_e)$ shape.

Figure 4 shows changes of in-phase and quadrature terms of the dispersion signal of PC in cis-PA at different B_1 values and in P3OT at different temperatures. The relaxation times of adiabatically saturated PC in these and other π-conductive polymers can be determined separately from the analysis of the u_i terms of Eq. 7 as[18]

$$T_1 = \frac{3\omega_m(1 + 6\Omega)}{\gamma_e^2 B_{10}^2 \Omega(1 + \Omega)} \quad (8a)$$

$$T_2 = \frac{\Omega}{\omega_m} \quad (8b)$$

(here $\Omega = u_3/u_2$, B_{10} is the polarizing field at which the condition $u_1 = -u_2$ is valid) at $\omega_m T_1 > 1$ and

$$T_1 = \frac{\pi u_3}{2\omega_m u_1} \quad (9a)$$

$$T_2 = \frac{\pi u_3}{2\omega_m(u_1 + 11u_2)} \quad (9b)$$

at $\omega_m T_1 < 1$.

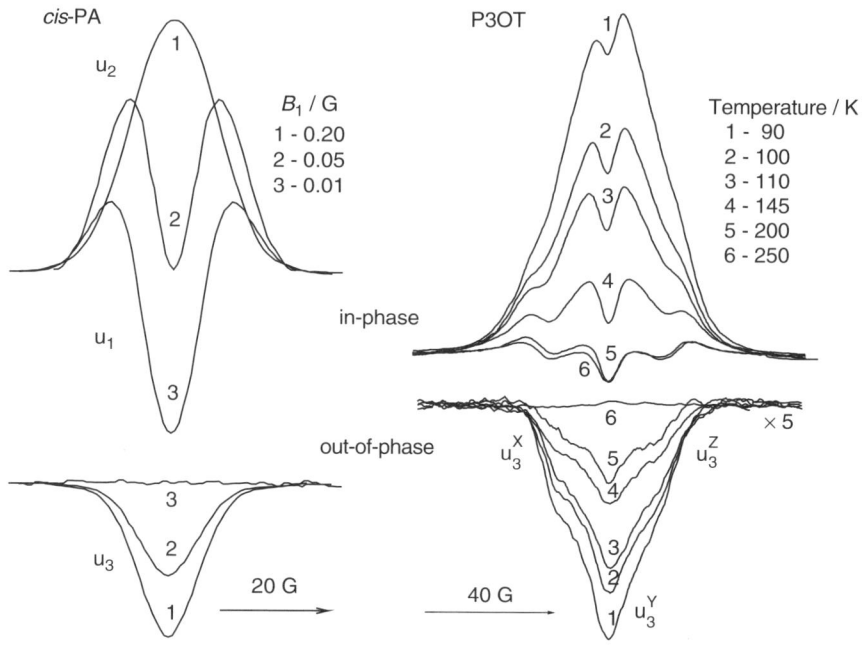

Fig. 4. The D-band in-phase and $\pi/2$-out-of-phase dispersion ESR spectra of *cis*-PA registered at different B_1 values and room temperature, and for P3OT registered at different temperatures and $B_1 = 0.2$ G (6). [From Refs. 12(b) and 30(b) with permission.

The inequalities $\omega_m T_1 > 1$ and $\omega_m T_1 < 1$ hold for *cis*- and *trans*-PA, respectively,[59] therefore their T_1 and T_2 values can be calculated by using Eq. 8 and 9, respectively. Fig. 5 exhibits these parameters for solitons in *trans*-PA with randomly oriented and with partly stretch-oriented polymer chains as a function of temperature and the angle ψ between the external magnetic field B_0 and the stretching directions. Figure 5 shows the field sensitivity of the relaxation parameters of solitons that can be explained by the depinning of their motion in *trans*-PA film.

Such spin diffusion along and between polymer chains, respectively, with coefficients D_{1D} and D_{3D}, (where 1D = one dimensional and 3D = three dimensional) induces an additional magnetic field onto other electron and nuclear spins and, therefore, should accelerate both the relaxation times of the whole spin reservoir. Dipole–dipole interaction between equal spins prevails in conductive polymers, so the relaxation rates can be calculated as[22,60]

$$T_1^{-1} = \langle \Delta\omega^2 \rangle [2J(\omega_e) + 8J(2\omega_e)] \quad (10a)$$

$$T_2^{-1} = \langle \Delta\omega^2 \rangle [3J(0) + 5J(2\omega_e)] \quad (10b)$$

where $\langle \Delta\omega^2 \rangle$ is the averaged constant of dipole–dipole spin interaction in a powder-like sample and $J(\omega_e) = (2D_{1D}\omega_e)^{-1/2}$ at $D_{3D} \leq \omega_e \leq D_{1D}$ or $J(\omega_e) = (2 D_{1D} D_{3D})^{-1/2}$

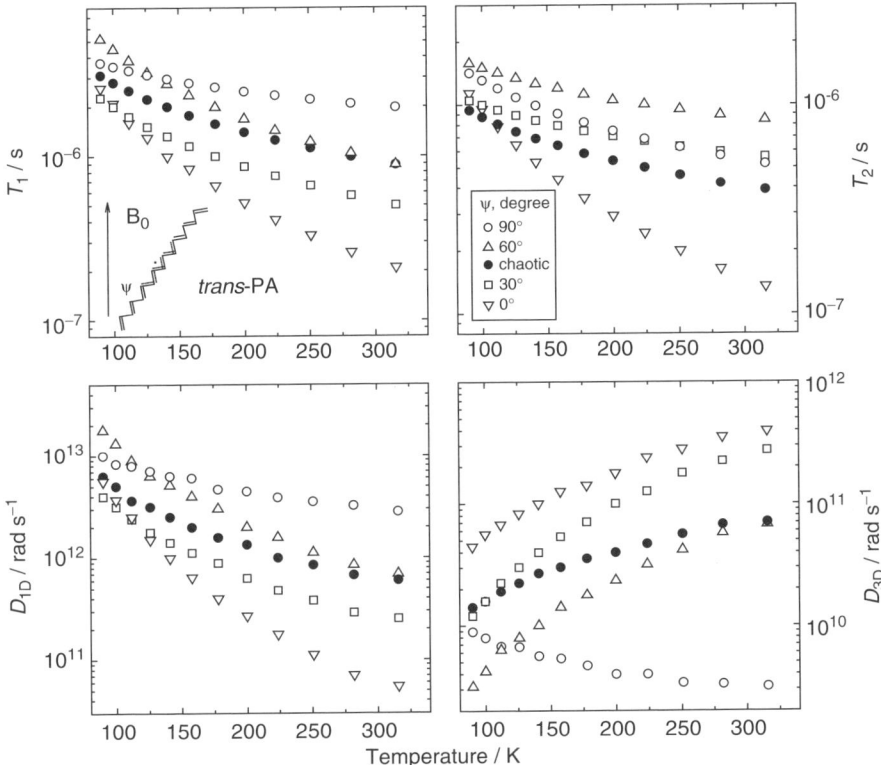

Fig. 5. Temperature dependence of effective spin–lattice T_1 and spin–spin T_2 relaxation times calculated from Eqs. 8 as well as of intrachain D_{1D} and interchain D_{3D} diffusion rates calculated from Eq. 10 for neutral solitons diffusing in *trans*-PA along and between randomly oriented chains and chains oriented with their c-axis with respect to an external magnetic field by $\psi = 90°, 60°, 30°,$ and $0°$. [From Refs. 12(b) and 61 with permission.]

at $\omega_e \leq D_{3D}$ is the spectral density function for a spin motion in a Q1D system. These equations become more complex in the case of an oriented polymer system.

Fig. 5 displays the temperature dependencies of the D_{1D} and D_{3D} coefficients calculated for the soliton diffusion in *trans*-PA with randomly oriented and partly stretch-oriented chains. This figure shows that, as in the case of relaxation parameters, both the 1D and 3D diffusion rates are sensitive to the orientation of the polymer chains in an external magnetic field due to soliton diffusion in this Q1D system. The $D_{1D}(\psi)$ function for an oriented sample is opposite to the $D_{3D}(\psi)$ one. Since the main c-axes are arbitrarily oriented in an initial *trans*-PA, these values are averaged over angle ψ. Moreover, the averaged D_{1D} value is well described by the equation $\langle D_{1D} \rangle = D_{1D}^{\perp} \cos^2\psi + D_{1D}^{\parallel} \sin^2\psi$, where $D_{1D}^{\perp}$ and $D_{1D}^{\parallel}$ are the extremes of the $D_{1D}(\psi)$ function. Thus, $D_{1D}^{\perp} \gg D_{1D}^{\parallel}$ inequality displays spin delocalization over N_s soliton sites equal to $\sqrt{D_{1D}^{\perp}/D_{1D}^{\parallel}}$.[61] The analysis of the data presented in Fig. 5 gives $N_s(T) \propto T^{1.0}$ temperature dependence for the soliton width and $2N_s = 14.8$ cell units at room temperature. This value is in good agreement with that theoretically predicted[4] and experimentally obtained.[62] Extrapolation to the low temperature range allows us to suppose the lowest $T_0 \approx 60$ K, where the soliton width starts to increase. Krinichnyi et al.[61] have shown that a soliton in *trans*-PA is transferred according to the Burgers–Korteweg–de-Vries theory, which described a dynamics of solitary wave in solids.

The ac conductivity of a conductive polymer due to dynamics of N_s spin charge carriers can be calculated from the modified Einstein relation

$$\sigma_{1,3D}(T) = \frac{N_s e^2 D_{1,3D} c_{1,3D}^2}{k_B T} \quad (11)$$

where c_{iD} is the appropriate lattice constant.

Figure 6 exhibits the conductivity due to intra- and interchain soliton diffusion in slightly iodine-doped *trans*-PA. The intrachain charge transfer was analyzed[63] in terms of phonon-assisted spin hopping between soliton sites, in the framework of the Kivelson phenomenological model,[64] with predicted conductivity

$$\sigma_{ac}(T) = \sigma_0 \omega_e T^{-1} \left[\ln \frac{k_1 \omega_e}{T^{n+1}} \right]^4 \quad (12)$$

where k_1 is constant. The fitting of experimental data by Eq. 12 confirms the applicability of the Kivelson theory for the explanation of charge transfer by soliton in *trans*-PA.

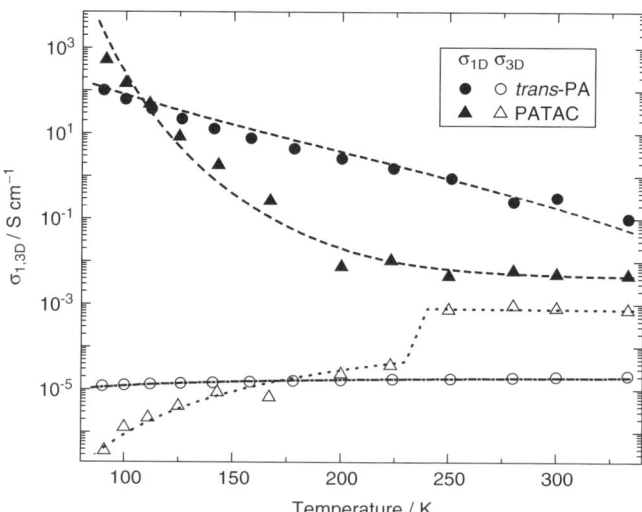

Fig. 6. Temperature dependence of conductivity σ_{1D} and σ_{3D} due to soliton motion in slightly doped *trans*-PA and polaron motion in laser-modified PATAC. The dependences calculated from Eq. 12 with $\sigma_0 = 2.8 \times 10^{-13}$ S s K cm^{-1}, $k_1 = 6.6 \times 10^{14}$ s K$^{14.2}$, and $n = 13.2$ (upper dashed line), Eq. 15 with $\sigma_0 = 2.9 \times 10^{-9}$ S K cm^{-1} and $\hbar\omega_{ph} = 0.18$ eV (lower dashed line), Eq. 14 with $\sigma_0 = 6.5 \times 10^{-20}$ S s (K cm)$^{-1}$ (dash–dotted line), and Eq. 13 with $\sigma_0 = 3.3 \times 10^{-18}$ S s (cm K)$^{-1}$ ($T < 230$ K), $\sigma_0 = 5.2 \times 10^{-17}$ S s (cm K)$^{-1}$ ($T > 230$ K) and $E_a = 0.055$ eV (dotted line) are shown as well. [From Refs. 12(b) and 21(b) with permission.]

One can expect that the interchain charge transfer be caused by thermal activation of a soliton to the band tails; therefore[65]

$$\sigma_{ac}(T) = \sigma_0 T \omega_e^\gamma \exp\left(-\frac{E_a}{k_B T}\right) \quad (13)$$

Here, γ is a constant that lies normally near 0.3–0.8 for solids and E_a is the energy for activation of charge carrier to extended states. It was deduced[66] that these two values of, for example, poly(3-methylthiophene), correlate as $\gamma = 1 - \alpha k_B T/E_a$ with $\alpha = 1$ and $E_a = 0.18$ eV.

Analysis of data obtained for *trans*-PA has shown, however, that its conductivity follows Eq. 13 at a too small value of E_a. For a better explanation of experimental data, the pair approximation[67] for Q3D charge hopping appeared more suitable. According to this approach, the charge tunneling between random energy electron potential wells governs the ac conductivity,[68]

$$\sigma_{ac}(T) = \frac{1}{3}\pi e^2 k_B T n^2 k_B(\varepsilon_F) \langle L \rangle^5 \omega_e \ln^4\left(\frac{\omega_0}{\omega_e}\right) = \sigma_0 T \quad (14)$$

where $\langle L \rangle$ is the averaged length of charge wave localization function and ω_0 is a hopping frequency.

The comparison of the dependences experimentally obtained and calculated by using Eq. 14 provide evidence for charge tunneling transfer in *trans*-PA (Fig. 6).

The relaxation and electron dynamics properties of this polymer are changed sufficiently as a hydrogen atom is replaced by a sulfur-based alkyl group. Such substitution leads to the stronger temperature dependence of relaxation times in laser-modified PATAC as compared with *trans*-PA.[21]

Figure 6 also shows the contribution of both 1D and 3D polaron motions to the ac conductivity of the laser-modified PATAC. Its intrachain conductivity σ_{1D} was interpreted in terms of the model of the charge-carrier scattering on the lattice optical phonons proposed by Kivelson and Heeger for metal-like clusters in conjugated polymers,[69]

$$\sigma_{ac}(T) = \sigma_0 T \left[\sinh\left(\frac{\hbar\omega_{ph}}{k_B T}\right) - 1\right] \quad (15)$$

where ω_{ph} is the angular frequency of the optical phonon.

The $\sigma_{1D}(T)$ and $\sigma_{3D}(T)$ dependences obtained for PATAC are fairly well fitted by equation 15 with $\hbar\omega_{ph} = 0.18$ eV and Eq. 14 with $E_a = 0.061$ eV, respectively (Fig. 6). The break in the $\sigma_{3D}(T)$ curve can be attributed to a change in the conformation of the system. The energy necessary for the charge-carrier activation hopping between polymer chains is close to the energy of activation of the polymer chain librations (see below). This fact is evidence of the interference of charge transfer and macromolecular dynamics processes in the polymer.

Figure 7 shows the temperature dependence of the relaxation times of polarons in P3OT determined from its dispersion spectrum terms presented in Fig. 4; of the spin diffusion constants D_{1D} and D_{3D}; and of the conductivity due to polaron mobility along σ_{1D} and between σ_{1D} polymer chains.[30]

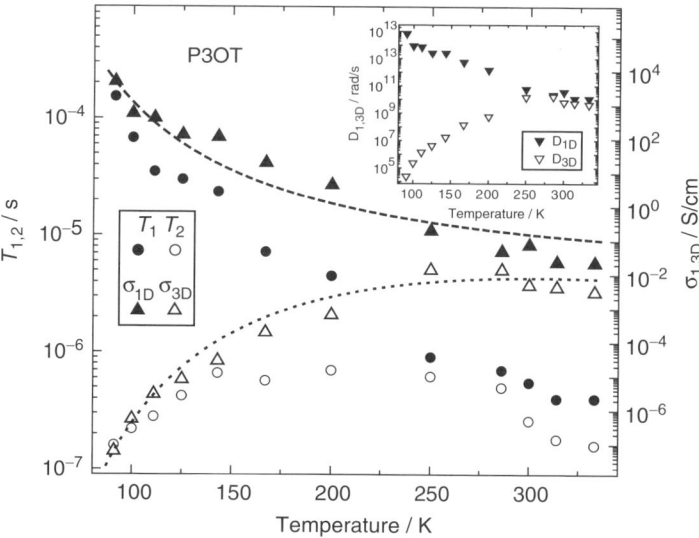

Fig. 7. Temperature dependence of the spin–lattice T_1, spin–spin T_2 relaxation times, intrachain D_{1D} and interchain D_{3D} diffusion rates (inset) of polarons in P3OT and appropriate terms of its ac conductivity. The dashed line shows the dependences calculated from Eq. 15 with $\sigma_0 = 5.5 \times 10^{-8}$ S cm^{-1} K^{-1} and $\hbar\omega_{ph} = 0.13$ eV. The dotted line shows the dependence calculated from Eq. 13 with $\sigma_0 = 9.1 \times 10^{-11}$ S s$^\gamma$ cm^{-1} K^{-1}, $\alpha = 2.1$, $E_a = 0.18$ eV. [From Ref. 30(b) with permission.]

The D_{1D} value exceeds the lower limit of the spin motion D_{1D}^0 by more than one to two orders of magnitude; however, the D_{3D} value obtained lies near $D \approx 2.1 \times 10^{10}$ s^{-1} evaluated from the charge-carrier mobility in slightly doped P3OT.[70] The anisotropy of spin dynamics D_{1D}/D_{3D} increases from 6 at 300 K up to 2500 at 200 K and up to 3.8×10^8 at 100 K. This leads to the respective increase in the anisotropy of conductivity σ_{1D}/σ_{3D} from 15 up to 6.8×10^3, and then up to 1.0×10^9.

As for the PATAC sample, the charge transfer along the chains in P3OT is characterized by the strong temperature dependence, especially in the low-temperature region (Fig. 7). Such behavior is usually associated with the scattering of charge carriers on the optical lattice phonons. Indeed, the intramolecular ac conductivity is in good agreement and follows Eq. 15 with $\hbar\omega_{ph} = 0.13$ eV. This value is close to that obtained for laser-modified PATAC and also for PANI-ES (see below). On the other hand, the interchain conductivity σ_{3D} increases with the temperature increase at $T \leq 200$ K, and then slightly decreases at higher temperatures. This is typical for the systems with a strong coupling of the charge with the lattice phonons. The γ value in Eq. 13 was shown[66] to depend on the activation energy, E_a, of polaron transfer between P3AT chains, so one can use this approach for the explanation of the 3D conductivity in the P3OT sample as well.

Figure 7 shows that the σ_{3D} term of conductivity of P3OT is well fitted by Eq. 13 with $\alpha = 2.1$ and $E_a = 0.18$ eV. The latter value is close to that determined for poly(3-methylthiophene)[66] and also to the energy of lattice phonons in P3OT determined above.

Relaxation and dynamics properties of charge carriers depend not only on the polymer structure, but also on its doping level. The temperature dependence of effective D_{1D} and D_{3D} calculated from Eqs. 10 for PC in PTTF samples with different structure and iodine content are shown in Fig. 8.[12,24,25] The D_{1D} parameter increases by two orders of magnitude at the transition from PTTF-Et-Ph to PTTF-THA and to PTTF-Me-Ph samples, due probably to a more planar conformation in the chain monomer units or crystallinity in this series.

The σ_{1D} and σ_{3D} conductivity terms of these PTTF samples are presented in Fig. 9 as a function of temperature. It is seen that the most acceptable intrachain charge transfer can be suggested in terms of the Kivelson–Heeger theory of spin–phonon interaction. Figure 9 shows that ac conductivity of the PTTF-Et-Ph and both of the PTTF-THA samples are really fitted by Eq. 15. The σ_{3D} value of the PTTF-Me-Ph$_{0.02}$ sample monotonically decreases with a temperature decrease. On the other hand, this value of PTTF-Me-Ph with $y \geq 0.05$, somewhat increases at the sample cooling from room temperature down to $T_c \approx 180$ K and then decreases on further cooling of the sample (Fig. 9). Such dependence can be explained by the thermally activated interchain polaron hopping in conduction band tails with activation energy E_a. Temperature dependences of conductivity due to such spin motion calculated from Eq. 13 are shown by dotted lines in Fig. 9. The σ_{3D} value increases in PTTF when the Ph group is replaced by THA where THA=tetrahydroanthracene (i.e., PTTF–THA is PTTF in which THF units are linked via tetrahydroanthracene bridges). This fact allows us to conclude that the structure of a polymer governs the polaron mobility and anisotropy of its transfer in PTTF.

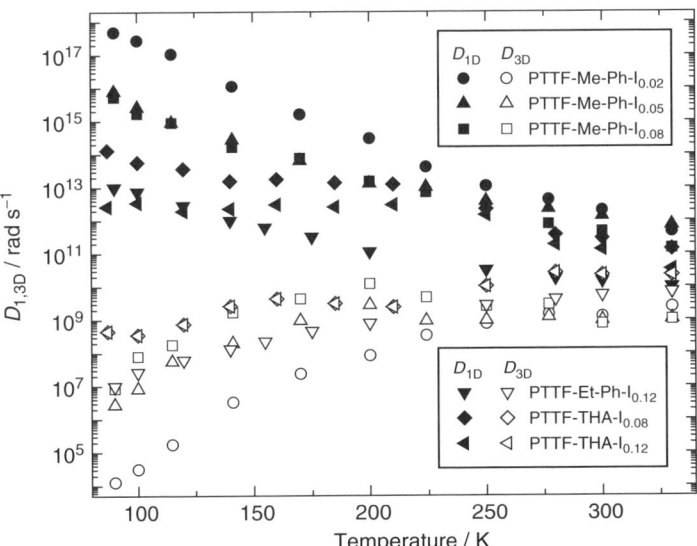

Fig. 8. Temperature dependence of the effective intrachain D_{1D} and interchain D_{3D} polaron diffusion rates in iodine-doped PTTF samples with different doping levels, determined from Eqs. 10. [From Refs. 12(b) and 24 with permission.]

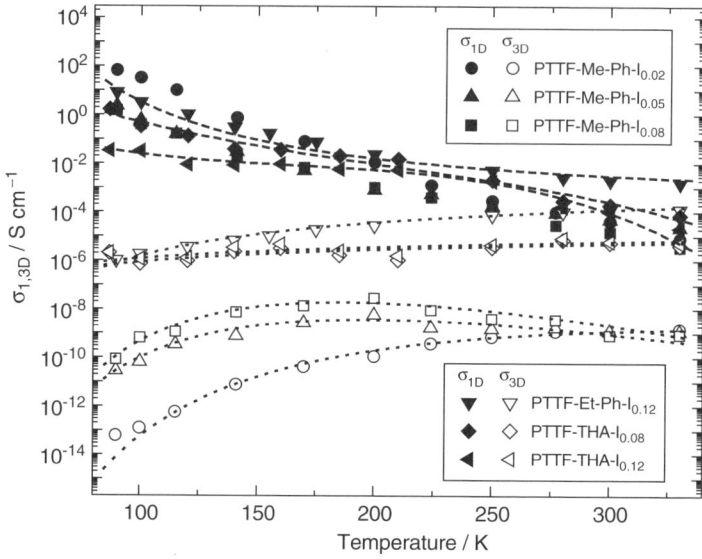

Fig. 9. Temperature dependence of ac conductivity of iodine-doped PTTF due to intrachain and interchain spin motion, calculated from Eq. 11 using data presented in Fig. 8. From the top down, dashed lines show $\sigma_{1D}(T)$ calculated from Eq. 15 with $\sigma_0 = 4.2 \times 10^{-7}$ S cm^{-1} and $\hbar\omega_{ph} = 0.11$ eV, $\sigma_0 = 2.0 \times 10^{-5}$ S cm^{-1} and $\hbar\omega_{ph} = 0.025$ eV, and $\sigma_0 = 4.1 \times 10^{-5}$ S cm^{-1} and $\hbar\omega_{ph} = 0.024$ eV. Top to down dotted lines show $\sigma_{3D}(T)$ calculated from Eq. 13 with $\sigma_0 = 2.7 \times 10^{-17}$ S K^{-1}cm^{-1}, $\alpha = 0$, $E_a = 0.052$ eV; $\sigma_0 = 1.9 \times 10^{-19}$ S K^{-1}cm^{-1}, $\alpha = 0$, $E_a = 0.007$ eV; $\sigma_0 = 1.6 \times 10^{-19}$ S K^{-1}cm^{-1}, $\alpha = 0$, $E_a = 0.09$ eV; $\sigma_0 = 1.1 \times 10^{-13}$ S K^{-1}cm^{-1}, $\alpha = 3.2$, $E_a = 0.14$ eV; $\sigma_0 = 5.4 \times 10^{-16}$ S K^{-1}cm^{-1}, $\alpha = 2.1$, $E_a = 0.12$ eV; $\sigma_0 = 1.6 \times 10^{-18}$ S K^{-1}cm^{-1}, $\alpha = 1.1$, $E_a = 0.16$ eV as well. [From Refs. 12(b) and 24 with permission.]

The Fermi velocity v_F in PTTF samples is near 2×10^7 cm s^{-1}. The mean free path l_i of a charge was determined for the PTTF samples to be $l_i = v_{1D}c_{1D}^2v_F^{-1} = 10^{-2}$–$10^{-4}$ nm at room temperature, and l_i is less than the lattice constant a; therefore the charge transfer is incoherent in such a Q1D metal-like polymer.

The temperature dependences of effective D_{1D} and D_{3D} calculated from Eqs. 10 for both types of PC in the PANI-EB and some slightly doped PANI were also determined (Fig. 10).[25,44,45,48] It seems reasonable that the anisotropy of the spin dynamics, being maximal in PANI-EB, decreases as y increases. For $y \geq 0.21$ the dimensionality of PANI-ES grows and the spin motion tends to become almost isotropic. This result disagrees with the results[71] obtained at lower ω_e, which concluded that there was high anisotropy of the spin dynamics in highly HCA-doped PANI even at room temperature.

The strong temperature dependence of the D_{1D} of PC in PANI-EB is a result of multiphonon charge hopping processes due to strong spin–lattice interaction. In contrast to undoped *trans*-PA, a small number of charge carriers exists even in the emeraldine base form of PANI. For this reason, a charge dynamics in the polymer can

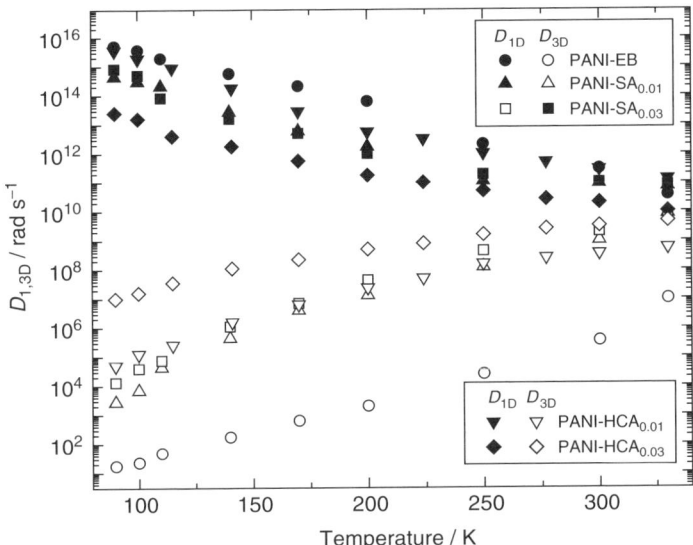

Fig. 10. Temperature dependence of effective coefficients for intrachain D_{1D} and interchain D_{3D} polaron diffusion in the PANI-EB and slightly SA- and HCA-doped PANI samples. [From Refs. 12(b), 44(b), and 45(a) with permission.]

also be considered in terms of the Kivelson formalism[64] of isoenergetic interchain charge transfer. Figure 11 shows that the experimental data for σ_{1D} of the initial PANI sample is well fitted by Eq. 12 with $n = 8.5$. This approach is not evident for PANI doped up to $0.01 \leq y \leq 0.03$, with a flatter temperature dependence. The model of charge-carrier scattering on optical phonons in metal-like domains seems to be more convenient for explaining the conductivity behavior.

As seen in Fig. 11, the $\sigma_{1D}(T)$ dependence, for example, the slightly doped PANI-SA, is also fairly well fitted by Eq. 15 with $\hbar\omega_{ph} \approx 0.1$ eV. This value lies near the energy (0.19 eV) of the polaron pinning in heavily doped PANI-ES.[72] The strong temperature dependence σ_{3D} of the initial sample can then be interpreted in the framework of the model of the activation charge transfer between the polymer chains described by Eq. 13, with E_a different for low- and high-temperature regions (Fig. 11). The respective interchain charge transfer in these polymers occurs with $E_a = 0.1$ eV (Fig. 11).

Spin–lattice and spin–spin relaxation times measured by the saturation method at X-band ESR for the PANI-PTSA with $y = 0.5$ are $T_1 = 1.1 \times 10^{-7}$ s and $T_2 = 1.6 \times 10^{-8}$ s.[48] For polarons, diffusion along and between polymer chains in this highly doped polymer, $D_{1D} = 8.1 \times 10^{11}$ rad s^{-1} and $D_{3D} = 2.3 \times 10^8$ rad s^{-1}, should be evaluated from Eqs. 10. The value $D_{1D}/D_{3D} \approx 10^4$ substantially exceeds the value $D_{1D}/D_{3D} \approx 50$ obtained for highly doped PANI-HCA.[71] The corresponding terms of conductivity due to possible polaron mobility calculated from Eq. 12 are $\sigma_{1D} = 29$ S cm^{-1} and $\sigma_{3D} = 2.4 \times 10^{-3}$ S cm^{-1}.

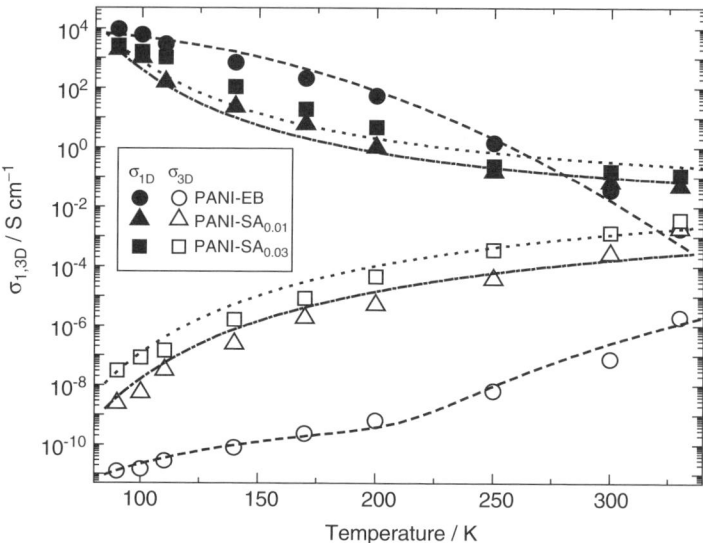

Fig. 11. Temperature dependence of the ac conductivity due to polaron motion along σ_{1D} and between σ_{3D} polymer chains in PANI-EB and slightly SA-doped PANI samples. The lines show the dependence calculated from Eq. 12 with $\sigma_0 = 4.5 \times 10^{-11}$ S K^{-1}s cm^{-1}, $k_1 = 3.1 \times 10^{12}$ s K$^{-9.5}$, and $n = 8.5$ (upper dashed line), from Eq. 15 with $\sigma_0 = 3.95 \times 10^{-6}$ S cm^{-1} K^{-1} and $\hbar\omega_{ph} = 0.12$ eV (upper dash–dotted line), $\sigma_0 = 1.75 \times 10^{-6}$ S cm^{-1} K^{-1} and $\hbar\omega_{ph} = 0.11$ eV (upper dotted line), and from Eq. 13 with $\sigma_0 = 8.2 \times 10^{-21}$ S K^{-1}s$^{0.8}$cm^{-1} and $E_a = 0.033$ eV (low-temperature region) and $\sigma_0 = 3.9 \times 10^{-12}$ S K^{-1}s$^{0.8}$cm^{-1} and $E_a = 0.41$ eV (high-temperature region) (lower dashed line), $\sigma_0 = 4.5 \times 10^{-14}$ S K^{-1}s$^{0.8}$cm^{-1} and $E_a = 0.10$ eV (lower dash–dotted line), $\sigma_0 = 3.1 \times 10^{-13}$ S K^{-1}s$^{0.8}$cm^{-1} and $E_a = 0.10$ eV (lower dotted line). [From Ref. 44(b) with permission.]

Note, however, that the steady-state saturation method used to obtain relaxation parameters of PC in lightly doped polymers reveals serious limitations for the study of spin dynamics in highly doped polymers, as seen below.

4. CHARGE TRANSFER IN HIGHLY DOPED CONDUCTIVE POLYMERS

The saturation of spin packets in highly doped polymers decreases significantly due to the increase in spin–spin and spin–lattice interactions. In the ESR spectra of such samples, the Dysonian term normally appears due to the formation of a skin layer with thickness δ. In contrast with the classic ESR signal, the Dyson-like spectrum shape "feels" both the spin polarons and the spinless bipolarons diffusing through a skin layer. It is possible to determine the intrinsic conductivity σ_{ac} of the sample directly from its Dysonian ESR spectrum. If the skin–layer is formed on the surface of a spherical powder particle with radius R, the coefficients A and D in Eq. 5 can be determined from Eqs. 16,[73]

$$\frac{4A}{9} = \frac{8}{p^4} - \frac{8(\sinh p + \sin p)}{p^3(\cosh p - \cos p)} + \frac{8 \sinh p \sin p}{p^2(\cosh p - \cos p)^2}$$
$$+ \frac{(\sinh p - \sin p)}{p(\cosh p - \cos p)} - \frac{(\sinh^2 p - \sin^2 p)}{(\cosh p - \cos p)^2} + 1 \quad (16a)$$

$$\frac{4D}{9} = \frac{8(\sinh p - \sin p)}{p^3(\cosh p - \cos p)} - \frac{4(\sinh^2 p - \sin^2 p)}{p^2(\cosh p - \cos p)^2}$$
$$+ \frac{(\sinh p + \sin p)}{p(\cosh p - \cos p)} - \frac{2\sinh p \sin p}{(\cosh p - \cos p)^2} \quad (16b)$$

where $p = 2R/\delta$, $\delta = \sqrt{2/\mu_0 \omega_e \sigma_{ac}}$, and μ_0 is the magnetic permeability for vacuum.

Figure 12 exhibits the temperature dependence of the ac conductivity of some doped PANI samples determined from their Dysonian D-band ESR spectra using Eqs. 5 and 16. The shape of the temperature dependence demonstrates nonmonotonous temperature dependence with a characteristic point $T_c \approx 170$–200 K. Such a temperature dependence can be a result of two parallel processes: the above-mentioned tunneling of charge carriers at $T \leq T_c$ (the semiconducting regime), and their interaction with lattice phonons at $T \geq T_c$ (the metallic regime), as described by Eqs. 14 and 15, respectively.

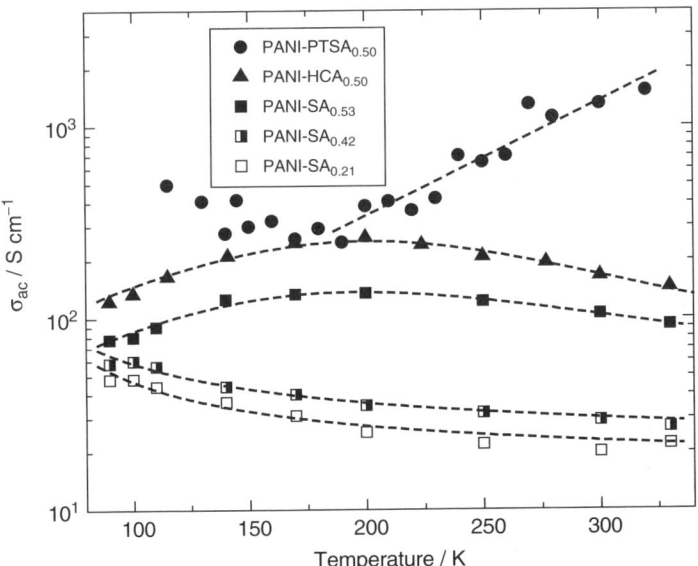

Fig. 12. Temperature dependence of ac conductivity determined from Dysonian ESR spectra of doped PANI-ES. Top to down dashed lines show the dependence calculated from Eq. 14 with $\sigma_0 = 24.9$ S (K cm)$^{-1}$ and from a combination of Eqs. 14 and 15 with, respectively, $\sigma_{01} = 1.47$ S cm^{-1} K^{-1}, $\sigma_{02} = 2.1 \times 10^{-2}$ S cm^{-1} K^{-1}, and $\hbar\omega_{ph} = 0.12$ eV; $\sigma_{01} = 0.86$ S cm^{-1} K^{-1}, $\sigma_{02} = 4.3 \times 10^{-2}$ S cm^{-1} K^{-1}, $\hbar\omega_{ph} = 0.087$ eV; $\sigma_{01} = 0.48$ S cm^{-1} K^{-1}, $\sigma_{02} = 8.5 \times 10^{-3}$ S cm^{-1} K^{-1}, $\hbar\omega_{ph} = 0.049$ eV; $\sigma_{01} = 0.41$ S cm^{-1} K^{-1}, $\sigma_{02} = 5.7 \times 10^{-3}$ S cm^{-1} K^{-1}, $\hbar\omega_{ph} = 0.052$ eV. [From Refs. 12(b), 44(b), and 45(a) with permission.]

Figure 12 shows that the experimental σ_{ac} values obtained for PANI are well fitted by a combination of Eqs. 14 and 15. The energy of lattice phonons $\hbar\omega_{ph}$ determined for highly doped PANI-ES lies near (0.066 eV) evaluated from data obtained by Wang et al.[74] The constant of charge carrier diffusion along the chains of PANI-SA and PANI-HCA, $D_{1D} = \sigma_{ac}/e^2 n(\varepsilon_F) c_{1D}^2 = (0.5–1.1) \times 10^{14}$ rad s^{-1}, exceeds at least by an order of magnitude the D_{1D} determined above for slightly doped samples at room temperature. The mean free path was calculated for the highly doped PANI-ES to be 0.5–6 nm. This value is smaller than that estimated for oriented *trans*-PA,[69] but holds, however, for extended electron states in these polymers as well.

An intrinsic conductivity determined for the highly doped PANI-PTSA sample at $T \geq 170$ K is described by Eq. 15 with $\hbar\omega_{ph} = 0.022$ eV (Fig. 12). The value calculated at 300 K, 1.3×10^3 S cm^{-1}, exceeds the σ_{1D} and σ_{3D} values assessed above, and indicates that the saturation method cannot be applied for the study of spin dynamics in highly doped polymers.

The conductivity of heavily doped PANI-SA, PANI-HCA, and PANI-PTSA estimated from their Dysonian ESR spectra is much smaller than that of $\approx 10^7$ S cm^{-1} calculated theoretically;[75,76] however, it is near that obtained for metal-like domains in PANI-ES at 6.5 GHz.[76]

The data obtained allows us to conclude that highly doped PANI-SA and PANI-HCA are Fermi glasses with electronic states localizes at the Fermi energy due to disorder, whereas PANI-TPSA and PANI-CSA are disordered metals on the metal–insulator boundary, so that the metallic quality of the emeraldine form of PANI grows in the series PANI-HCA → PANI-SA → PANI-PTSA → PANI-CSA.

5. HIGH–FIELD SATURATION TRANSFER ESR METHOD IN THE STUDY OF CONDUCTIVE POLYMERS

An additional advantage of D-band ESR is that it offers the opportunity to investigate the macromolecular mobility in conductive polymers with PC interacting with heteroatoms and, therefore, possessing anisotropic magnetic parameters. Such motion is *a priori* realized with correlation times in the range of $10^{-3} > \tau_c > 10^{-7}$ s.[77] The most sensitive to such molecular motions are the $\pi/2$-out-of-phase first term of the first harmonic of dispersion and second harmonic of absorption spectra. According to the saturation-transfer ESR (ST–ESR) method,[79] if all radicals rotate about their own *x*-axis, the conditions of adiabatic saturation are fulfilled for some of the radicals oriented by the *x*-axis along the external magnetic field B_0 and cannot be realized for radicals with other orientations. This results in the elimination of the saturation for radicals whose *y* and *z* axes are oriented parallel to the field B_0 and in the decrease of their contribution to the total ST–ESR spectrum.

It was demonstrated[11,12,78] that all terms of anisotropic magnetic parameters of stable organic radicals are registered separately at D-band ESR. In contrast with the nitroxide radical usually used as a spin probe or label in condensed systems, the native polaron with anisotropic magnetic parameters becomes itself a stable spin label. Therefore, the nearest environment of such PC remains undisturbed and the results obtained on structure or spin and molecular motion become more accurate and complete. Besides, the sensitivity of the method increases with ω_e,[79,80] so it can be used more efficiently in the study of macromolecular dynamics in conductive polymers at D-band ESR. By measuring ESR at this frequency, where both dispersion terms of PC are localized on the polymer chain, it is possible to determine separately the relaxation times (see above) and the correlation time of anisotropic activation motion with a chain near, for example, the main molecular x-axis from the following simple equation,[12]

$$\tau_c^x = \tau_{c0}^x \left(u_3^x / u_3^y \right)^\alpha = \tau_{c0}^x \exp\left(\frac{E_a}{k_B T} \right) \quad (17)$$

where α is a constant determined by an anisotropy of the g-factor and E_a is an activation energy of the radical motion near a specific molecular axis. The preexponential factor τ_{c0}^x is the lowest limit for the correlation times in a respective polymer matrix.

From Fig. 4, it is seen how the relative intensity of the $\pi/2$-out-of-phase term of D-band ESR dispersion spectra of P3OT changes with temperature. The increase of the spectral X-component u_3^x with increasing temperature shows the appearance of such

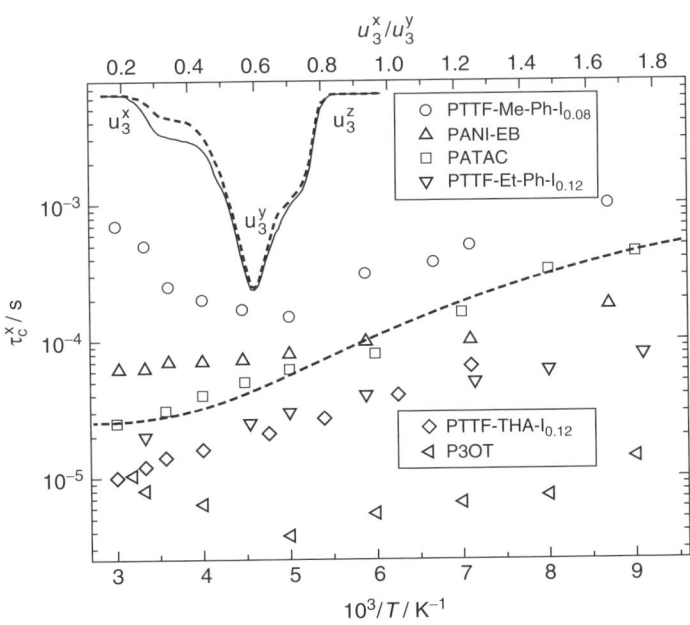

Fig. 13. Arrhenius dependence of correlation times for superslow librations near the polymer main x-axis of polarons R_1 localized on polymer chain segments in some conductive polymers evaluated from their ST–ESR spectra (inset). In the inset, the typical $\pi/2$-out-of-phase dispersion spectra registered at 100 K (solid line) and 200 K (dashed line) are shown. The dependence calculated from Eq. 17 is shown by a dotted line. [From Refs. 12(b), 21(b), 24, 30(b), and 45(a) with permission.]

saturation transfer over the spectrum due to superslow macromolecular libration dynamics. The Arrhenius dependence of τ_c^x determined from Eq. 17 for such motion in different conductive polymers is presented in Fig.13.

The value deduced for P3OT decreases with a temperature increase up to $T_c \approx 150$ K and then increases above this critical temperature. Note that Masubuchi et al.[81] observed the proton nuclear magnetic resonance (^{1}H NMR) T_1 temperature dependence with the same critical temperature. This result was attributed to the motion of the alkyl chain end groups. The dependences obtained can be interpreted in the frame of the superslow 1D libration of the polymer chains together with polarons near their main x-axis at low temperatures for $T \leq T_c$, whereas their high-temperature part can be explained by the collective 2D motion at $T \geq T_c$ with $E_a = 0.069$ eV. Treatment of this polymer by annealing and by both recrystallization and annealing leads to lower (0.054 eV) and higher (0.073 eV) E_a values, respectively.[30]

The energy of activation also depends on the structure and the doping level of a polymer. For example, the values obtained for PANI–EB, laser-modified PATAC, iodine doped PTTF-Me-Ph, PTTF-Et-Ph, and PTTF–THA from the data summarized in Fig. 13 and Eq. 17 are 0.015, 0.043, 0.036, 0.021, and 0.041 eV, respectively. Slight doping of an initial PTTF sample doubles the activation energy of macromolecular librations. The E_a values obtained at D-band ESR are comparable with those determined for interchain charge transfer in doped PTTF at a lower measuring frequency,[26] indicating the interaction of pinned and mobile polarons in this polymer matrix. The upper limit for the correlation time registered by the ST–ESR method is 4×10^{-4} s at 66 K for P3OT and 1×10^{-4} s for PTTF-Me-Ph and PANI-EB at 75 and 125 K, respectively.

6. HIGH-FIELD SPIN PROBE METHOD IN THE STUDY OF CONDUCTIVE POLYMERS

In some conductive polymers, pairs of polarons can merge into spinless bipolarons.[8] In this case, the method of spin–label and probe,[82,83] especially at D-band ESR,[11] seems to be more effective for the study of their structure and dynamics.

The X- and D-band absorption ESR spectra of a nitroxide radical, 2,2,6,6-tetramethyl-1-oxypiperid-4-yl acetic acid, introduced as both a probe and dopant in PP and as a probe in a frozen nonpolar model system, are shown in Fig. 14.[84] The X-band ESR single line of PC (R) stabilized in PP overlaps the lines of nitroxide radical rotating with correlation time $\tau_c > 10^{-7}$ s^{-1}.

The D-band ESR spectra of these model and modified polymer systems are more informative (Fig. 14). At this waveband, all terms of g- and A-tensors of the probe are completely resolved. Nevertheless, the asymmetric spectrum of radicals R stabilized in PP with magnetic parameters $g_\parallel^R = 2.00380$, $g_\perp^R = 2.00235$, and $\Delta B_{pp} = 5.7$ G is registered near the z component of the probe spectrum. In nonpolar toluene, the probe is characterized by $g_x = 2.00987$, $g_y = 2.00637$, $g_z = 2.00233$, $A_x = A_y = 6.0$ G, and $A_z = 33.1$ G. The difference $\Delta g^R = 1.45 \times 10^{-3}$ corresponds to an excited electron configuration in R with $\Delta E_{\sigma\pi^*} = 5.1$ eV. In conducting PP, the g_x value of the probe decreases to 2.00906 and its x- and y-components become broadened by 40 G (Fig. 14). In addition, the shape of the probe spectrum shows the localization of PC

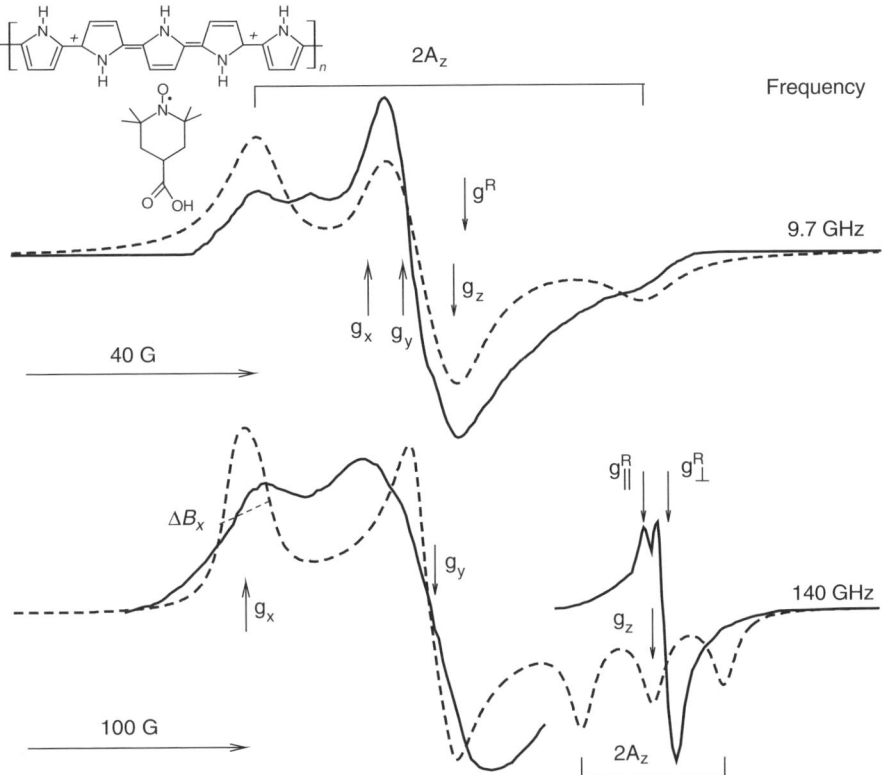

Fig. 14. Both X- and D-band ESR absorption spectra of a nitroxide radical in frozen (120 K) toluene as a spin probe (dotted line) and in conductive PP (solid line) as a dopant and spin probe. The anisotropic spectrum of localized PC marked by the symbol R and measured at a smaller amplification is also shown in the lower spectrum. The bipolaron spreads out over a larger number of units than shown schematically in this figure. [From Refs. 12(b) and 84 with permission.]

R in the polymer pocket of 1 nm-size, that is, the charge in this polymer is really transferred by spinless bipolarons.

The fragments with a considerable dipole moment are *a priori* absent in an initial PP. Besides, the dipole–dipole interactions between the radicals can be neglected due to a low concentration of the probe and PC localized on the chain. Therefore, the change in the probe's magnetic resonance parameters may be caused by a Coulombic interaction of the probe active fragment with mobile spinless bipolarons. The effective electric dipole moment of bipolarons diffusing near the probe was determined from the shift of the g_x component to be $\mu_v = 2.3$ D. The shift of g_x may be calculated based on the electrostatic interaction of the probe and bipolaron dipole by using the potential of the electric field induced by the bipolaron near the probe site,[84,85]

$$\Delta g_x = \frac{6 \cdot 10^{-3} e r_{NO} k_B T}{I \mu_u} (x \coth x - 1) \qquad (18)$$

where r_{NO} is the distance between N and O atoms in the probe active fragment, I is the resonant overlapping integral of the C=C bond, μ_u is the dipole moment of the probe, $x = 2\mu_u\mu_v(\pi\varepsilon\varepsilon_0 k_B T r^3)^{-1}$, ε and ε_0 are the dielectric constants for PP and vacuum, respectively, r is the distance between the nitroxide radical and the bipolaron. The value of $r = 0.92$ nm is obtained by using $\mu_u = 2.7$ D,[86] $\mu_v = 2.3$ D, and $r_{NO} = 0.13$ nm.[49]

The effective spin–spin relaxation rate $T_2^{-1} = T_{2(0)}^{-1} + T_{2(D)}^{-1}$ is defined by relaxation of radicals not interacting with the environment, $T_{2(0)}^{-1}$, and the increment due to dipole–dipole interactions, $T_{2(D)}^{-1} = \gamma_e\delta(\Delta B_{pp}^{x,y})$. The characteristic time τ_c of such an interaction was calculated from the broadening of the spectral lines using Eq. 10b with $J(\omega_e) = 2\tau_c/(1 + \omega_e^2\tau_c^2)$ to be $\tau_c = 8.1 \times 10^{-11}$ s. The value is close to the time of polaron hopping between chains, $\tau_{3D} \cong 1.1 \times 10^{-10}$ s estimated for slightly doped PP.[87] Taking into account that the average time between the translating jumps of charge carriers is defined by the diffusion coefficient D and by the average jump distance equal to a product of lattice constant c_{1D} on the half-width of the bipolaron $N_{bp}/2$, $\tau_c = 1.5 \langle c_{1D}^2 N_{bp}^2 \rangle/D$, and by using $D = 5 \times 10^{-3}$ cm^{2}s^{-1} typical for conductive polymers, it is possible to determine $\langle c_{1D}N_{bp} \rangle = 3$ nm, equal approximately to four pyrrole rings.

By measuring at D-band, all anisotropic magnetic parameters of organic PC allows us to determine the subtle features of structure, conformation, dynamics, and polarity of the PC microenvironment in different condensed systems.[11,12,78] At this waveband, the correlation times of radical rotation near the main axis can be defined separately from the broadening of corresponding spectral components $\delta(\Delta B_i)$. Krinichnyi et al.[88] showed that the method of spin macroprobe can be used for more detailed investigations of the polymer systems in solution. The method is based on the analysis of simultaneous rotation in the system under study of both the nitroxide radical as a spin microprobe and a single microcrystal of a suitable ion-radical salt as a spin macroprobe. The macroprobe tends to orient the main crystal axis along the direction B_0 to attain the minimum energy of spin interaction with the magnetic field. Such a reorientation process is easily registered by its ESR line shift to lower magnetic fields.

The inset in Fig. 15 shows a D-band ESR spectrum of a frozen nujol/*tert*-butylbenzene mixture in whom a nitroxide radical is solved as a spin microprobe and a dibenzotetrathiafulvalene$_3$PtBr$_6$, (DBTTF$_3$PtBr$_6$) single microcrystal with the characteristic size of $\approx 10^{-4}$ mm^3 is introduced as a spin macroprobe. The change in the nitroxide radical spectrum shape below the T_g of the matrix glass transition is caused by its Brownian diffusion rotation. The correlation time $\tau_c = 1/\gamma_e\delta(\Delta B_i)$ of the spin microprobe reorientation was determined [78] to be $4.7 \times 10^{-20} \exp(0.57 \text{ eV}/k_B T)$ s. The spin macroprobe was preliminary oriented by its d crystallographic axis along the magnetic field at a temperature significantly exceeding T_g. Then, the spin-modified sample is frozen down to $T \leq T_g$, turned by 90^0 (upper spectrum in the inset), and slowly warmed to $T \cong T_g$. The ESR line of the spin macroprobe is shifted from an initial position $B(t = 0)$ at $t = 0$ to a lower magnetic field $B(t)$, as shown on the lower spectrum in the inset. This line shift follows an exponential law, and attains the maximal value, $\delta B \approx 200$ G for this sample at $t \to \infty$ (Fig. 15).

The dynamics of the spin macroprobe depends on many parameters of the matrix and can be described by the differential equation for a one-point-fixed oscillator

$$\dot{\varphi} + \tau_m^{-1}\varphi = 0 \qquad (19)$$

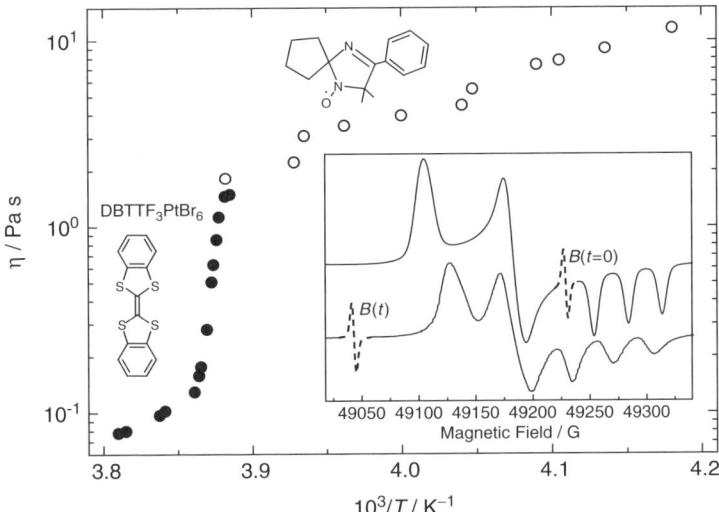

Fig. 15. The Arrhenius dependence of the dynamic viscosity η in the mixture obtained by both spin microprobe (open symbols) and spin macroprobe (filled symbols) methods. Inset: D-band ESR spectra of nitroxide radical (solid line) and single-crystal DBTTF$_3$PtBr$_6$ (dashed lines) introduced into a frozen and heat softened nujol/*tert*-butylbenzene mixture (1:10) at $t = 0$ (upper spectrum) and arbitrary t at $T \leq T_g$ (lower spectrum) time period. (From Ref. 89 with permission.]

where $\varphi(t) = \arccos[\delta B(t)/\delta B(t \to \infty)] = \varphi_0 \exp(-t/\tau_m)$, $\tau_m = 6\pi\eta r^3/(NV\mu_0\mu_B B_0)$ is the mechanical relaxation time, η is the coefficient of dynamic viscosity of the matrix, r and V are the characteristic size and volume of the crystal, respectively, and N is the PC bulk concentration. This allows us to determine the dynamic viscosity of the system under consideration, by using the Stokes equation:

$$\eta(T) = \frac{3k_B T}{4\pi r^3} \tau_c \qquad (20)$$

Figure 15 depicts the temperature dependence of the viscosity of the model system, defined by the spin micro- and macroprobe methods. As expected, $\eta(T) \propto T\exp(E_a/k_B T)$ at $T_g \approx 256$ K. At these temperatures, a spin probe reflects the motion of the radical glass-like microenvironment. The change in viscosity seems to have an activation character at $T \geq T_g$.

7. CONCLUDING REMARKS

The data presented in this chapter show the variety of electronic processes that take place in the low-dimensional organic conductive polymers, caused by the structure, conformation, packing, and degree of ordering of polymer chains, and also by the amount and type of the dopant introduced into the polymer matrix. The most important deductions are the following.

Spin and spinless nonlinear excitations exist as charge carriers in organic conductive polymers. The number of these carriers depends on the various properties of the polymer and the dopant introduced. With increasing doping level some polaron pairs merge into diamagnetic bipolarons. However, such a process can be restricted in some polymers due to their structural–conformational peculiarities. The doping of a polymer initiates changes in the charge-transfer mechanism. Conductivity in neutral or weakly doped samples is defined mainly by isoenergetic charge tunneling, which is characterized by a high interaction of spins with several phonons of a lattice and leads to the correlation of Q1D spin motion and interchain charge transfer. This mechanism ceases to dominate with the increase of doping level, and the charge can be transferred by its thermal activation from widely separated localized states in the gap to close localized states in the tails of the valence and conducting bands. Therefore, complex quasiparticles, namely, the molecular-lattice polarons, are formed in some polymers due to libron–phonon interactions analogous to that in organic molecular crystals. Note that as conductive polymers have *a priori* a lower dimensionality compared to molecular crystals, the dynamics of charge carriers is more anisotropic. In heavily doped samples, the dominant mechanism is the interchain charge transport, characterized by a strong interaction of charge carriers with lattice phonons.

The higher spectral resolution at D-band provides a higher accuracy of all magnetic measured magnetic parameters, with the g-factor of organic free radicals an important informative characteristic. As in the case of other organic solids, this allows us to establish the correlation between the structure of organic radicals and their g-tensor principal values, and to identify the PC in conductive polymers. High-field ESR allows us to obtain qualitatively new information on the spin carrier and on molecular dynamics, as well as on the magnetic and relaxation properties of polymer systems. Further progress in the study of conductive polymers is the use of different pulse methods at D-band ESR, for example, light-induced ESR (LESR) and spin–echo. Such studies are currently in progress in our laboratory.

ACKNOWLEDGMENTS

Support by the Russian Foundation of Basic Research, grants 97-03-33707, 01-03-33255, 03-03-04005, 05-03-33148, by the UK Royal Society, grant 638072.P885/bll, and by the German Foundation Deutsche Forschungsgemeinschaft (DFG), grant 436 RUS 113/734/1-1, is gratefully acknowledged. The author expresses his gratitude to E. Fanghänel for the gift of polytetrathiafulvalenes, to G. Hinrichsen, A.P. Monkman and B. Wessling for the gift of polyanilines, to S.D. Chemerisov and N.N. Denisov for assistance in the ESR experiments, and to S.A. Brazovskii, H.-K. Roth, K. Lüders, and Y.S. Lebedev for fruitful discussions.

REFERENCES

1. (a) *Encyclopedia of Polymeric Materials*, Salamone, J.C., Ed.; CRC Press: Boca Raton, Fl., 1996. (b) *Handbook of Organic Conductive Molecules and Polymers*, Nalwa, H.S., Ed.; John Wiley & Sons, Inc.: Chichester, New York, 1997; Vols. 1–4.

REFERENCES

2. (a) *Handbook of Conducting Polymers*, Scotheim, T.E.; Elsenbaumer, R.L.; Reynolds, J.R., Eds.; Marcel Dekker: New York, 1997. (b) *Handbook of Conducting Polymers*, Skotheim, T.; Elsenbaumer, R.; Reynolds, J., Eds.; Marcel Dekker, Inc., 1998. (c) *Conducting Polymers, Fundamentals and Applications: A Practical Approach*, Chandrasekhar, P., Ed.; Kluwer Academic Publishers: Boston, MA, 1999. (d) *Encyclopedia of Polymer Science and Technology, Part 3*, Mark, H.F., Ed.; Wiley-Interscience: New York, 2004. (e) *One-Dimensional Metals: Conducting Polymers, Organic Crystals, Carbon Nanotubes*, Roth, S.; Carroll, D., Eds.; Wiley-VCH: Weinheim, 2004.

3. Peierls, R.E. *Quantum Theory of Solids*, Oxford University Press: London, 1955.

4. Su, W.P.; Schrieffer, J.R.; Heeger, A.J. *Phys. Rev. B* **1980**, *22*, 2209.

5. Brédas, J.L. In *Handbook of Conducting Polymers*, Scotheim, T.E., Ed.; Marcel Dekker, Inc.: New York, 1986; Vol. 2, Chapt. 25, pp. 859–913.

6. Chance, R.R.; Boudreaux, D.S.; Brédas, J.L.; Silbey, R. In *Handbook of Conducting Polymers*, Scotheim, T.E., Ed.; Marcel Dekker, Inc.: New York, 1986; Vol. 2, Chapt. 24, pp. 825–857.

7. Devreux, F.; Genoud, F.; Nechtschein, M.; Villeret, B. In *Electronic Properties of Conjugated Polymers*, Kuzmany, H.; Mehring, M.; Roth, S., Eds.; Springer-Verlag: Berlin, 1987; Vol. 76, pp. 270–276.

8. (a) Bernier, P. In *Handbook of Conducting Polymers*, Scotheim, T.E., Ed.; Marcel Deccer, Inc.: New York, 1986; Vol. 2, Chapt. 30, pp. 1099–1125. (b) Mizoguchi, K.; Kuroda, S. In *Handbook of Organic Conductive Molecules and Polymers*, Nalwa, H.S., Ed.; John Wiley & Sons, Inc.: Chichester, New York, 1997; Vol. 3, Chapt. 6, pp. 251–317.

9. (a) Thomann, H.; Dalton, L.R. In *Handbook of Conducting Polymers*, Scotheim, T.E., Ed.; Marcel Dekker, Inc.: New York, 1986; Vol. 2, Chapt. 32, pp. 1157–1190. (b) Clarke, T.C.; Scott, J.C. In *Handbook of Conducting Polymers*, Skotheim, T.A., Ed.; Marcel Dekker, Inc.: New York, 1986; Vol. 2, Chapt. 31, pp. 1127–1156.

10. Altshuler, S.A.; Kozirev, B.M. *Electron Paramagnetic Resonance* (Russ), Fizmatgiz: Moscow, 1961.

11. (a) Krinichnyi, V.I. *J. Biochem. Biophys. Meths.* **1991**, *23*, 1. (b) Krinichnyi, V.I. *Appl. Magn. Reson.* **1991**, *2*, 29, and references cited therein.

12. (a) Krinichnyi, V.I. *2-mm Wave Band EPR Spectroscopy of Condensed Systems*, CRC Press: Boca Raton, FL, 1995. (b) Krinichnyi, V.I. *Synth. Met.* **2000**, *108*, 173. See also website: http://hf-epr.sitesled.com/publications.htm

13. Poole, C.P. *Electron Spin Resonance*, International Science Publishers: London, 1967.

14. Altshuler, S.A.; Kozirev, B.M. *Electron Paramagnetic Resonance of Compounds of Elements of Intermediate Groups* (Russ), 2 ed.; Nauka: Moscow, 1972.

15. (a) Krinichnyi, V.I. *Russ. Chem. Rev.* **1996**, *65*, 81. (b) Krinichnyi, V.I. *Russ. Chem. Rev.* **1996**, *65*, 521. (c) Krinichnyi, V.I. *Phys. Solid State* **1997**, *39*, 1.

16. Chien, J.C.W. *Polyacetylene: Chemistry, Physics and Material Science*, Academic Press: Orlando, FL, 1984.

17. Krinichnyi, V.I.; Tkachenko, L.I.; Kozub, G.I. *Khimich. Fiz.* **1989**, *8*, 1282.

18. (a) Krinichnyi, V.I.; Pelekh, A.E.; Brezgunov, A.Y.; Tkachenko, L.I.; Kozub, G.I. *Mater. Sci. Eng.* **1991**, *17*, 25. (b) Pelekh, A.E.; Krinichnyi, V.I.; Brezgunov, A.Y. ; Tkachenko, L.I.; Kozub, G.I. *Vysokomol. Soedin. A* **1991**, *33*, 1731.

19. Krinichnyi, V.I.; Pelekh, A.E.; Lebedev, Y.S.; Tkachenko, L.I.; Kozub, G.I.; Barra, A.L.; Brunel, L.C.; Robert, J.B. *Appl. Magn. Reson.* **1994**, *7*, 459.

20. Goldberg, I.B.; Crowe, H.R.; Newman, P.R.; Heeger, A.J.; MacDiarmid, A.G. *J. Chem. Phys.* **1979**, *70*, 1132.
21. (a) Roth, H.-K.; Krinichnyi, V.I.; Schrödner, M.; Stohn, R.-I. *Synth. Met.* **1999**, *101*, 832. (b) Krinichnyi, V.I.; Roth, H.-K.; Schrödner, M. *Appl. Magn. Reson.* **2002**, *23*, 1.
22. Carrington, F.; McLachlan, A.D. *Introduction to Magnetic Resonance with Application to Chemistry and Chemical Physics*, Harper & Row: New York, 1967.
23. Traven, V.F. *Electronic Structure and Properties of Organic Molecules (Russ)*, Khimija: Moscow, 1989.
24. (a) Krinichnyi, V.I.; Pelekh, A.E.; Roth, H.-K.; Lüders, K. *Appl. Magn. Reson.* **1993**, *4*, 345. (b) Krinichnyi, V.I.; Denisov, N.N.; Roth, H.-K.; Fanghänel, E.; Lüders, K. *Polym. Sci. A* **1998**, *40*, 1259.
25. Roth, H.-K.; Krinichnyi, V.I. *Makromolec. Chem. Macromolec. Symp.* **1993**, *72*, 143.
26. (a) Roth, H.-K.; Gruber, H.; Fanghänel, E.; Trinh, V.Q. *Progr. Colloid Polym. Sci.* **1988**, *78*, 75. (b) Roth, H.-K.; Brunner, W.; Volkel, G.; Schrödner, M.; Gruber, H. *Makromolec. Chem. Macromolec. Symp.* **1990**, *34*, 293.
27. Nechtschein, M. In *Handbook of Conducting Polymers*, Skotheim, T.A.; Elsenbaumer, R.L.; Reynolds, J.R., Eds.; Marcel Dekker: New York, 1997; Chapt. 5, pp. 141–163.
28. *Thiophene and its Derivatives*, Gronowitz, S., Ed.; John Wiley & Sons, Inc.: New York, 1991.
29. Krinichnyi, V.I.; Grinberg, O.Y.; Nazarova, I.B.; Kozub, G.I.; Tkachenko, L.I.; Khidekel, M.L.; Lebedev, Y.S. *Bull. Acad. Sci. USSR, Div. Chem.* **1985**, *34*, 425.
30. (a) Roth, H.-K.; Krinichnyi, V.I. *Synth. Met.* **2003**, *137*, 1431. (b) Krinichnyi, V.I.; Roth, H.-K. *Appl. Magn. Reson.* **2004**, *26*, 395. (c) Krinichnyi, V.I.; Roth, H.-K.; Konkin, A.L. *Physica B* **2004**, *344*, 430.
31. Owens, J. *Phys. Stat. Solidi B* **1977**, *79*, 623.
32. Houze, E.; Nechtschein, M. *Phys. Rev. B* **1996**, *53*, 14309.
33. Molin, Y.N.; Salikhov, K.M.; Zamaraev, K.I. *Spin Exchange*, Springer: Berlin, 1980.
34. Patzsch, J. Ph.D. Dissertation, Technische Hochschule Leipzig, 1991.
35. Elliott, R.J. *Phys. Rev.* **1954**, *96*, 266.
36. Williams, J.M.; Ferraro, J.R.; Thorn, R.J.; Carlson, K.D.; Geiser, U.; Wang, H.H.; Kini, A.M.; Whangboo, M.-H. *Organic Superconductors (Including Fullerenes): Synthesis, Structure, Properties, and Theory*, Prentice-Hall, Inc., Englewood Cliffs: NJ, 1992.
37. Trivedi, D.C. In *Handbook of Organic Conductive Molecules and Polymers*, Nalwa, H. S., Ed.; John Wiley & Sons, Inc.: Chichester, 1997; Vol. 2, Chapt. 12, pp. 505–572.
38. Ginder, J.M.; Richter, A.F.; MacDiarmid, A.G.; Epstein, A.J. *Solid State Commun.* **1987**, *63*, 97.
39. Epstein, A.J.; MacDiarmid, A.G. *J. Molec. Electr.* **1988**, *4*, 161.
40. (a) Zuo, F.; Angelopoulos, M.; MacDiarmid, A.G.; Epstein, A.J. *Phys. Rev. B* **1987**, *36*, 3475. (b) MacDiarmid, A.G.; Epstein, A.J. *Faraday Disc.* **1989**, *88*, 317. (c) Epstein, A.J.; MacDiarmid, A.G. *Synth. Met.* **1991**, *41*, 601.
41. Stafstrom, S.; Brédas, J.L.; Epstein, A.J.; Woo, H.S.; Tanner, D.B.; Huang, W.S.; MacDiarmid, A.G. *Phys. Rev. Lett.* **1987**, *59*, 1464.
42. Saprigin, A.V.; Brenneman, K.R.; Lee, W.P.; Long, S.M.; Kohlman, R.S.; Epstein, A.J. *Synth. Met.* **1999**, *100*, 55.
43. Lubentsov, B.Z.; Timofeeva, O.N.; Saratovskikh, S.L.; Krinichnyi, V.I.; Pelekh, A.E.; Dmitrenko, V.V.; Khidekel, M.L. *Synth. Met.* **1992**, *47*, 187.

44. (a) Lux, F.; Hinrichsen, G.; Krinichnyi, V.I.; Nazarova, I.B.; Chemerisov, S.D.; Pohl, M. *Synth. Met.* **1993**, *55*, 347. (b) Krinichnyi, V.I.; Roth, H.-K.; Hinrichsen, G.; Lux, F.; Lüders, K. *Phys. Rev. B* **2002,** *65*, 155205. (c) Krinichnyi, V.I.; Roth, H.-K.; Hinrichsen, G. *Synth. Met.* **2003,** *135–136*, 431.
45. (a) Krinichnyi, V.I.; Chemerisov, S.D.; Lebedev, Y.S. *Phys. Rev. B* **1997**, *55*, 16233. (b) Krinichnyi, V.I.; Chemerisov, S.D.; Lebedev, Y.S. *Synth. Met.* **1997,** *84*, 819.
46. Krinichnyi, V.I. *Russ. Chem. Bull.* **2000,** *49*, 207.
47. (a) Krinichnyi, V.I.; Konkin, A.L.; Devasagayam, P.; Monkman, A.P. *Synth. Met.* **2001,** *119*, 281. (b) Konkin, A.L.; Shtyrlin, V.G.; Garipov, R.R.; Aganov, A.V.; Zakharov, A.V.; Krinichnyi, V.I.; Adams, P.N.; Monkman, A.P.; *Phys. Rev. B* **2002,** *66*, 075203.
48. (a) Krinichnyi, V.I.; Tokarev, S.V.; Roth, H.-K.; Schrödner, M.; Wessling, B. *Synth. Met.* **2005,** *152*, 165. (b) Krinichnyi, V.I.; Tokarev, S.V. *Polym. Sci. A* **2005,** *47*, 261.
49. Buchachenko, A.L.; Vasserman, A.M. *Stable Radicals (Russ)*, Khimija: Moscow, 1973.
50. Long, S.M.; Cromack, K.R.; Epstein, A.J.; Sun, Y.; MacDiarmid, A.G. *Synth. Met.* **1994,** *62*, 287.
51. (a) Goldenberg, L.M.; Pelekh, A.E.; Krinichnyi, V.I.; Roshchupkina, O.S.; Zueva, A.F.; Lyubovskaja, R.N.; Efimiv, O.N. *Synth. Met.* **1990,** *36*, 217. (b) Goldenberg, L.M.; Pelekh, A.E.; Krinichnyi, V.I.; Roshchupkina, O.S.; Zueva, A.F.; Lyubovskaja, R.N.; Efimiv, O.N. *Synth. Met.* **1991,** *43*, 3071.
52. Dyson, F.J. *Phys. Rev. B* **1955,** *98*, 349.
53. Vonsovskii, S.V. *Magnetism* (Russ), Nauka: Moscow, 1971.
54. Iida, M.; Asaji, T.; Inoue, M.B.; Inoue, M. *Synth. Met.* **1993,** *55*, 607.
55. (a) Kahol, P.K.; Pinto, N.J.; McCormick, B.J. *Solid State Commun.* **1994,** *91*, 21. (b) Lee, K.; Heeger, A.J.; Cao, Y. *Synth. Met.* **1995,** *69*, 261.
56. Blakemore, J.S. *Solid State Physics*, Cambridge University Press: Cambridge, 1985.
57. Lee, K.H.; Heeger, A.J.; Cao, Y. *Phys. Rev. B* **1993,** *48*, 14884.
58. Gullis, P.R. *J. Magn. Reson.* **1976,** *21*, 397.
59. Pelekh, A.E.; Krinichnyi, V.I.; Brezgunov, A.Y.; Tkachenko, L.I.; Kozub, G.I. *Vysokomolekul. Soedin. A* **1991,** *33*, 1731.
60. Abragam, A. *The Principles of Nuclear Magnetism*, Clarendon Press: Oxford, 1961.
61. Krinichnyi, V.I.; Pelekh, A.E.; Tkachenko, L.I.; Kozub, G.I. *Synth. Met.* **1992,** *46*, 13.
62. Nechtschein, M.; Devreux, F.; Genoud, F.; Guglielmi, M.; Holczer, K. *Phys. Rev. B* **1983,** *27*, 61.
63. Krinichnyi, V.I.; Pelekh, A.E.; Tkachenko, L.I.; Kozub, G.I. *Synth. Met.* **1992,** *46*, 1.
64. Kivelson, S. *Phys. Rev. B* **1982,** *25* , 3798.
65. Epstein, A.J. In *Handbook of Conducting Polymers*, Scotheim, T.E., Ed.; Marcel Dekker: New York, 1986; Vol. 2, Chapt. 29, pp. 1041–1097.
66. Parneix, J.P.; El Kadiri, M. In *Electronic Properties of Conjugated Polymers*, Kuzmany, H.; Mehring, M.; Roth, S., Eds.; Springer-Verlag: Berlin, 1987; Vol. 76, pp. 23–26.
67. Long, A.R. *Adv. Phys.* **1982,** *31*, 553.
68. Mott, N.F.; Davis, E. A. *Electronic Processes in Non-Crystalline Materials*, Clarendon Press: Oxford, 1979.
69. Kivelson, S.; Heeger, A.J. *Synth. Met.* **1988,** *22*, 371.
70. Kunugi, Y.; Harima, Y.; Yamashita, K.; Ohta, N.; Ito, S. *J. Mater. Chem.* **2000,** *10*, 2673.

71. Mizoguchi, K.; Nechtschein, M.; Travers, J.-P.; Menardo, C. *Phys. Rev. Lett.* **1989**, *63*, 66.
72. Javadi, H.H.S.; Cromack, K.R.; MacDiarmid, A.G.; Epstein, A.J. *Phys. Rev. B* **1989**, *39*, 3579.
73. Chapman, A.C.; Rhodes, P.; Seymour, E.F.W. *Proc. Phys. Soc.* **1957**, *B 70*, 345.
74. (a) Wang, Z.H.; Li, C.; Scherr, E.M.; MacDiarmid, A.G.; Epstein, A.J. *Phys. Rev. Lett.* **1991**, *66*, 1745. (b) Wang, Z.H.; Scherr, E.M.; MacDiarmid, A.G.; Epstein, A.J. *Phys. Rev. B* **1992**, *45*, 4190.
75. Lim, H.Y.; Jeong, S.K.; Suh, J.S.; Oh, E.J.; Park, Y.W.; Ryu, K.S.; Yo, C.H. *Synth. Met.* **1995**, *70*, 1463.
76. Joo, J.; Oh, E. J.; Min, G.; MacDiarmid, A.G.; Epstein, A.J. *Synth. Met.* **1995**, *69*, 251.
77. Rånby, B.; Rabek, J.F. *ESR Spectroscopy in Polymer Research*, Springer-Verlag: Berlin, Heidelberg, New York, 1977.
78. Dubinski, A.A.; Grinberg, O.Y.; Kurochkin, V.I.; Oransky, L.G.; Poluektov, O.G.; Lebedev, Y.S. *Teor. Eksper. Khim.* **1981**, *17*, 231.
79. Hyde, J.S.; Dalton, L.R. In *Spin Labeling II: Theory and Application*, Berliner, L.J., Ed.; Academic Press: New York, 1979; Vol. 2, Chapt. 1, pp. 1–70.
80. Hustedt, E.J.; Beth, A.H. *Biophys. J.* **2004**, *86*, 3940.
81. Masubuchi, S.; Imai, R.; Yamazaki, K.; Kazama, S.; Takada, J.; Matsuyama, T. *Synth. Met.* **1999**, *101*, 594.
82. Wasserman, A.M.; Kovarski, A.L. *Spin Labels and Probes in Physics and Chemistry of Polymers (Russ)*, Nauka: Moscow, 1986.
83. *Spin Labeling: The Next Millenium*, Berliner, L.J., Ed.; Plenum Press: New York, 1998.
84. Pelekh, A.E.; Goldenberg, L.M.; Krinichnyi, V.I. *Synth. Met.* **1991**, *44*, 205.
85. Buchachenko, A.L. *Complexes of Radicals and Molecular Oxygen with Organic Molecules (Russ)*, Nauka: Moscow, 1984.
86. Reddoch, A.; Konishi, S. *J. Chem. Phys.* **1979**, *70*, 2121.
87. Kanemoto, K.; Yamauchi, J. *Phys. Rev. B* **2000**, *61*, 1075.
88. Krinichnyi, V.I.; Lebedev, Y.S.; Grinberg, O.Y. *Appl. Magn. Reson.* **1997**, *13*, 259.

INDEX

Acrylates, ESR analysis:
  chain-transfer reaction, 111–119
  penultimate unit effect, dimeric model radicals, 124–128
Acrylic polymers:
  nitroxide kinetics and HALS stabilization, 272–276
  photoinitiation rate measurements, 267–269
Acrylonitrile-butadiene-styrene (ABS) polymers, electron spin resonance imaging:
  FTIR *vs.*, 249–250
  photodegradation, 241–243
  spatially resolved degradation:
    aged heterophasic polymers, 238–243
    basic principles, 230–233
    experimental protocols, 233–235
    hindered amine stabilizers, 235–238
    thermal aging, 238–241
Activation rate constants, ESR analysis, monomeric and dimeric alkyl bromides, 128–129
Alkyl bromides, ESR analysis, penultimate unit effect:
  dimeric model radicals, 124–128
  monomeric model radicals, 122–124
Alumina interactions, dendrimer structures, spin labeled nitroxides, 295–296

Amino acids, dendrimer structures, spin labeled nitroxides, 295–296
Amphiphilic probes, strategic applications, 170–171
Anisotropic properties:
  conductive polymers, high-field ESR, spin probe method, 332–333
  electron spin echo envelope modulation, 44–49
  electron spin resonance, basic principles, 5–9
  ESR spectrum, 54–55
  hyperfine interactions, g-tensor and, 10–12
  nitroxide magnetic tensors, 55–60
Anisotropic viscosity model, spin-labeled polymer orientation, 68
Arrhenius behavior:
  conductive polymers:
    saturation-transfer ESR, 329–330
    spin probe method, 332–333
  poly(2-hydroxyethyl methacrylate) segmental mobility, 158–161
  polymer dynamics, NMR and fluorescence techniques, 134–135
  polystyrene segmental mobility, 150–155
  segmental mobility of polymer radicals, 143–147
A-tensor anisotropy, nitroxide rotational dynamics, 136–140

Advanced ESR Methods in Polymer Research, edited by Shulamith Schlick.
Copyright © 2006 John Wiley & Sons, Inc.

Atom-transfer radical addition (ATRA):
　ESR analysis of radical polymerization,
　　104–105
　metallo-dendrimer structures, 301
Atom transfer radical polymerization (ATRP),
　ESR analysis:
　basic principles, 102–105
　chain length dependence, 105–111
　　controlled chain length models, 106–111
　　tert-butyl methacrylate, 105–106
　chain transfer reaction, 111–119
　　chain length-dependent changes, 117–119
　　controlled chain lengths, 113–117
　　tert-butyl acrylate, 111–113
　penultimate unit effect, 119–129
　　activation rate constants, 128–129
　　dimeric model radicals, 124–128
　　monomeric model radicals, 122–124
Attenuated total reflectance (ATR) FTIR,
　acrylonitrile-butadiene-styrene
　polymers, 249–250
Automotive paint technology, polymer coating
　stabilization, 256–258
Azobisisobutyronitrile (AIBN), photoinitiation
　rate measurements, 265–269

Basis sets, stochastic Liouville equation:
　calculation, 73–74
　pruning, 78–79
　symmetrization, 74
　truncation parameters, 74–78
Best-fit values:
　poly(2-hydroxyethyl methacrylate) segmental
　　mobility, 156–161
　polystyrene segmental mobility, 150–155
Bioreduction rate, dendrimer structure, ESR
　analysis, nitroxide spin labeling,
　291–296
Bjerrum length, polyelectrolyte-counterion
　interactions, 173–178
Bohr magneton, double resonance ESR, dipole-
　dipole coupling, 28–29
Boltzmann relation, spin-labeled polymer
　orientation, 65–68
Burgers-Korteweg-de-Vries theory, high-field
　electron spin resonance, conductive
　polymers, charge carrier relaxation and
　dynamics, 320–326

Catalyst-coated membranes (CCMs), in situ ESR
　experiments, fuel cell technology,
　222–224
[2]Catenanes:
　coconformation and motional modes, 178–182

mesoscopic interfaces and nanostructure,
　182–192
Cationic nitroxide spin probes:
　dendrimer spin labeling, 291–296
　dendrimer structure, ESR analysis, 284–289
Cetyl-trimethylammonium bromide (CTAB),
　dendrimer structure, spin probe
　analysis, 286–289
Chain conformation, polyelectrolyte-counterion
　interactions, 174–178
Chain-end radicals, Nafion and Dow ionomers,
　membrane-derived fluorinated radicals,
　217–221
Chain length dependence:
　dendrimer structure, ESR analysis, spin probe
　　techniques, 284–289
　ESR analysis of radical polymerization,
　　105–111
　　chain transfer reactions, 117–119
　　controlled chain length models, 106–111
　　tert-butyl methacrylate, 105–106
Chain transfer reaction:
　ESR analysis, radical polymerization,
　　111–119
　　chain length-dependent changes, 117–119
　　controlled chain lengths, 113–117
　　tert-butyl acrylate, 111–113
　polymer coatings, photoinitiation rate
　　measurements, 268–269
Charge carriers, conductive polymers, high-field
　ESR:
　magnetic parameters, 310–317
　relaxation and dynamics, 317–326
Charge transfer reactions, doped conductive
　polymers, 326–328
Cobalt complexes, metallo-dendrimer structures,
　300–301
Coherence:
　electron-electron double resonance, 32–35
　experimental techniques, 33–35
　electron spin echo envelope modulation,
　　46–49
Coil expansion, segmental mobility of polymer
　radicals, 146–147
Conductive polymers, high-field electron spin
　resonance:
　basic principles, 307–310
　carrier relaxation and dynamics, 317–326
　charge carrier magnetic parameters, 310–317
　highly-doped charge transfer, 326–328
　saturation transfer, 328–330
　spin probes, 330–333
Cone distributions, spin-labeled polymer
　orientation, 67–68

Continuous-wave electron spin resonance
(CW-ESR):
   anisotropic hyperfine interaction and g-tensor,
      10–12
   electron-electron double resonance, 32–33
   macromolecular materials:
      mesoscopic interfaces and nanostructure,
         182–187
         ionomers, 187–192
      in solution, 171–182
         [2]catenane coconformation and motional
            modes, 178–182
         frozen solutions, 172–173
         polyelectrolyte-counterion interactions,
            173–178
         solvent and temperature effects,
            171–172
      spin labeling:
         research overview, 166–167
         strategies for, 167–168
      spin probing:
         amphiphilic probes, 170–171
         hydrogen-bonding probes, 169
         hydrophobic probes, 169
         ionic probes, 169–170
         strategies for, 168–171
   multifrequency and high-field electron spin
      resonance, 18
   nitroxide line shape analysis, 138–140
   nitroxide magnetic tensors, 55–60
   paramagnetic quencher accessibility, 14–15
   pulsed methods, 18–22
Controlled chain length, ESR analysis of radical
      polymerization:
   model radicals, 106–111
   tert-butyl acrylate radicals, 113–119
Copolymerization:
   ESR analysis, penultimate unit effect, dimeric
      model radicals, 128
   mesoscopic ionomers, 188–192
   polystyrene segmental mobility, 148–155
Copper complexes:
   dendrimer structures, additions to, 296–299
   membrane-derived fluorinated radicals in
      Nafion, 215–221
Counterion-polyelectrolyte interactions,
      solution-based macromolecules,
      173–178
Counterions, Nafion and Dow ionomers,
      membrane-derived fluorinated radicals,
      215–221
Cross-linked coatings:
   nitroxide kinetics and HALS stabilization,
      272–276

   photoinitiation rate measurements:
      nonweathered coatings, 267–269
      weathered coatings, 270–272
Crystalline polymers, microtoming sensitivity,
      250–251

Data analysis, electron spin echo envelope
      modulation, 47–49
Davies electron-nuclear double resonance,
      experimental techniques, 41–42
D-band electron spin resonance, conductive
      polymers:
   charge carrier relaxation and dynamics, 317–326
   magnetic parameters, 312–317
   saturation transfer, 328–330
   spin probe method, 330–333
Degradation processes, electron spin resonance
      imaging of polymeric systems, 94–96
Dendrimer structures, ESR analysis:
   basic properties, 279–282
   metallo-dendrimers, 299–301
   nitroxide spin labeling, 289–296
   paramagnetic metal ion additions, 296–299
   spin probes, 283–289
Density functional theory (DFT), membrane-
      derived fluorinated radicals in Nafion,
      220–221
Deuterated spin labels, nitroxide rotational
      dynamics, 138–139
Diblock copolymers, mesoscopic ionomers,
      188–192
Dibutyl phthalate (DBP), segmental mobility,
      149–155
Dielectric permittivity, polyelectrolyte-
      counterion interactions, 174–178
Diffusion-limited oxidation (DLO) regime:
   acrylonitrile-butadiene-styrene polymers:
      FTIR vs ESRI, 249–250
      thermal aging, 239–241
   electron spin resonance imaging:
      polymeric systems, 95–96
      spatially resolved degradation, 231–233
   heterophasic propylene-ethylene copolymers,
      thermal aging, 243–244
Diffusion models:
   electron spin resonance imaging of polymeric
      systems, 93–96
   slow-motion ESR spectra, 60–63
Dilute solutions, nitroxide spin label dynamics,
      140–141
Dimeric radicals, ESR analysis:
   activation rate constants, 128–129
   chain length dependence, 117–119
   penultimate unit effect, 124–128

Dimyristoylphosphatidylcholine (DMPC), dendrimer structure, spin probe analysis, 286–289
Dioctyl phthalate (DOP), segmental mobility, 149–155
Dipole-dipole (DD) coupling:
 conductive polymers, high-field ESR:
  charge carrier relaxation and dynamics, 318–326
  magnetic parameters, 313–317
  spin probe method, 331–333
 double resonance ESR methods, 27–29
 electron spin resonance principles, 8–9
 solution-based macromolecules, frozen solutions, 172–173
Direct computation techniques, electron-electron double resonance, distance distributions, 38–39
Direct electron spin resonance, radical intermediate detection, 204–206
Direct methanol fuel cell, membrane stability, 201–204
Director frame:
 slow-motion ESR spectra, diffusion models, 61–63
 spin-labeled polymer orientation, 63–68
Dispersion mechanisms, mesoscopic interfaces and nanostructure, 183–187
Distance distributions, electron-electron double resonance, direct computation, 38–39
DMPO solution, *in situ* ESR experiments, fuel cell technology, 224–226
Dodecyl-trimethylammonium bromide (DTAB), dendrimer structure, spin probe analysis, 286–289
Doped conductive polymers:
 charge carrier relaxation and dynamics, 317–326, 321–326
 charge transfer reactions, 326–328
d orbital spin density, double resonance ESR, 31
Double electron-electron resonance (DEER):
 [2]catenane coconformation and motional modes, 179–182
 experimental techniques, 33–35
 mesoscopic ionomers, 189–192
 polyelectrolyte-counterion interactions, 176–178
Double resonance ESR methods:
 basic principles, 26–27
 electron-electron double resonance, 31–39
  basic principles, 31–33
  distance distribution computation, 38–39
  experimental techniques, 33–35
  scattering, form factor and structure factor, 35–38

 electron-nuclear double resonance, 39–43
  basic principles, 39–40
  experimental techniques, 40–42
  high-field ENDOR, 42–43
 electron spin echo envelope modulation, 44–49
  basic principles, 44
  data analysis, 47–49
  experimental techniques, 44–47
 spin-spin couplings, 27–31
  dipole-dipole coupling, 27–29
  exchange coupling, 30
  Fermi contact interaction, 30–31
  spin density, p and d orbitals, 31
Dow ionomers, fluorinated radicals, 215–221
DOXYL nitroxides:
 amphiphilic probes, 170–171
 mesoscopic interfaces and nanostructure, polymer lattices, 185–187
 spin labeling and probing, 168
Dynamic electrostatic attachment, polyelectrolyte-counterion interactions, 175–178
Dynamic light scattering (DLS), poly(2-hydroxyethyl methacrylate) segmental mobility, 159–161
Dysonian ESR spectrum, doped conductive polymers, charge transfer reactions, 326–328

Echo-detected electron spin resonance:
 anisotropic hyperfine interaction and g-tensor, 10–12
 mesoscopic ionomers, 192
Electron-electron double resonance (ELDOR), 31–39
 basic principles, 31–33
 dipole-dipole coupling, 27–29
 distance distribution computation, 38–39
 experimental techniques, 33–35
 pulsed ESR methods, 19–22
 scattering, form factor and structure factor, 35–38
 weak coupling separation, 26
Electron-nuclear double resonance (ENDOR), 39–43
 basic principles, 39–40
 conductive polymers, 310
 dipole-dipole coupling, 27–29
 experimental techniques, 40–42
 high-field ENDOR, 42–43
 isotropic hyperfine analysis, 12
 metallo-dendrimer structures, 300
 multifrequency and high-field electron spin resonance, 17–18

INDEX

pulsed ESR methods, 19–22
weak coupling separation, 26
Electron spin echo envelope modulation
    (ESEEM), 44–49
  basic principles, 44
  conductive polymers, 310
  data analysis, 47–49
  dipole-dipole coupling, 27–29
  double-resonance techniques, 27
  experimental techniques, 44–47
  multifrequency and high-field electron spin
      resonance, 17–18
  polyelectrolyte-counterion interactions,
      175–178
  pulsed ESR methods, 19–22
Electron spin resonance (ESR):
  anisotropic hyperfine interactions and g-tensor,
      10–12
  basic principles, 3–9
  environmental effects, 12–13
  imaging studies:
    basic principles, 85–87
    polymeric systems, 93–96
    spatially resolved degradation, 230–251
      aged heterophasic polymers, 238–251
      experimental protocols, 233–235
      hindered amine stabilizers, 235–238
    system components, 87–93
      hardware requirements, 88–90
      magnetic field gradients, 87–88
      one-dimensional intensity profiling, 90–92
      two-dimensional line shape profiling, 92–93
  isotropic hyperfine analysis, 12
  line shape analysis, tumbling nitroxide
      radicals, 15–16
  multifrequency and high-field methods,
      16–18
  paramagnetic quencher accessibility, 13–15
  pulsed methods, 18–22
Electron Zeeman (EZ) interaction, basic
    principles, 4–9
Elliot mechanism, conductive polymers, high-
    field ESR, magnetic parameters,
    314–317
Elongated spot distribution, spin-labeled polymer
    orientation, 67–68
Energy potential, spin-labeled polymer
    orientation, 65–68
Environmental effects, g-tensor and hyperfine
    interaction, 12–13
EPRLL program:
  available packages, 81–82
  microscopic order-macroscopic disorder
      model, 68–70, 69–70

nonlinear least-squares analysis, 80–81
slowly relaxing local structure model (SRLS)
    model, 70–73
slow-motional ESR analysis, 55
  diffusion models, 62–63
  nitroxide magnetic tensors, 59–60
  spin-labeled polymer orientation, 63–68
  stochastic Liouville equation basis sets, 73–78
  pruning, 78–79
Ether bridges, fuel cell membranes, 207–210
Euler angles:
  nitroxide diffusion models, 62–63
  spin-labeled polymer orientation, 64–68
Exchange coupling, dipole-dipole interactions, 30

Far-infrared (FIR) electron spin resonance
    (ESR), polystyrene segmental mobility,
    147–155
Fast internal motion (FIM) model, slow-motion
    ESR spectra, 71–73
Fast-motional spectra, nitroxide line shape
    analysis, 138–139
Fenton media, fuel cell membrane stability,
    202–204
  direct ESR and spin trapping detection, radical
      intermediates, 204–206
Fermi contact interaction, double resonance ESR,
    30–31
Fermi velocity, conductive polymers, high-field
    ESR, charge carrier relaxation and
    dynamics, 324–326
Ferromagnetic exchange coupling, electron spin
    responance principles, 8–9
Fluorescence analysis, polymer dynamics,
    134–136
Fluorinated alkyl radicals:
  Nafion membrane in Fenton media, 214–215
  UV-irradiated Nafion and Dow monomers,
      215–221
Fourier transform (FT):
  electron-electron double resonance:
    distance distributions, 38–39
    form and structure scattering, 35–38
  one-dimensional electron spin resonance
      imaging, intensity profiling, 91–92
Fourier transform infrared (FTIR)
    spectroscopy:
  fuel cell membrane stability, 203–204
  spatially resolved degradation, electron spin
      resonance imaging, 233, 248–250
Free induction decay (FID), pulsed ESR
    methods, 20–22
Free ions, polyelectrolyte-counterion
    interactions, 173–178

Free-radical reactions, polymer coating photooxidation and stabilization, 257–260
  nitroxide decay experiments, 263–265
Fremy's salt dianion, polyelectrolyte-counterion interactions, 174–178
Frozen solutions, solution-based macromolecules, 172–173
Fuel cells:
  basic elements, 198–201
  membrane stability, 201–204
    future research issues, 224–226
    Nafion membranes in Fenton media, 210–215
      fluorinated alkyl radicals, 214–215
      oxygen-centered radicals, 211–213
    oxygen radical reactions, 206–210
    in situ ESR experiments, 222–224
    UV-irradiated Nafion and Dow ionomers, 215–221
      fluorinated radical fragments, 217–221
    radical intermediate detection, direct ESR and spin trapping, 204–206

Gas voltaic battery, fuel cell technology, 198–201
Gaussian line shape analysis, nitroxide magnetic tensors, 59–60
Gauss-Newton method, nonlinear least-squares analysis, 80–81
Genetic algorithm, one-dimensional electron spin resonance imaging, intensity profiling, 92
Gibbs free energy, fuel cell membrane stability, 203–204
Glass transition temperature:
  mesoscopic interfaces and nanostructure, polymer lattices, 183–187
  solution-based macromolecules, frozen solutions, 172–173
  spin labeling and, 167
Gradient coils, electron spin resonance imaging, 89–90
g-tensor:
  anisotropic hyperfine interactions, 10–12
  conductive polymers, high-field ESR, magnetic parameters, 311–317
  electron spin resonance, basic principles, 4–9
  environmental effects, 12–13
  nitroxide magnetic tensors, 55–60
  nitroxide rotational dynamics, 136–140

Hahn echo experiments, pulsed ESR methods, 20–22
Hamiltonian equations, electron spin resonance principles, 4–9

Hardware requirements, electron spin resonance imaging, 88–90
Heterophasic polymers, electron spin resonance imaging, spatially resolved degradation, 230–251
  aged heterophasic polymers, 238–251
    poly(acrylonitrile-butadiene-styrene) systems, 238–243
    propylene-ethylene copolymers, 243–248
    crystalline polymers microtome sensitivity, 250–251
  experimental protocols, 233–235
  FTIR comparison, 248–250
  hindered amine stabilizers, 235–238
    nitroxides derived from hindered amine stabilizers, 235–238
Heterophasic propylene-ethylene copolymers (HPEC):
  photodegradation, 244–248
  spatially resolved degradation:
    basic properties, 233–235
    ESRI *vs.* FTIR, 249–250
  thermal aging, 243–244
High-field electron-nuclear double resonance (ENDOR), basic principles, 42–43
High-field electron spin resonance:
  basic principles, 16–18
  conductive polymers:
    basic principles, 307–310
    carrier relaxation and dynamics, 317–326
    charge carrier magnetic parameters, 310–317
    highly-doped charge transfer, 326–328
    saturation transfer, 328–330
    spin probes, 330–333
High-spin polycarbenes, dendrimer spin labeling, 290–296
Hindered amine stabilizers (HAS):
  electron spin resonance imaging:
    nitroxide derivation, 235–238
    polymeric formation from HAS, 94–96
    spatially resolved degradation, 230–233
  heterophasic propylene-ethylene copolymers, thermal aging, 243–244
  polymer coating photooxidation and stabilization:
    basic properties, 257–258
    ESR-based nitroxide concentration, 260–262
    free-radical reactions, 259–260
    nitroxide kinetics, 272–276
    nitroxide measurements, 264
    photoinitiation rate measurements:
      nonweathered coatings, 269
      weathered coatings, 272

Houzé-Nechtschein model, conductive polymers, high-field ESR, magnetic parameters, 313–317
Hydrogen bonding:
  local polarity, 13
  solution-based macromolecules, solvent and temperature effects, 172
Hydrogen peroxide:
  fuel cell membrane stability, 202–204
    direct ESR and spin trapping detection, 204–206
  Nafion membranes in Fenton media, 210–215
  toluene sulfonic acid reaction, 207–210
  UV-irradiated Nafion and Dow ionomers, 215–221
Hydroperoxyl radicals (HOO•):
  fuel cell membrane stability, 202–204
  Nafion membranes in Fenton media, 210–215
    fluorinated alkyl radicals, 214–215
  polymer coating photooxidation and stabilization, 258–260
  toluene sulfonic acid reaction, 207–210
Hydrophobic probes, strategic applications, 169
2-Hydroxyethyl methacrylate (HEMA), segmental mobility, 155–161
Hydroxyl radicals (HO•):
  fuel cell membrane stability, 202–204
    direct ESR and spin trapping detection, 204–206
  toluene sulfonic acid reaction, 206–210
Hyperfine interactions:
  anisotropic parameters and g-tensor, 10–12
  dendrimer spin labeling, 289–296
  double resonance ESR, 26–27
  electron spin echo envelope modulation, 44–49
  electron spin resonance, basic principles, 5–9
  ESR analysis, radical polymerization:
    chain length dependence, 117–119
    penultimate unit effect, 123–124
  isotropic analysis, 12
  Nafion and Dow ionomers, membrane-derived fluorinated radicals, 217–221
  nitroxide magnetic tensors, 55–60
  pulsed ESR methods, 18–22
Hyperfine sublevel correlation (HYSCORE) experiments:
  double-resonance techniques, 27
  electron spin echo envelope modulation, 47–49
  metallo-dendrimer structures, 300–301
  pulsed ESR methods, 19–22

Imaging studies with ESR:
  basic principles, 85–87
  polymeric systems, 93–96

system components, 87–93
  hardware requirements, 88–90
  magnetic field gradients, 87–88
  one-dimensional intensity profiling, 90–92
  two-dimensional line shape profiling, 92–93
"Immobilized" nitroxides, polymer coating photooxidation and stabilization, 261–262
Infrared spectroscopy, photoinitiation rate measurements, weathered coatings, 271–272
In situ electron spin resonance, fuel cell membrane stability, 222–224
Intensity profiling, one-dimensional electron spin resonance imaging, 90–92
Intercluster distance, mesoscopic ionomers, 188–192
Interpulse delay, electron spin echo envelope modulation, 45–49
Ionic probes:
  polyelectrolyte-counterion interactions, 174–178
Ionomers:
  Dow ionomers, membrane-derived fluorinated radicals, 215–221
  mesoscopic interfaces and nanostructure, 187–192
Iron complexes:
  metallo-dendrimer structures, 300–301
  Nafion and Dow ionomers, membrane-derived fluorinated radicals, 215–221
Isotropic values:
  electron spin resonance:
    basic principles, 5–9
    hyperfine analysis, 12
  hyperfine analysis, 12
  membrane-derived fluorinated radicals in Nafion, 218–221

J couplings, dipole-dipole interactions, 30

Ketone formation, polymer coating photooxidation and stabilization, photoinitiation rate measurements, 269
Kinetic models, polymer coatings, nitroxide kinetics and HALS stabilization, 272–276
Kivelson theory, high-field electron spin resonance, conductive polymers, charge carrier relaxation and dynamics, 320–326
Kramers' theory:
  polymer dynamics, NMR and fluorescence techniques, 134–135
  polystyrene segmental mobility, 149–155

K values, stochastic Liouville equation truncation parameters, 76–78

Least-squares analysis, nitroxide magnetic tensors, 57–60
Levenberg-Marquardt algorithm, nonlinear least-squares analysis, 80–81
Ligand-to-metal charge transfer (LMCT), Nafion membrane radicals in Fenton media, 210–215
Linear charge density, polyelectrolyte-counterion interactions, 174–178
Line shape analysis:
  [2]catenanes, 181–182
  conductive polymers, high-field ESR, magnetic parameters, 311–317
  ESR spectrum calculations, 55
  nitroxide magnetic tensors, 59–60
  nitroxide rotational dynamics, 136–140
    fast-motional spectra, 138–139
    slow-motional spectra and ordering, 139–140
  polystyrene segmental mobility, 150–155
  tumbling nitroxide radicals, 15–16
  two-dimensional spectral-spatial ESRI, 92–93
Liquid crystalline structures in dendrimers, 299
Longitudinal relaxation:
  paramagnetic quencher accessibility, 14–15
  pulsed ESR methods, 20–22
Lorentzian line shapes:
  [2]catenanes, 181–182
  conductive polymers, magnetic parameters, 315–317
  nitroxide rotational dynamics, 138–139
  polystyrene segmental mobility, 149–155
L values, stochastic Liouville equation truncation parameters, 76–78

Macrocycle motion, [2]catenane coconformation and motional modes, 179–182
Macromolecular materials:
  dendrimer structures, ESR analysis:
    basic properties, 279–282
    metallo-dendrimers, 299–301
    nitroxide spin labeling, 289–296
    paramagnetic metal ion additions, 296–299
    spin probes, 283–289
  mesoscopic interfaces and nanostructure, 182–187
  ionomers, 187–192
  solution-based ESR, 171–182
    [2]catenane coconformation and motional modes, 178–182
    frozen solutions, 172–173

polyelectrolyte-counterion interactions, 173–178
solvent and temperature effects, 171–172
spin labeling:
  research overview, 166–167
  strategies for, 167–168
spin probing:
  amphiphilic probes, 170–171
  hydrogen-bonding probes, 169
  hydrophobic probes, 169
  ionic probes, 169–170
  strategies for, 168–171
Magnetic frame, nitroxide spin labels, 61–62
Magnetic parameters:
  conductive polymers, high-field ESR:
    charge carriers, 310–317
    spin probe method, 332–333
  dendrimer structures, 302–303
  electron spin resonance imaging, 87–88
  membrane-derived fluorinated radicals in Nafion, 218–221
  one-dimensional electron spin resonance imaging, intensity profiling, 90–92
Magnetic tensors, nitroxide labels, 55–60
Magnetic tilt angles, nitroxide magnetic tensors, 56–60
Manganese complexes, dendrimer structure, ESR analysis, paramagnetic metal ion additions, 297–299
Manning parameter, polyelectrolyte-counterion interactions, 173–178
Maximum dipolar time evolution, electron-electron double resonance, 32–33
Membrane electrode assembly (MEA), fuel cell technology, 198–201
Membrane stability, fuel cell technology, 201–204
  future research issues, 224–226
  Nafion membranes in Fenton media, 210–215
    fluorinated alkyl radicals, 214–215
    oxygen-centered radicals, 211–213
  oxygen radical reactions, 206–210
  in situ ESR experiments, 222–224
  UV-irradiated Nafion and Dow ionomers, 215–221
    fluorinated radical fragments, 217–221
Mesoscopic interfaces and nanostructure, macromolecular materials, 182–192
  ionomers, 187–192
  polymer lattices, 182–187
Metallo-dendrimers, ESR analysis, 299–301
Methacrylates, ESR analysis:
  chain-transfer reaction, 111–119
  penultimate unit effect, dimeric model radicals, 124–128

INDEX **347**

Micelle concentrations, dendrimer structure, spin probe analysis, 285–289
Microscopic order-macroscopic disorder (MOMD) model:
  nitroxide rotational dynamics, slow-motional spectra, 140
  poly(2-hydroxyethyl methacrylate) segmental mobility, 156–161
  polystyrene segmental mobility, 147–155
  slowly relaxing local structure model, 71–73
  slow-motion ESR spectra, 68–70
  stochastic Liouville equation truncation parameters, 75–78
Microtoming sensitivity, crystalline polymers, 250–251
Mims electron-nuclear double resonance, experimental techniques, 41–42
Minimum truncation set (MTS), stochastic Liouville equation, 75–78
  basis set pruning, 78–79
Molybdenum complexes, metallo-dendrimer structures, 301
Monomeric radicals, ESR analysis:
  activation rate constants, 128–129
  chain length dependence, 117–119
  penultimate unit effect, 119–124
Monte Carlo procedures, one-dimensional electron spin resonance imaging, intensity profiling, 91–92
Motional modes:
  [2]catenanes, 178–182
  Nafion membrane radicals in Fenton media, 211–215
Multifrequency electron spin resonance, basic principles, 16–18

Nafion:
  fluorinated radicals in UV-irradiated compounds, 215–221
  fuel cell technology, 198–201
  radical intermediate detection, 205–206
  radical species in Fenton media, 210–215
Nanomaterials:
  [2]catenane coconformation and motional modes, 178–182
  mesoscopic interfaces and nanostructure: ionomers, 187–192
  polymer lattices, 182–187
Nelder-Mead search, nonlinear least-squares analysis, 80–81
Nickel catalysts, metallo-dendrimer structures, 301
Nitroxide compounds:
  anisotropic hyperfine interaction and g-tensor, 10–12

electron spin resonance imaging, 87–88
  hindered amine stabilizers, 235–238
  spatially resolved degradation, 231–233
environmental effects on g-tensor and hyperfine interactions, 12–13
heterophasic propylene-ethylene copolymers:
  photodegradation, 244–248
  thermal aging, 243–244
line shape analysis:
  rotational effects, 136–140
  fast-motional spectra, 138–139
  slow-motional spectra and ordering, 139–140
  tumbling radicals, 15–16
microscopic order-macroscopic disorder model, 69–70
polymer coating photooxidation and stabilization:
  decay experiments, 263–264
  ESR-based concentration, 260–262
  free-radical decay kinetics and photoinitiation rates, 264–265
  HALS-stabilized sample preparation, 264
  kinetics, 272–276
  photoinitiation rate measurements:
    nonweathered coatings, 265–269
    weathered coatings, 269–272
  UV exposures, 262–263
radical intermediate detection, spin trapping techniques, 205–206
segmental dynamics, 140–141
spin labeling and probing:
  TEMPO and PROXYL labels, 168–169
Non-Brownian diffusion models, slow-motional ESR analysis, 62–63
Nonlinear least-squares analysis (NLLS):
  poly(2-hydroxyethyl methacrylate) segmental mobility, 156–161
  polystyrene segmental mobility, 148–155
  slow-motional ESR, 79–81
Nonweathered coatings, photoinitiation rate measurements, 265–269
Nuclear frequency spectrum, electron spin echo envelope modulation, 45–49
Nuclear Overhauser effect (NOE) enhancements, polymer dynamics, 134
Nuclear Zeeman (NZ) interaction, basic principles, 6–9

Odijk-Skolnick-Fixman theory, polyelectrolyte-counterion interactions, 174–178
Oligomeric radicals, ESR analysis, controlled chain lengths, 109–111

One-dimensional electron spin echo envelope modulation, data analysis, 48–49
One-dimensional electron spin resonance imaging:
 basic components, 87–88
 intensity profiling, 90–91
 polymeric systems studied, 94–96
 spatially resolved degradation, 232–233
  aged heterophasic polymers, 238–251
Optimization techniques, one-dimensional electron spin resonance imaging, intensity profiling, 90–92
Orientational ordering:
 nitroxide rotational dynamics, 139–140
 poly(2-hydroxyethyl methacrylate) segmental mobility, 156–161
 polystyrene segmental mobility, 151–155
 spin-labeled polymer orientation, 63–68
Oxygen centered radicals:
 ESR analysis, controlled chain lengths, 116–119
 Nafion membrane radicals in Fenton media, 211–215

Pair distance distribution function, electron-electron double resonance, 35–38
Pake pattern, double resonance ESR, dipole-dipole coupling, 29
PANI conductive polymer:
 charge carrier relaxation and dynamics, 321–326
 charge transfer reactions, 327–328
 magnetic parameters, 314–317
 saturation-transfer ESR, 329–330
Paramagnetic parameters:
 conductive polymers, high-field ESR, 310
 dendrimer structures, ion additions, 296–299
 electron spin resonance principles, 8–9
 ESR imaging studies, 85–87
 quencher accessibility, 13–15
Pauli susceptibility, high-field electron spin resonance, conductive polymers, magnetic parameters, 315–317
Penultimate unit effect, ESR analysis, radical polymerization, 119–129
 activation rate constants, 128–129
 dimeric model radicals, 124–128
 monomeric model radicals, 122–124
Persistence length, polyelectrolyte-counterion interactions, 174–178
Phenoxyl radicals, toluene sulfonic acid reaction, 207–210
Photoacoustic (PA) FTIR spectroscopy, acrylonitrile-butadiene-styrene polymers, 249–250

Photodegradation:
 acrylonitrile-butadiene-styrene polymers, 241–243
 heterophasic propylene-ethylene copolymers, 244–248
 FTIR spectroscopy, 249–250
Photoinitiation rate (PIR), polymer coating photooxidation and stabilization:
 free-radical reactions, 258–260
 nonweathered coatings, free-radical decay kinetics, 265–269
 weathered coatings, 269–272
Photooxidation, polymer coating stabilization:
 basic principles, 256–258
 free-radical processes, 258–260
  nitroxide decay, 264–272
 nitroxide concentration by ESR, 260–262
 nitroxide decay sample preparation, 263–264
 nitroxide kinetics and HALS stabilization, 272–277
 ultraviolet exposures, 262–263
π-conjugated polymers, high-field electron spin resonance, 308–310
π radicals, isotropic hyperfine analysis, 12
Platinum catalysts, in situ electron spin resonance experiments, fuel cell membrane stability, 222–224
Poisson-Boltzmann equation:
 electron-electron double resonance, form factor and structure factor in scattering, 36–38
 polyelectrolyte-counterion interactions, 173–178
Polar angles, spin-labeled polymer orientation, 64–68
Polaron motion:
 conductive polymers, high-field ESR:
  magnetic parameters, 312–317
  spin probe method, 331–333
 high-field electron spin resonance, conductive polymers, charge carrier relaxation and dynamics, 321–326
Polyacetylene (PA), high-field electron spin resonance, 308–310
Polyacrylates, ESR analysis, controlled chain lengths, 116–119
Poly(acrylic acid) (PAA), segmental mobility, 141–147
Polyamidoamine (PAMAM), dendrimer structure, ESR analysis, 280–282
 nitroxide spin labeling, 291–296
 paramagnetic metal ion additions, 296–299
 spin probe techniques, 284–289

Polybutadiene (PB), electron spin resonance imaging, spatially resolved degradation, 233–235
Poly(butylacrylate) (P(BA)), mesoscopic interfaces and nanostructure, polymer lattices, 183–187
Poly(butylmethacrylate) (PBA), mesoscopic interfaces and nanostructure, 183–187
Poly(diallyldimethylammonium chloride) (PDADMAC), polyelectrolyte-counterion interactions, 174–178
Polyelectrolytes:
  counterion interactions, solution-based macromolecules, 173–178
  segmental mobility, 141–147
Poly(2-hydroxyethyl methacrylate) (PHEMA), segmental mobility, 141, 155–161
Polymer coatings, photooxidation and stabilization:
  basic principles, 256–258
  free-radical processes, 258–260
    nitroxide decay, 264–272
  nitroxide concentration by ESR, 260–262
  nitroxide decay sample preparation, 263–264
  nitroxide kinetics and HALS stabilization, 272–277
  ultraviolet exposures, 262–263
Polymeric systems, electron spin resonance imaging, 93–96
Polymer lattices, mesoscopic interfaces and nanostructure, 182–187
Poly(methacrylic acid) (PMA), segmental mobility, 141–147
Poly(methyl methacrylate) (polyMMA) radicals:
  ESR analysis, controlled chain lengths, 109–111
  segmental mobility, 141–142
Polynucleotides, dendrimer structures, spin labeled nitroxides, 293–296
Poly(propyleneimine) (PPI) dendrimer:
  paramagnetic metal ion additions, 298–299
  spin labeling with nitroxides, 289–296
  spin probe analysis, 283–289
Polypropylenepolyamine dendrimer, spin probe analysis, 283–289
Polypropylene (PP), spatially resolved degradation, basic properties, 233–235
Polystyrene (PS), segmental mobility, 141, 147–155
Polythiophene, conductive polymers, high-field ESR, magnetic parameters, 312–317
Population distributions, spin-labeled polymer orientation, 65–68
p orbital spin density, double resonance ESR, 31

Probe ordering, spin-labeled polymer orientation, 63–68
Projection slice algorithm, two-dimensional electron spin resonance imaging, line shape profiling, 92–93
Proton exchange membrane (PEM):
  direct ESR and spin trapping detection, radical intermediates, 205–206
  fuel cell technology, 198–201
    future research issues, 224–226
  oxygen radical reactions, 210–215
  stability in, 201–204
PROXYL nitroxides, spin labeling and probing, 168–169
  dendrimer structures, 289–296
Pruning, stochastic Liouville equation basis sets, 78–79
Pseudo-spherical tensor components, nitroxide magnetic tensors, 59–60
Pulsed electron-electron double resonance:
  experimental techniques, 33–35
  form factor and structure factor in scattering, 35–38
  mesoscopic ionomers, 189–192
Pulsed electron spin resonance:
  basic principles, 18–22
  dendrimer spin labeling, 290–296
    metallo-dendrimer structures, 300–301
  electron-electron double resonance, 33–35
    DEER sequences, 33–35
Pump pulse, electron-electron double resonance, 31–33
bis[2-(2-Pyridyl)ethyl]-amine (PY2) ligands, dendrimer structures, paramagnetic metal ion additions, 298–299
α-(4-Pyridyl-1-oxide)-N-tert butylnitrone (POBN), in situ ESR experiments, fuel cell technology, 223–225

Quadruople interaction, basic principles, 6–9
Quartet ESR spectra, Nafion and Dow ionomers, membrane-derived fluorinated radicals, 216–217
Quasi-one-dimensional (Q1D) systems, high-field electron spin resonance, conductive polymers, 308–310
  charge carrier relaxation and dynamics, 319–326
  magnetic parameters, 315–317
Quenchers, paramagnetic, accessibility to, 13–15
Quintet ESR spectra, Nafion and Dow ionomers, membrane-derived fluorinated radicals, 215–221

Radial distribution function, polyelectrolyte-counterion interactions, 176–178
Radical intermediates:
  direct ESR and spin trapping detection, 204–206
  ESR imaging, 230–233
Radical polymerization. See also Free-radical formation
  ESR analysis:
    basic principles, 102–105
    chain length dependence, 105–111
      controlled chain length models, 106–111
      tert-butyl methacrylate, 105–106
    chain transfer reaction, 111–119
      chain length-dependent changes, 117–119
      controlled chain lengths, 113–117
      tert-butyl acrylate, 111–113
    penultimate unit effect, 119–129
      activation rate constants, 128–129
      dimeric model radicals, 124–128
      monomeric model radicals, 122–124
Radical-radical collisions, dipole-dipole interactions, 30
Relative anisotropy, line shape analysis, tumbling nitroxide radicals, 15–16
Relaxation times:
  dendrimer structures, 302–303
  electron-electron double resonance, 32–33
  high-field electron spin resonance, conductive polymers, charge carrier relaxation and dynamics, 321–326
  mesoscopic ionomers, 191–192
Resonant radio frequency irradiation, electron-nuclear double resonance, 40–42
Rhombic distortion:
  poly(2-hydroxyethyl methacrylate) segmental mobility, 156–161
  polystyrene segmental mobility, 151–155
Rotational correlation time (τ):
  dendrimer structures:
    paramagnetic metal ion additions, 298–299
    spin probe analysis, 284–289
  electron-electron double resonance, 33–35
  electron spin echo envelope modulation, 45–49
  line shape analysis, tumbling nitroxide radicals, 15–16
  nitroxide line shape analysis, 136–140
  nitroxide spin label dynamics, 140–141
  polyelectrolyte-counterion interactions, 175–178
  poly(2-hydroxyethyl methacrylate) segmental mobility, 157–161
  polystyrene segmental mobility, 152–155

segmental mobility of polymer radicals, 145–147
solution-based macromolecules, solvent and temperature effects, 171–172
Rotational diffusion frame:
  nitroxide spin labels, 61–62
  polyelectrolyte-counterion interactions, 174–178
  slowly relaxing local structure model, 71–73
Ruthenium complexes, dendrimer structure, spin probe analysis, 288–289

Saturation recovery, mesoscopic ionomers, 191–192
Saturation-transfer ESR (ST-ESR), conductive polymers, 328–330
Scanning electron microscopy (SEM), heterophasic propylene-ethylene copolymers, 234–235
Schneider-Freed programs, nitroxide rotational dynamics, slow-motional spectra, 139–140
Segmental mobility, solution-based polymer dynamics:
  NMR and fluorescence techniques, 134–135
  spin-label electron spin resonance, 140–162
    nitroxide spin labels, 140–141
    poly(2-hydroxyethyl methacrylate), 155–161
    poly(methacrylic acid) and poly(acrylic acid), 142–147
    poly(methyl methacrylate), 142
    polystyrene, 147–155
Semiquinone radicals, hydroxyl radicals, 207–210
Single crystal samples:
  anisotropic hyperfine interaction and g-tensor, 10–12
  nitroxide magnetic tensors, 56–60
Site-bound ions, polyelectrolyte-counterion interactions, 173–178
Slowly relaxing local structure model (SRLS) model, slow-motion ESR spectra, 70–73
Slow-motion ESR spectra:
  nitroxide parameters, 55–73
    diffusion models, 60–63
    magnetic tensors, 55–60
    microscopic order-macroscopic disorder model, 69–70
    slowly relaxing local structure model, 70–73
    spin-labeled polymer orientation, 63–68
  nitroxide rotational dynamics, 139–140
  nonlinear least-squares analysis (NLLS), 79–81

INDEX

program availability, 81–82
stochastic Liouville equation basis sets, 73–79
　calculation, 73–74
　pruning, 78–79
　symmetrization, 74
　truncation parameters, 74–78
Small angle neutron scattering (SANS), electron-electron double resonance, 35–38
Small angle X-ray scattering (SAXS):
　electron-electron double resonance, 35–38
　mesoscopic interfaces and nanostructure, ionomers, 187–192
　poly(2-hydroxyethyl methacrylate) segmental mobility, 159–161
Solution-based macromolecules, 171–182
　[2]catenane coconformation and motional modes, 178–182
　frozen solutions, 172–173
　polyelectrolyte-counterion interactions, 173–178
　solvent and temperature effects, 171–172
Solution-based polymer dynamics, spin-labeled ESR analysis:
　basic principles, 134
　nitroxide rotational effects, ESR line shapes, 136–140
　　fast-motional spectra, 138–139
　　slow-motional spectra and ordering, 139–140
　NMR and fluorescence analysis, 134–136
　　segmental mobility, 134–135
　segmental mobility, 140–162
　　nitroxide spin labels, 140–141
　　poly(2-hydroxyethyl methacrylate), 155–161
　　poly(methacrylic acid) and poly(acrylic acid), 142–147
　　poly(methyl methacrylate), 142
　　polystyrene, 147–155
Solvent effects:
　polymer coating stabilization, 256–258
　polystyrene segmental mobility, 151–155
Spatially resolved degradation, electron spin resonance imaging, 230–251
　aged heterophasic polymers, 238–251
　　poly(acrylonitrile-butadiene-styrene) systems, 238–243
　　propylene-ethylene copolymers, 243–248
　　crystalline polymers microtome sensitivity, 250–251
　experimental protocols, 233–235
　FTIR comparisons, 248–250
　hindered amine stabilizers, 235–238
Spatial resolution, electron spin resonance imaging, 87–88

Spectral diffusion, pulsed ESR methods, 21–22
Spin density, double resonance ESR, 31
Spin diffusion, conductive polymers, charge carrier relaxation and dynamics, 318–326
Spin labeling:
　dendrimer structures, 289–296
　macromolecular materials:
　　research overview, 166–167
　　strategies for, 167–168
　orientational ordering, 63–68
　solution-based polymer dynamics:
　　basic principles, 134
　　nitroxide rotational effects, ESR line shapes, 136–140
　　　fast-motional spectra, 138–139
　　　slow-motional spectra and ordering, 139–140
　　NMR and fluorescence analysis, 134–136
　　　segmental mobility, 134–135
　　segmental mobility, 140–162
　　　nitroxide spin labels, 140–141
　　　poly(2-hydroxyethyl methacrylate), 155–161
　　　poly(methacrylic acid) and poly(acrylic acid), 142–147
　　　poly(methyl methacrylate), 142
　　　polystyrene, 147–155
Spin-orbit coupling, environmental effects on g-tensor and hyperfine interactions, 13
Spin probing:
　conductive polymers, high-field ESR, 330–333
　dendrimer structures, ESR analysis, 283–289
　macromolecular materials:
　　amphiphilic probes, 170–171
　　hydrogen-bonding probes, 169
　　hydrophobic probes, 169
　　ionic probes, 169–170
　　strategies for, 168–171
　mesoscopic interfaces and nanostructure, polymer lattices, 183–187
Spin-spin couplings:
　conductive polymers, high-field ESR:
　　magnetic parameters, 314–317
　　spin probe method, 331–333
　double resonance ESR methods, 27–31
　　dipole-dipole coupling, 27–29
　　exchange coupling, 30
　　Fermi contact interaction, 30–31
　　spin density, p and d orbitals, 31
Spin trapping:
　radical intermediate detection, 204–206
　in situ ESR experiments, fuel cell technology, 224
Stabilization kinetics, polymer coatings:

basic principles, 256–258
free-radical processes, 258–260
  nitroxide decay, 264–272
  nitroxide concentration by ESR, 260–262
  nitroxide decay sample preparation, 263–264
  nitroxide kinetics and HALS stabilization, 272–277
  ultraviolet exposures, 262–263
Starburst dendrimers, ESR analysis, 280–282
Steric hindrance, dendrimer structures, ESR analysis, 280–282
Stochastic Liouville equation (SLE):
  basis set:
    calculation, 73–74
    pruning, 78–79
    symmetrization, 74
    truncation parameters, 74–78
  ESR spectrum calculations, 54–55
  nitroxide rotational dynamics, slow-motional spectra, 139–140
  nonlinear least-squares analysis, 80–81
  slow-motion ESR spectra, diffusion models, 60–63
Structure factor scattering, electron-electron double resonance, 35–38
Styrene-acrylonitrile (SAN):
  ESRI vs. FTIR, 249–250
  spatially resolved degradation, 233–235
  thermal aging, 241
Sulfonic acid groups:
  fuel cell membrane stability, 202–204
  oxygen radical reactions, 206–210
Superoxide radical anions, toluene sulfonic reaction, 207–210
Supramolecular dendrimer complexes, spin probe analysis, 284–289
Swept-field calculations, stochastic Liouville equation basis set pruning, 78–79
Symmetrization:
  poly(2-hydroxyethyl methacrylate) segmental mobility, 156–161
  polystyrene segmental mobility, 150–155
  stochastic Liouville equation basis sets, 74

Teflon surfaces:
  ESR imaging, spatially resolved degradation studies, 230–233
  in situ ESR experiments, fuel cell technology, 223–224
Temperature dependence:
  conductive polymers, high-field ESR:
    charge carrier relaxation and dynamics, 319–326
    magnetic parameters, 312–317

ESR analysis of radical polymerization, controlled chain lengths, 108–111
poly(2-hydroxyethyl methacrylate) segmental mobility, 156–161
polystyrene segmental mobility, 151–155
solution-based macromolecules, 171–172
TEMPO-4-carboxylate spin probe, strategic applications, 170
TEMPOL probes, hydrogen-bond acceptors, 169
TEMPO nitroxides:
  dendrimer structures:
    spin labeling, 291–296
    spin probe analysis, 284–289
  hyprophobic probes, 169
  mesoscopic interfaces and nanostructure:
    ionomers, 187–192
    polymer lattices, 184–187
  polymer coating photooxidation and stabilization, decay experiments, 263–264
  spin labeling and probing, 168–169
TEMPO-4-phosphonooxylate spin probe, strategic applications, 170
Territorially bound ions, polyelectrolyte-counterion interactions, 173–178
tert-Butyl acrylate radicals, ESR analysis:
  chain length dependence, 117–119
  controlled chain lengths, 113–119
  penultimate unit effect, 119–129
tert-Butyl methacrylate (tBMA), ESR analysis:
  chain length dependence, 105–106
  controlled chain lengths, 106–111
7,7,8,8-Tetracianoquinodimethane (TCNQ), spin probe analysis, 283–289
Tetrathiafulvalene (TTF), dendrimer spin labeling, 291–296
Thermal aging:
  acrylonitrile-butadiene-styrene polymers, 238–241
  heterophasic propylene-ethylene copolymers, 243–244
Tikhonov regularization, electron-electron double resonance, distance distribution computation, 38–39
TIN 440/770/292 compounds, nitroxide kinetics and HALS stabilization, 273–276
Titanium complexes, Nafion membrane radicals in Fenton media, 210–215
  fluorinated alkyl radicals, 214–215
Toluene, segmental mobility of polymers in, 149–155
Toluene sulfonic acid, oxygen radical reactions, 206–210

INDEX

Trajectory method, ESR spectrum calculations, 54–55
Transition metal ions:
  dendrimer structures, additions to, 296–299
    metallo-dendrimer structures, 299–301
  electron spin resonance principles, 7–9
  fuel cell membrane stability, 202–204
  Nafion and Dow ionomers, membrane-derived fluorinated radicals, 215–221
Triarylmethyl (TAM) trianion radical, polyelectrolyte-counterion interactions, 178
Truncation parameters, stochastic Liouville equation basis sets, 74–78
Turning points, nitroxide magnetic tensors, 57–60
Twisted boat configuration, nitroxide magnetic tensors, 57–60
Two-dimensional electron spin resonance imaging:
  line shape profiling, 92–93
  spatially resolved degradation, 232–233
  aged heterophasic polymers, 238–251
Two-dimensional electron-spin transient nutation (2D-ESTN), dendrimer spin labeling, 290–296
Two-pulse echo decay, mesoscopic ionomers, 191–192

Ultraviolet (UV) radiation:
  electron spin resonance imaging, spatially resolved degradation, 232–233
  Nafion, membrane-derived fluorinated radicals, 215–221
  polymer coating stabilization:
    basic principles, 256–258
    experimental protocols, 262–263
    free-radical reactions, 259–260
Urethane coatings, photoinitiation rate measurements, 271–272

Vanadium catalysts, Nafion membrane radicals in Fenton media, 211
Vanadyl complexes, dendrimer structures, paramagnetic metal ion additions, 298–299
Viscosity dependence, polystyrene segmental mobility, 151–155

Weathered polymer coatings, photoinitiation rate measurements, 269–272
Weighting factors, stochastic Liouville equation basis set pruning, 78–79
Wigner rotation matrix, polystyrene segmental mobility, 148–155

X-ray photoelecron spectroscopy (XPS), Nafion membrane stability, 201
X-ray powder diffraction (XRD), Nafion membrane stability, 201

Zeeman interaction:
  basic principles, 4–9
  dipole-dipole coupling, 27–29
  electron-nuclear double resonance, 40
  electron spin echo envelope modulation, 44–49
  electron spin resonance imaging, 89–90
  high-field electron-nuclear double resonance, 43
Zero-field splitting (ZFS):
  dendrimer structure, ESR analysis:
    paramagnetic metal ion additions, 297–299
  electron-electron double resonance, 31–33
  electron spin responance principles, 8–9
  multifrequency and high-field electron spin resonance, 17–18
Zwitterionic diblock copolymers, mesoscopic ionomers, 189–192